JNTU
ANANTHAPUR
Edition

TEXTBOOK OF ENGINEERING CHEMISTRY

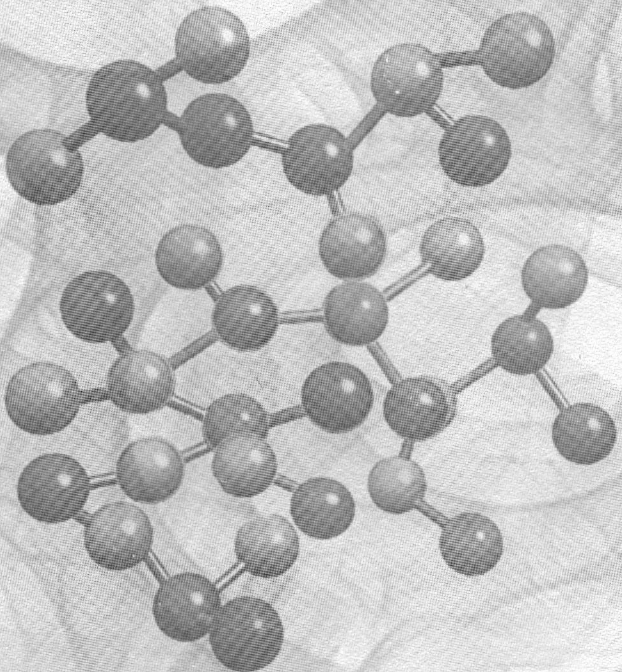

JNTU
ANANTHAPUR
Edition

TEXTBOOK OF ENGINEERING CHEMISTRY

C.Parameswara Murthy
M.Sc., Ph.D.
Professor of Chemistry,
Osmania University,
Hyderabad.

Andra Naidu
M.Sc., Ph.D.
Professor of Chemistry,
JNTU College of Engineering,
Kukatpally, Hyderabad.

BSP **BS Publications**
4-4-309, Giriraj Lane, Sultan Bazar,
Hyderabad - 500 095 - A.P.
Phone : 040 - 23445605, 23445688

Published by :

BSP BS Publications
4-4-309, Giriraj Lane, Sultan Bazar,
Hyderabad - 500 095 - A.P.
Phone : 040 - 23445605, 23445688
e-mail : contactus@bspublications.net
www.bspublications.net

Printed at

Adithya Art Printers
Hyderabad.

Price : Rs. 210.00

ISBN : 978-81-7800-215-6

Preface

You see things; and you say ***"Why?"*** *But I dream things that never were; and I say* ***"Why not?"***

— George Bernard Shaw, Great Dramatist and Nobel Laureate.

Chemistry has an important role to play in achieving a sustainable civilization on earth.

— Dr. Terry Collins, Professor of Chemistry, Carnegie Mellon University.

Most of the environmental problems of past centuries and decades, such as the biological contamination of drinking water, were solved only when the methods of science in general—and chemistry in particular—were applied to them. The phenomenal rise in human life expectancy and in the material quality of life that has come about in recent decades is due in no small measure to chemicals and chemistry.

— Colin Baird, Environmentalist.

Chemistry is the study of matter, its properties, the changes that it may undergo and how these properties and changes are affected by its composition. It is important for engineers to have knowledge of chemistry, since they can expect to find problems in fields as diverse as the design and development of new materials, quality control and environmental engineering that are basically chemical in nature. The standing and firmness of a structure is decided by the strength of foundation laid, although it is never expressed to the external world. Similarly the applied and engineering sciences derive strength from basic sciences and their growth is decided by inputs given by basic sciences.

Chemistry is the back bone in designing and understanding the nature of various engineering materials. Many advances in engineering either produce a new chemical demand as in the case of polymers or wait upon chemical developments for their application as in the case of implants and alloys. Currently, the electronics and computers engineers are waiting for suitable biopolymers and nano molecules for use in miniature super computers; the electrical engineers are in search of proper conducting polymers; the mechanical engineers are on lookout for micro fluids and the civil engineers are looking for materials that are environment friendly, economical but long lasting.

Engineering Chemistry is a maze of topics of basic chemistry and applied chemistry. In developing this book, an attempt is made to mix both the facets of the subject to reach proper consequence and meet the demands of teachers and taught.

This book is intended as a customized textbook to suit the JNTU, Ananthapur syllabus. The contents in each chapter includes both theoretical concepts & practical procedures required for undergraduate Engineering & Technology students.

This book is an amalgamation of expertise of

Prof. C. Parameswara Murthy, the author of *College Chemistry,* three volumes of *University Chemistry* and *Chemistry of Engineering Materials*.

Prof. A. Naidu's teaching experience in Chemistry of Engineering Materials.

We are confident that this book too would get acceptance of teacher and student communities like our previous books.

To give a logical end, the applied topics are preceded by the corresponding basic topics. Thus the fuel related topics are preceded by thermodynamics and thermo chemistry, the corrosion is dealt after giving sufficient information of electrochemistry, solid state and phase rule precede alloys, material science and so on. Several numerical problems, multiple choice questions and short & essay type questions are included at the end of chapters to help the students to prepare for exams and self test their knowledge.

No book is perfect without the inputs of constructive criticisms from the readers. So we would appreciate receiving comments and suggestions for improving the content of text, additions and deletions to be made, errors crept and the same would be acknowledged in the future editions. You can use the email ID *cpmurthy@hotmail.com* for this purpose.

We take this opportunity to express our deep sense of gratitude to the authorities of Osmania University and Jawaharlal Nehru Technological University for permitting us to take up this venture. We are highly thankful to our colleagues for their discussions and suggestions. In particular we thank our colleagues Prof. P. K. Dubey, Prof. A. S. R. Swamy, Prof. S. Sondu, Prof. G. Venkat Reddy, Dr. P. Yadagiriswamy, Dr. B. Satyanarayana, Dr. P. Veera Somaiah, Prof. Arun Kumar, Dr. Gouri shankar for providing various inputs to get a best book.

The good students help in shaping their teachers with their inquisitive questions and observations and their role is acknowledged through this work. We are thankful to Mr. Nikhil Shah & Anil Shah proprietors of B. S. Publications for their cooperation. We are also thankful to Mr. Lakshminarayan Jakkani and Mr. Naresh for good DTP work. We are thankful to the management of Tara Publications for their encouragement for writing this book.

Our family members are the under-current of our energies and inspiration and we deeply acknowledge their patience and encouragement.

- Authors

Contents

JAWAHARLAL NEHRU TECHNOLOGICAL UNIVERSITY ANANTHAPUR

I Year B.Tech.

ENGINEERING CHEMISTRY

(Common to all Branches)

UNIT I:

Water: Sources of Water, Types of impurities in Water, Hardness of Water – Temporary and Permanent hardness. Units. Estimation of hardness by EDTA Method. Analysis of Water - Dissolved Oxygen. Disadvantages of Hard Water. Problems on hardness of water. Methods of Treatment of Water for Domestic Purpose – Sterilisation: Chlorination, Ozonisation.

Water for Industrial purpose - Water for Steam Making, Boiler Troubles – Carry Over (Priming and Foaming), Boiler Corrosion, Scales and Sludge, Caustic Embrittlement. Water Treatment: - Internal Treatment – Colloidal, Phosphate, Calgon, Carbonate, Sodium aluminates Conditioning of Water. External Treatment - Ion- Exchange Process; Demineralization of Brakish Water – Reverse Osmosis.

UNIT II:

Science of Corrosion: Definition, Types of corrosion: Dry Corrosion, (Direct Chemical attack), Wet Corrosion, Theories of Corrosion and Mechanism, Electro Chemical Theory of Corrosion. Galvanic Series, Galvanic Corrosion, Concentration Cell Corrosion, Oxygen absorption type. Factors Influencing Corrosion. Control of Corrosion – Cathodic Protection – Sacrificial anode and Impressed Current. Uses of Inhibitors. Electro Plating, and Electro less plating (copper and nickel)

UNIT III:

Polymers: Polymerization Reactions – Basic concepts. Types of Polymerization – Addition and Condensation Polymerization. Plastics –Thermosetting and Thermoplastics. Composition, Properties and Engineering Uses of the Following: Teflon, Bakelite, Nylon. Rubber – Processing of Natural Rubber and Compounding. Elastomers – Buna S, Buna N, Polyurethane Rubber; Silicone Rubber. Conducting Polymers, Synthesis and applications of Polyacetylene and Poly aniline. Liquid Crystals definition, properties, suitable examples and Engineering Applications

UNIT IV: **Chemistry of nano materials:** Nano materials definition, properties and applications;

Explosives and Propellants: Explosives, Classification, precautions during storage, blasting fuses, important explosives. Rocket propellants, classification of propellants.

Lubricants : Principles and function of lubricants - Classification and properties of lubricants – Viscosity, flash and fire points, cloud and pour points, aniline point, Neutralisation Number and Mechanical Strength.

UNIT V:

Electro Chemistry: Conductance – Equivalent Conductance – Molecular Conductance, Conductometric Titrations – Applications of Conductivity Measurements.

Electrochemical Cells: Measurement of EMF, Standard electrode potential, concentration cells, batteries (Ni–Cd cell), Lithium batteries. Fuel cell: hydrogen oxygen fuel cell and methanol fuel cell

Insulators – Definition, Properties and Characteristics of Insulating Materials; Engineering Applications.

UNIT VI:

Phase rule: Definition, Terms involved in Phase Rule and Phase rule equation. Phase diagrams – one component system (water system), two component system (lead- silver system) Eutectics, heat treatment based on iron-carbon phase diagram, hardening, annealing.

UNIT VII:

Fuels and Combustion: Definition and Classification of fuels. Solid, liquid & gaseous fuels, Characteristics of a good fuel. Metallurgical Coke – Characteristics & Manufacture (Otto-Halfmann). Petroleum – Refining – Synthetic Petrol. Calorific Value & its determination (Bomb Calorimeter – Junker's Gas Calorimeter). Combustion: Flue gas analysis by Orsat's apparatus.

UNIT VIII:

Building Materials: Cement: composition of Portland cement, analysis, setting and hardening of cement (reactions).

Refractories : Definition, Classification With Examples; Criteria of a Good Refractory Material; Causes for the failure of a Refractory Material

1 Water and Water for Industries

1 Water and Water for Industries

1.1 Sources of Water

Although the problem of water supply is one of the most important of the problems that we face, it continues to draw very little attention from most people because many things are taken for granted. When the engineer is called upon to select a site for the location of a factory, he is sometimes apt to consider all other factors except that of water supply. Yet, it may be a deciding factor in the working of a factory. In some cases, the water requirement is very great. For example, about 30 million litres of water are needed per 1,000 tonnes of steel produced and about half a million litres per tonne of paper and over one and a half lakh litres per automobile. It may be noted that the largest requirement of water is for municipal supply. Besides the quantity of water, the nature and amount of impurities present in the water are of no less importance. Therefore, in selecting a proper site for a factory the following factors have to be taken in to account regarding the water supply: 1. quality and quantity of water available, 2. effect of impurities on the process and 3. quantity of water needed.

Generally, the water available contains a wide range of impurities but only a few of them may affect a process, so one is concerned with the removal, or reduction of concentration of only the few impurities and the rest may be allowed to remain as such, e.g., bacteria, if present, need not be removed from water for use in boilers. It may be noted that the water supplied for a particular industry is generally not 'consumed' in the operation in the sense of the term as electricity is consumed in a process. Almost the total quantity of water appears in one form or another and the engineer has also to deal with the disposal or re-use of waste water. The nature of the impurities in the outgoing water may be such that it is not safe or wise to let it flow out in the river stream or into the sources of water supply. Thus, connected with the water supply is the problem of the disposal of the waste or sewage. Again, in the same industry, water for several purposes is needed, e.g., steam generation, cooling, general needs of the processes and plant services, and for each purpose the specifications may be different.

Even *rain water,* which is the purest form of water found in nature, contains a considerable amount of dissolved gases and solids – both of organic and inorganic origin. Rain water either flows in to rivers or lakes and ultimately into the sea and is known as *surface water,* or it sinks underground and is available as well or spring water, when it is known as *underground water.* In general, the greater the contact that water has with the soil, or the more soluble the earthy matter with which it has come in contact, the greater is the amount of dissolved impurities. A sufficient quantity of flowing water of low hardness is available from many rivers. Even in any particular river the amount of dissolved mineral matter varies 2 to 3 times during the year. The amount of suspended and organic matter is still more variable. For cooling purposes, river water does not need any special treatment unless the amount of suspended matter is very great. Sometimes even *sea water,* in spite of its causing corrosion, can be used for cooling purposes. Sea water contains about 3.5 percent of dissolved salts, principally sodium chloride (2.7%), and magnesium chloride and sulphate (0.6%). It is an important source of these and other substances.

Lake water has a more constant chemical analysis and the seasonal changes can be predicted quite accurately. It is softer than well water and a big lake can be taken to be a dependable source of water supply, but attention should be paid to the organic matter present. The quality of water of some of the big lakes is so good that hardly any treatment is needed for industrial uses, e.g., the water of Lake Superior, U S A contains only 2 mg/litre of turbidity and 60 mg/litre of total dissolved solids.

Usually well water is still more constant in its composition. It is clear but contains more of dissolved matter and is harder than surface water. It is more suitable for human consumption because of clarity and freedom from organic matter, harmful bacteria, unpleasant taste and odour. Deep well water is cool and is available at a more constant temperature throughout the year but contains more of dissolved minerals. Usually it is of high organic purity.

1.1.1 Types of Impurities in Water

The common impurities present in natural waters may be classified as follows.

1. *Dissolved minerals* – mostly comprise of carbonates, bicarbonates, sulphates and chlorides of calcium, magnesium, sodium and potassium.
2. *Dissolved gases* – mostly air and carbon dioxide.
3. *Suspended matter* – mostly mineral matter, giving turbid or muddy water. Organic matter may also be present.
4. *Microscopic matter* – consists mostly of plant and bacterial life giving colour, taste and odour.

In general, the removal of impurities from water of classes 1, 2 and 3 form the chief problem for industrial usage, and 3 and 4 for municipal supplies.

1.2 Hardness of Water

From the standpoint of an engineer the dissolved salts of calcium and magnesium are of special interest as they are responsible for some of the common and major boiler troubles and therefore will be considered in greater detail. A simple method of detecting these is by the use of soap (not detergent). A sample of water that reacts with soap to form a white scum and does not produce lather is known as *hard water*. *Soft water* lathers freely with soap. Calcium and magnesium salts are present mostly as bicarbonates, sulphates and also as chlorides. The difficulty in producing lather arises because of the fact that soaps are sodium and potassium salts of higher fatty acids and are soluble in water but when soap is used in water containing calcium and magnesium salts, insoluble soaps of calcium and magnesium are formed and precipitated, thus destroying the soap. A typical reaction may be illustrated as follows:

$$\underset{\substack{\text{Soap (Sodium stearate)} \\ \text{Soluble}}}{2C_{17}H_{35}COONa} + CaCl_2 \rightarrow \underset{\substack{\text{(Precipitate)} \\ \text{Insoluble}}}{(C_{17}H_{35}COO)_2\,Ca} + 2NaCl$$

Such precipitates are not formed with the newer detergents, which are sulphonates of long chain alcohols. Titration of cold water will give total hardness of water sample when the water sample is boiled for 5 min; cooled to room temperature and titrated to the end point it will give the permanent hardness as temporary hardness is removed by boiling.

1.2.1 Types of Hardness – Temporary & Permanent Hardness

Calcium and magnesium salts are not the only salts that cause hardness. They are mentioned because they are most commonly present. As a general rule, all soluble salts of heavy metals cause hardness because they form precipitates with soap. The bicarbonates of calcium and magnesium are decomposed by boiling the water when carbonates are formed which are less soluble. Thus the hardness due to the bicarbonates can be removed by boiling. The hardness due to these salts is called ***temporary hardness***. As magnesium carbonate is only partially precipitated the term ***cabonate hardness*** is now used instead of temporary hardness. Sulphates and chlorides of calcium and magnesium are not removed by boiling and hardness due to these is called ***permanent or non carbonate hardness.*** The corresponding chemical reactions are given in 1.9.3.

1.2.2 Estimation of Hardness by EDTA Method

In a boiler house, the degree of hardness of feed water forms a simple and important guide to its suitability for boiler feed. In a simple test, usually performed for water to be used in a low pressure boiler, 70 ml of water is taken in a 300 ml bottle. A standard soap solution is added from a burette and the bottle is shaken. The volume of soap required to produce lather stable for 5 minutes is noted and hardness is calculated. It is an empirical method but after some experience fairly accurate results can be obtained. It is estimated that if about 2 kg of common soap is added to 100 litres of water of hardness 100 mg per litre it would get precipitated by the salts present in the water. Lather would be produced only if more than this amount of soap is used.

For more accurate results the soap method has been replaced by the *complexometric method.* This depends upon the fact that when ethylenediamine tetra acetic acid (abbreviated as EDTA) or its sodium salt solution is added to a solution containing certain metals, chelated soluble complex is formed. If the pH value of the solution is made 10 and a dye, known as *Chrome Black T or Eriochrome Black T or Erio-T* is added in presence of calcium and magnesium ions the solution becomes light red in colour. If EDTA is added to such a solution the calcium and magnesium ions will form complex with EDTA. When the formation of complex with free ions is completed, EDTA extracts the ions from the dye and the solution turns blue. This is the end point. If certain other metallic ions are present, they interfere. This interference can be removed by adding certain salts.

The reaction taking place on the addition of Eriochrome black T to a solution containing Ca^{++} and Mg^{++} may be represented thus:

$$Ca^{++} + \text{Eriochrome black T} \rightarrow \underset{\text{light red complex}}{(Ca^{++}\text{ chrome black T})}$$

On the addition of EDTA the free Ca^{++} and Mg^{++} form complexes with it. When they have been complexed the red complex is disrupted and a more stable complex is formed as follows:

$$\underset{\text{light red}}{(Ca^{++}\text{ chrome black T})} + \text{EDTA} \rightarrow \underset{\text{blue}}{\text{chrome black T}} + (\text{Ca. EDTA})$$

A very useful engineering use of EDTA is in the prevention of scale formation. It is dissolved in water which is to be used in the circulatory system or for cooling, as in the radiators of automobiles. EDTA forms stable complexes with calcium and magnesium ions and prevents their deposition. It can also be used to remove some of the deposits.

The EDTA solution is also called the Versenate solution and sometimes the method is called Versenate method.

1.2.3 Expression of Hardness and Units

At this stage the units in which hardness is usually expressed may be considered. The weights of all the substances are expressed in terms of a weight equivalent to the weight of $CaCO_3$. For example, 162 parts by weight of Ca $(HCO_3)_2$ would react with the same amount of soap as 100 parts by weight of $CaCO_3$. Therefore, the weight of Ca $(HCO_3)_2$, is multiplied by 100/162 to give weight in terms of $CaCO_3$. Similarly, weights of $CaSO_4$, $CaCl_2$, $MgSO_4$, Mg $(HCO_3)_2$ and $MgCl_2$ actually present are converted in terms of weight of $CaCO_3$ by multiplying x with

$$\frac{100}{136};\ \frac{100}{111};\ \frac{100}{120};\ \frac{100}{146};\ \text{and}\ \frac{100}{95}\ \text{respectively.}\ \left(x \times \frac{100}{\text{mol.wt of the compound}}\right)$$

The common unit of expression at present is in terms of parts of the hardness causing impurities in terms of weight of $CaCO_3$ per million parts (are particle parts) of water. For example, if 6 kg of $MgSO_4$, are present in 100,000 kg of water, then in 1,000,000 kg of water 60 kg of it would be present. Therefore, the weight in terms of $CaCO_3$, would be $60 \times \frac{100}{120} =$ 50 kg. So, the hardness of this sample of water, is 50 parts per million parts of water, which is written as 50 ppm (parts per million). In the French system it is expressed in terms of parts per 100,000 parts. It is said to be degree French of °F (one part in 10^5 parts). In the English system, hardness is expressed in terms of grains (1/7000 lb) per gallon (10 lbs) of water, i.e., parts per 70,000 parts. This is also called degree Clark, e.g., 14° Clark means impurities equivalent to 14 grains in terms of $CaCO_3$ are present per gallon of water or 14 parts per 70,000 parts of water. Therefore on ppm scale it is $14 \times \frac{1,000,000}{70,000} = 200$ ppm

ppm = 1 part/106 parts. °Fr = 1 part/10^5 parts °cl = 1 part/70,000 parts

1 ppm = 0.1 °Fr = 0.07°Cl

It is becoming more common to express the results in terms of milligrams of the substance (expressed as $CaCO_3$) present in a litre of the solution (mg/l). Taking density of water as unity, milligrams per litre means 1/1000 g per 1000 ml of water (or solution), or 1/1000 g per 1000 g of water. This is the same as parts per 1,000,000 parts. Therefore, in common cases where the density of the solution is almost unity, parts per million parts is equal to milligrams per litre. Of course, for more concentrated solutions a correction has to be made. As the quantities of water (or solutions) are mostly expressed in terms of volumes, the engineer is able to make more direct calculations by using mg/l scale than by using ppm scale (although for most practical purposes both are the same).

A more recent trend is to report the various cations and anions separately in terms of their equivalents as $CaCO_3$. This has been extended to express the three forms of alkalinity, namely, bicarbonate, carbonate and caustic or hydroxide. This form of expression is now used to express even the mineral acidity (i.e., due to the sulphates, chlorides and nitrates) which might be shown as negative alkalinity. In spite of all this, free CO_2, Fe, Mn, SiO_2 and fluorides are generally expressed as these constituents. It is therefore, necessary that whenever the results are expressed in terms of equivalents of $CaCO_3$, it be clearly indicated by a statement like "expressed as p. p. m. of $CaCO_3$." It is a general practice to express the results in three significant figures; e.g., the figure 151.65 ppm is rounded off to 152 ppm.

The maximum permissible hardness of water for boilers operating at different pressures is as follows :

at	0 to 10 kg/cm^2	... ...	80 ppm
at	10 to 15 kg/cm^2	... ...	40 ppm
at	15 to 30 kg/cm^2	... ...	10 ppm
	over 30 kg/cm^2	... ...	2 ppm

1.3 Treatment of Water for Domestic Purpose

Perhaps the most important problem of water technology that of municipal water supply, because if the water supplied to the cities is not of the right kind, the results will be disastrous to human lives. The requirements for this purpose are very different from those of the industries. Municipalities have to supply potable water. i.e., water which is safe and good to drink. Potable water does not mean pure water. Distilled water, for example, is flat and tasteless. At the same time potability cannot be determined merely on the basis of its taste, odour and appearance. Water may have disease producing bacteria which cannot be detected by these means.

Briefly, water supplied by the municipalities should be free from disease-producing bacteria, suspended impurities and excess of dissolved salts or chemicals used for water treatment. It should be clear, colourless, odourless and pleasant to taste. A sample of clear water which has low dissolved salt content but has disease-producing bacteria can be used in a boiler, but would be unfit for drinking. On the other hand, a sample of clear water, free from bacteria, but containing a certain amount of dissolved salts may be unfit for boiler feed, yet fit to drink.

Municipalities draw their water from rivers, wells and lakes. Water so obtained contains dissolved matter which in most cases is harmless for drinking. In general, municipal water treatment is not directed towards removal of the dissolved salts present in water. Water, such as sea water which contains a large amount of dissolved matter is not commonly looked upon as a source of drinking water. But, as good quality water is not uniformly distributed on the surface of the earth, and demand for drinking water is increasing due to increase in population, a considerable amount of research is being done, especially in the U S A, on *conversion of sea water* into drinking water. Some experimental plants, each producing over half a crore litres per day of potable water from sea water, are functioning in the U S A. Two avenues of approach for obtaining potable water from sea water are distillation, and freezing out pure water as ice, leaving the salt in the mother liquor. Distillation is a well-known process but the salts present in sea water cause severe corrosion of the equipment. Freezing is more expensive but there is less corrosion and is likely to be successful in the cold climates.

The latest approach in converting sea water fit for drinking is the application of reverse osmosis technique. On account of low capital cost, low operating cost, reliability and simplicity of operation the reverse osmosis process is becoming the principal method for this purpose. Such plants have been mainly installed in the petroleum rich countries of the Middle East.

Municipalities in general do not soften the water. There are only a few municipalities in the world which partially soften the water because the water available to them is very hard.

The treatments more widely given include: 1. Sedimentation, 2. Filtration, and 3. Sterilization. See Fig. 1.1.

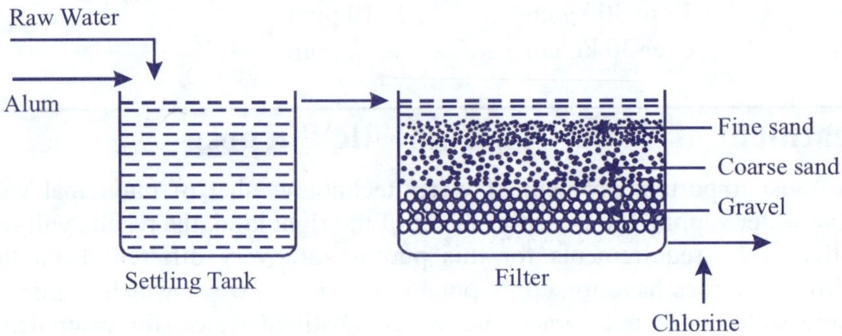

Fig. 1.1 Principle of municipal water treatment.

1. **Sedimentation**

   Sedimentation means simple settling. It is the process by which the force of gravity acts on particles heavier than water and causes them to move downwards and settle. For this purpose water is kept in huge tanks, called sedimentation tanks (or basins) in which water moves very slowly. The basins have to be designed carefully so that the formation of eddy currents (which cause mixing) or straight channeling of water is avoided.

   The time required for sedimentation depends upon the weight, size and shape of particles and the viscosity and frictional resistance of water. Because viscosity changes with temperature, the settling rate varies in different seasons. It has been calculated that the settling rate at 0° is only 45% of the rate at 30°. Consequently turbidity is more easily removed during summer than during winter. Water is kept in sedimentation tanks from 4 to 12 hours.

Finely divided suspended matter does not settle down easily. Generally, only 70% of the suspended matter settles down by plain sedimentation. For better clarification, chemical means have to be used. This is done by the addition of coagulants like alum, ferrous sulphate, ferric chloride, sodium aluminate, etc. Each of these works best within a particular pH range and has a certain function in addition to clarification. Generally about 95% of the suspended matter is removed by coagulation.

Coagulation reactions are complicated but a simplified explanation can be that the salt hydrolyses and produces a floc of the hydroxide, e.g., $Al(OH)_3$ and $Fe(OH)_3$. The flocculent precipitate so produced has an enormous surface area per unit volume. The floc removes the finely divided and colloidal matter by a variety of mechanisms, e.g., by adsorption, mechanical entanglement and electrical attraction of oppositely charged particles. Thus the smaller particles gather together to form bigger particles which quickly settle down. Some bacteria and colour are also removed. The dosage of the coagulant depends upon turbidity and may be 5 to 100 mg/l.

These years increasingly large quantities of *coagulant aids*, mostly polyelectrolytes, are being used. The polyelectrolytes are high molecular weight water soluble polymers. Some naturally occurring organic flocculants have been used. Polyacrylamides aid materially in the formation of rapidly settling flocs. Other organic polymeric materials which are strongly cationic have been used in place of aluminium or iron salts. These function through hydration. The true polyelectrolytes contain groups which undergo electrolytic dissociation giving highly charged large molecular weight ions which attract other charged particles and help coagulation. They may give anions, cations or may be non-ionic (i.e., provide both positive and negative charges in solution). Considerable research work is being done in this field but the mechanism of their activity has not been well understood.

Each coagulant gives best results in a pH range which is affected by the mineral characteristics or the water. For Example, in pure water, potash alum coagulates best at a pH of 5.5 but for most natural waters the range is 5.5 to 8.0. It should be noted that each ppm of added alum, consumes about 0.5 ppm of natural alkalinity thereby reducing the pH level. So lime or soda needs to be added. The presence of iron and manganese in water introduces additional factors affecting the choice and dosage of coagulant. It is therefore necessary that laboratory tests be made with different coagulants at varied pH levels and varied concentrations. Retention time of 2 to 6 hours is usually provided. Often coagulation can be combined with sedimentation followed by filtration. Potash alum and ferric alum are the most widely used coagulants.

2. Filtration

Filtration may be defined as 'the process of passing a liquid containing suspended matter through a suitable porous material in such a manner as to effectively remove the suspended matter from the liquid'. Generally, water is filtered after sedimentation and coagulation. In some cases where there is great shortage of space or the water is quite clear, it may be filtered directly thus saving large areas needed for the construction of sedimentation basins. Filtration is primarily a mechanical process. When water flows through a porous material many suspended particles are unable to pass through the interstices (or gaps) in the material because of their size. Thus these particles are removed and clarification takes place.

Filtration removes the suspended matter and to some extent bacteria. True colour is not removed by filtration. It can be removed by the use of coagulants.

Filters are of two types: (a) gravity filters and (b) pressure filters.

(a) *Gravity filters.* In these the water flows through the bed by gravity. These filters are of two types: (i) Slow sand filters and (ii) Rapid sand filters.

(i) *Slow sand filters* are still being used in India by many municipalities. They occupy large areas and the rate of filtration is slow. The filters are like large shallow tanks made of concrete. At the bottom there is an under-drain system through which the filtered water goes to the storage tank. Gravel is placed in the tank above the under-drain system. It consists of several layers of stone of sizes varying from 6 cm to 0.5 cm. On the gravel is placed a bed of coarse sand and above it a layer of fine sand is placed. See Fig. 1.1.

As water percolates through the filters, the suspended particles are retained on the top 5 to 10 cm layer of the fine sand. In fact, the layers of coarse sand and of gravel do not filter at all. They support the fine sand, the top layer of which retains the suspended matter. When the suspended matter has been deposited to such an extent that the rate of filtration becomes very slow, the portion of the top layer is scraped and the bed is used again. After continued use of the bed for a long time, it becomes necessary to replace the layer completely.

(ii) *Rapid sand filters* are coming into greater use primarily because their efficiency for filtration is many times that of slow sand filters, hence, they need much less space. The general arrangement of the sand bed is similar to that of slow sand filters, the distinguishing feature being the backwash system. The chief reason for a slow rate of filtration through the slow sand filter is that as the suspended matter is deposited, the gaps between the sand particles get filled up and the flow is restricted. To overcome this difficulty, the rapid sand filters are provided with backwash system. After filtering for some time the filtration is stopped and a portion of filtered water is *sent up* through the bed. This removes the suspended particles deposited in the pores and cleans the bed. When the suspended matter has been removed, the wash water supply is stopped and the filter bed which is as clean as a freshly made bed, is again realy for filtration. The cycle of filtration and washing is repeated. Thus high filter rates are obtained. The wash water is discarded. The rate of filtration may be 30 times that of the rate of filtration through slow sand filters.

(b) *Pressure filters* have essentially the same arrangement of the filter media as gravity filters. The filter material is kept in a closed cylinder and water is forced through the filter bed under pressure. Pressure filters can be installed in the water supply line, thus repumping of filtered water is avoided. As these are enclosed, they are more suitable for filtering hot water. If hot water is filtered through gravity filters, there is more loss of heat. The pressure filters occupy much less space than the other types of filters. Because of these advantages pressure filters are widely used for industrial water conditioning.

3. Sterilization and Disinfection

After filtration it becomes necessary to destroy or make inactive the residual bacteria and to prevent their further growth. This is achieved by sterilization and disinfection. Often ground water is free from disease-producing bacteria but the surface water, which is the main source of municipal water supply, is likely to be contaminated, especially because of the discharge of sewage into streams. Some epidemics, like cholera, plague and also typhoid can be controlled by taking better care of drinking water.

(i) *Chlorination :* Chlorination is mainly used for killing disease-producing organisms. Further, it reduces (or may even remove) objectionable tastes and odours from drinking water, and oxidizes iron, manganese and hydrogen sulphide, and thereby help their removal. It also causes oxidation, decolourization and destruction of organic matter. Because of wide range of action chlorine, besides its use in water treatment, is injected into sewage and industrial waste before being discharged into streams. In such cases it causes reduction in bacteria, destruction of chemicals that may react with oxygen, such as sulphides, sulphites, ferrous ions, etc. It helps coagulation by destroying organic matter, hence the practice of pre-chlorination. In water used for food processing industries chlorine should be used cautiously. It may spoil the flavour and even corrode the metal container.

Small volumes of water can be sterilized by boiling. For larger volumes chlorine is very widely used.

Bleaching powder is convenient for use on smaller scales. Its disinfecting action is due to the chlorine made available by it. When the chlorine requirement is over about 10 kg per day, it is more economical to use chlorine gas. The reactions of chlorine and water can be represented thus.

$$Cl_2 + H_2O \rightleftharpoons HOCl + H^+ + Cl^-$$

At higher pH values hypochlorous acid ionizes thus:

$$HOCl \rightleftharpoons H^+ + OCl^-$$

Earlier it was believed that the disinfecting properties of chlorine were due to the nascent oxygen liberated, which oxidizes harmful matter. More recent work has shown that the HOCl molecules (not hypochlorite ion, OCl^-) combine with an enzyme in the cells of the organisms and thereby destroy the organisms. Therefore, chlorine is more effective as a disinfectant at lower pH values (below 6.5). A part of the chlorine added is used in combining with or oxidizing some of the impurities. The amount of chlorine added is generally such that about 0.2 mg/l of chlorine is left behind after the reactions.

If the water is exposed to the atmosphere, especially to sunlight, chlorine is dissipated quickly. In order to overcame this disadvantage, ammonia may be added along with the chlorine when chloramines like NH_2Cl and $NHCl_2$, are formed. These are not as active as chlorine but are not so easily removed, and are more desirable when water is to be distributed over long distances.

Chloramines require higher concentrations and longer contact time for comparable bacterial reduction. Chloramine tablets are used by the army for sterilizing water. Water is collected in the bottles and chloramine tablets are added followed by sodium thiosulphate tablets which remove excess, of chlorine.

Break point chlorination is a specialized method of chlorination applied to highly polluted water and for the removal of ammonia. In the beginning as chlorine is added to such samples of water there is a proportionate increase in the amount of residual chlorine as is indicated in the Fig. 1.2. Ammonia is converted to monochloramine and then to dichloramine. Further chlorination gives trichloramine or oxidizes ammonia to free nitrogen. At this stage there is a marked decline in the residual chlorine but further addition of chlorine gives a proportionate increase of free available chlorine. Because of the peculiar character of the curve obtained on such chlorination, this practice is known as "break point chlorination".

Besides its use in the removal of organic matter from highly polluted water, break point chlorination is useful in removing ammonia from boiler feed water.

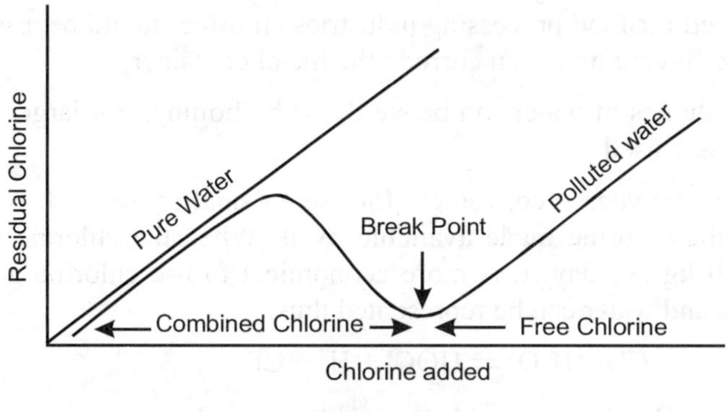

Fig. 1.2 Break point chlorination.

(ii) ***Ozonization:*** Disinfecting agents other than chlorine are also used. The use of ozone for disinfection has been suggested. It is expensive, hence is only used on a small scale.

In India well water is still used to a considerable extent. This can be disinfected by adding potassium permanganate solution to the well. Chlorine can also be added.

Most natural waters have pH values, between 6.0 and 8.0 but. 7.2 to 7.8 is the more common range. Water of pH less than 7.2 is not good for the human system. For boiler feed water, pH is adjusted up to about 10.5 to prevent corrosion and also to ensure the precipitation of scale forming salts during the chemical softening reactions. At lower pH values water has corrosive action on mild steel.

1.3.1 Measurement of pH Value

1. **Colorimetric Method.** The colorimetric measurement is based on the assumption that when a definite quantity of an indicator is added to a certain fixed volume of different solutions of the same pH value, the colours of these solutions will be the same. This assumption holds good for most of the solutions which are commonly come across, but there may be interference due to some ions. The working of the colorimetric method can be easily understood with the help of an illustration from acidimetric and alkalimetric titration. If methyl orange indicator is added to a sodium carbonate solution, the solution becomes yellow. If a small amount of an acid is now added to it the colour seems to be unchanged but near the endpoint the addition of each drop of acid brings about a definite change in the colour of the solution. If only a few drops of the acid are added in excess, the amount added can be guessed by observing the colour, but with a large excess of the acid no such guess can be made. This shows that over a small range the colour can indicate the acidic or alkaline nature of the solution. In other words, colour can indicate the hydrogen ion concentration.

   An indicator is a substance which gives different colours or different shades of a colour at different pH values. Therefore, by comparing the colours with the known standards, it is possible to tell the pH value. As the indication given is over a small range only, a number of indicators have to be used. Common indicators that cover the usual engineering range of pH are noted below :

Indicator	*pH range*	*Colour change*
Methyl red	4.4 to 6.2	Red to yellow
Brom thymol	6.0 to 7.6	Yellow to blue
Cresol red	72 to 8.8	Yellow to red
Phenolphthalein	8.2 to 10.0	Colourless to red

   Standard equipments for pH value determination are available on the market. The general procedure is that a certain prescribed volume (say, 10 ml) of the sample to be tested is taken in a tube and a fixed volume (say, 0.3 ml) of the indicator is mixed with it. The colour of the sample is then compared with the colours marked on a standard disc supplied by the manufacturer. These different colours correspond to certain pH values which are read on the instrument. Different colour discs have to be used with different indicators.

2. **Electrometric Method** : pH value can be more accurately determined by the electrometric method. Compact instruments called pH meters are available on the market. The pH meter has two electrodes which are dipped in the solution whose pH value is to he determined. One of the electrodes, known as calomel electrode, is a half cell (prepared by a combination of mercury and mercurus chloride in contact with a standard potassium chloride solution) and its potential is fixed. The potential of the other electrode varies with the pH value of the solution. This second electrode is a glass electrode. Its action is not well understood. It is like a tube, made of glass of a certain composition, having in it an electrolyte of a fixed concentration. When it is dipped in a solution, a potential is developed which depends upon the relative concentrations of H^+ on the inner and the outer surfaces of the glass.

Thus when these two electrodes (i.e., calomel electrode and glass electrode) are dipped in a solution, a potential is developed which depends upon the hydrogen ion concentration of the solution. As the potential is very small, a vacuum tube voltmeter is used. It is calibrated to read directly the pH value.

1.3.2 Alkalinity

Alkalinity in natural waters is mostly due to the presence of salts of weak acids. Carbon dioxide is formed in nature in considerable amounts and it produces bicarbonates by its action on basic materials in soils, this causes bicarbonates, to be present in large amounts, borates, silicates and phosphates present in small amounts, also contribute to alkalinity. Surface waters, which have algae, contain appreciable amounts of carbonate and hydroxide alkalinity. This happens because algae remove free as well as combined carbon dioxide from water, thereby raising the pH value. Water treated by lime soda process contains carbonates and excess hydroxide. The alkalinity of natural waters is related to its buffer capacity (i. e., capacity to receive acids without substantially lowering the pH) because the alkalinity is due to salts of weak acids.

In order of their effect on pH, the three major classes of materials may be classed as follows: (1) hydroxide, (2) carbonates and (3) bicarbonates.

Highly alkaline waters do not have a good taste. Alkalinity causes precipitates and sludges, to be deposited in the pipes and heating tubes and also makes the metal brittle.

In water-softening processes and in boiler water analysis, it is necessary to know the kinds and amounts of the various types of alkalinity present. Various formulas have been proposed to calculate these from the titration results. The formulas are mainly for the technicians who do not know chemistry, but the chemists should be able to calculate from the fundamental principles and memorizing the formulas need not be encouraged.

Any one of the following five conditions may arise in an alkaline solution:

1. *Hydroxide only.* These have pH above 10 and titration is done using phenolphthalein as indicator. In this case phenolphthalein alkalinity is equal to the hydroxide alkalinity or the total alkalinity.
2. *Carbonate only.* These have a pH of 9.5 or higher. The carbonate alkalinity is equal to the total alkalinity. The total titration, using methyl orange as indicator, is twice the titration to the phenolphthalein end point. Phenolphthalein becomes colourless at pH 8.3 when carbonate ion is converted to bicarbonate ions as follows:

$$CO_3^{--} + H^+ \rightleftharpoons HCO_3^-$$

Further addition of acid produces carbonic acid and the reaction is complete at pH 4.5 when methyl orange turns from yellow to orange red. It can be represented as follows:

$$HCO_3^- + H^+ \rightleftharpoons H_2CO_3 \rightleftharpoons H_2O + CO_2$$

3. *Bicarbonate only.* Solutions having only bicarbonates have a pH value of 8.3 or less. Methyl orange is used as indicator.

4. *Hydroxide and carbonate*. Solutions having both hydroxide and carbonate have pH above 10. The titration from the phenolphthalein end point to the methyl orange end point indicates only half of the carbonate alkalinity. The total titration to the methyl orange end point indicates the total alkalinity. Therefore, hydroxide and carbonate alkalinities may be easily calculated as follows:

   Carbonate alkalinity = 2 × (titration from phenolphthalein to methyl orange).

   Hydroxide alkalinity = Total alkalinity - Carbonate alkalinity.

5. *Carbonate and bicarbonate*. These solutions usually have pH between 8.3 and 11. Only half the carbonate alkalinity is indicated by the titration to the phenolphthalein end point. Methyl orange end point gives total alkalinity. The alkalinities may be calculated as follows:

   Carbonate alkalinity = 2 × (titration to phenolphthalein end point)

   Bicarbonate alkalinity = total alkalinity-carbonate alkalinity.

It is convenient to use N/50 acid solution for titration. A few drops of phenolphthalein indicator is added to the measured quantity of water and if pink colour is produced, it is titrated until just colourless. Let the number of milliliters of the acid used be P. A few drops of methyl orange are then added and titrated until the solution turns orange red. Let the total number of milliliters used for both the titrations be M.

From the foregoing discussion it is clear that M represents total alkalinity and P indicates all the hydroxides and half the carbonates. The results are summarized in the following Table 1.1.

Table 1.1

	Amount and type of alkalinity		
Titration result	**Hydroxide**	**Carbonate**	**Bicarbonate**
$P = 0$	nil	nil	M
$P = \frac{1}{2} M$	nil	2 P	nil
$P = M$	M	nil	nil
$P < \frac{1}{2} M$	nil	2 P	M – 2P
$P > \frac{1}{2} M$	2 P – M	2 (M – P)	nil

Alkalinity is usually expressed in terms of calcium carbonate. 1000 ml of N/50 acid solution is equivalent to 1g of $CaCO_3$ (equivalent weight of $CaCO_3$ being 50).

$$\therefore \quad 1 \text{ ml of N/50 acid solution} = \frac{1}{1000} \text{ g or 1 mg } CaCO_3$$

From the foregoing it is easy to calculate the alkalinity quantitatively and the result can be expressed as follows:

$$\text{Alkalinity} = \frac{\text{vol. of N/50 acid} \times 1000}{\text{vol. of sample for titration}} \text{ mg/l.}$$

1.4 Water for Industrial Purpose

Although the manufacturing industries need water for a variety of purposes the largest demand for water is for heat transfer. Water is the usual medium for extracting heat. Boilers, cooling towers, heat exchangers, condensers, coolers, refrigerators and air conditioning equipment illustrate the use of water for heat extraction. It is estimated that the large thermal power stations require 800 tons of water per ton of coal consumed.

Water is a *raw material* when it becomes a part of the product as in the beverage industry and some of the chemical industries. In a few cases it is a *reactant* as in the manufacture of acetylene from calcium carbide. Water being the cheapest liquid available is used widely for cleaning and washing in the homes as well as in industry. One is not conscious of its uses in *material handling,* in the pumping of materials like coal in the form of slurry for many kilometers. Harmful effluents from industries are *diluted* by large quantities of water. High pressure water jets for cleaning metal parts or the removal of bark from logs of wood illustrate the use of water for kinetic energy. Since water does not yield radio activities when exposed to neutron flux, it is used for *nuclear shielding,* or as a moderator in nuclear energy reactors.

It is obvious from the above examples that not only the quantity but also the quality of water needed for different uses vary widely. The high pressure boilers require water of very high purity for steam making necessitating very careful and elaborate treatment whereas for ordinary washing no treatment, other than simple settling for highly turbid water is needed.

As it is not possible in a small monograph to discuss the requirements of each industry separately, the treatment of water for steam making has been taken up by way of illustration showing how almost absolutely pure water can be obtained from the sea. Where the requirement is not that stringent the treatment may not be carried out to eliminate all the impurities.

1.4.1 Water for Steam Making

In water tube boilers steam is generated on the surface of the tubes which are heated by the hot gases and the flames produced by the combustion of the fuel. The steam travels through the tubes and is delivered to the boiler drums. At this stage the steam is mixed with the boiler water, foam and sludge. The problem of the production of clean-dry steam has been investigated by engineers for the last few decades. In recent times this has become a still more serious problem because boilers working at much higher pressures are being used. Therefore, all the boilers are provided with entrainment separators, i.e., devices for separating steam from the liquids and the solids. This equipment provides purification mainly by changing the direction of steam whereby the heavier particles, i.e., the liquid and the suspended matter, drop out. As a matter of fact, more heat is added to steam so that *super-heated* steam is obtained. The properties of superheated steam are almost those of a perfect gas, hence it does better work in steam engines and turbines and the steam consumption is lower. Super heated steam being at higher temperature does not have the problem due to the condensation of water in the pipelines, engines, etc., and so dry steam can be supplied over longer distances.

In water treatment complete elimination of all the impurities is not aimed at. Only those impurities which give rise to certain troubles are eliminated or kept within certain limits. In the modern high pressure boilers water purer than the distilled water used in laboratories has to be supplied. The major boiler troubles, due to the use of unsuitable water, may be classified as 1. Carry over, 2. Corrosion, 3. Scale formation, 4. Caustic embrittlement and 5. Turbine deposits. The total destruction of several large power houses has ultimately been traced to the use of unsuitable water which caused explosion of the boilers.

1.5 Boilers Troubles

Usage of hard water leads to many boiler troubles and heavy industrial losses due to its mal function.

1.5.1 Carry over

As steam emerges from the surface of the water, it is associated with tiny droplets of water. Such steam is called *wet steam*. The droplets may carry with them suspended matter, besides the dissolved matter present in the boiler water. Such carrying over of water by steam along with the suspended and the dissolved solids is called ***carry over***. It is mainly due to priming and foaming.

The liquid contamination in the steam is expressed in percentage, by weight, of the total mixture and is called ***steam quality***. If steam contains 0.3 per cent moisture steam quality is 99.7 percent. The impurity due to solids is represented by *steam purity* and is expressed in parts per million (ppm) or milligrams per litre (mg/l) of impurity. If steam contains 1 part by weight (or 1 mg) of solid contamination per million parts by weight of steam (or per litre of water), steam purity is 1 ppm (or 1 mg/l).

Priming is such rapid or violent boiling of water occurring in the boiler that water is carried into the steam outlet in the form of a spray. It may be caused by 1. very higher water level; 2. presence of excessive foam filling the steam space; 3. high steam velocity: 4. sudden steam demand leading to sudden boiling and 5. faulty boiler design.

Control of priming is fundamentally a problem of boiler design. It is controlled by providing proper evaporation and adequate heating surfaces, uniform distribution of fire and of working load, providing anti-priming pipes or dash plates, etc. In a given boiler priming can be minimized by 1. keeping the water level as low as it can safely be kept. Even a slight lowering of the water level greatly reduces priming; 2. avoiding sudden steam demands (by suddenly taking out a quantity of steam the pressure in the boiler is reduced, thus the boiling point of water is lowered and the superheated water is suddenly converted into steam); and 3. minimizing foaming.

Foaming is the formation of small but stable bubbles above the surface of water. It becomes a serious problem when the bubbles are persistent, i.e., do not break easily. These bubbles are carried over along with steam, leading to excessive priming.

Foam formation takes place when the concentration of solids in the surface layer is different from that in the mass of the liquid. This means that the solid must alter (whether increase or decrease) markedly the surface tension. Therefore all those solids which alter markedly the surface tension of water promote foaming. Presence of fatty oil is highly undesirable as it may form soap with the alkalinity. Soaps greatly reduce surface tension of water.

Pure water has little tendency to foam but almost all the impurities cause foaming, especially the dissolved impurities like soap, which reduce the surface tension of water. It may be noted that as the steam is being generated the concentration of the dissolved impurities, especially of sodium and potassium salts is increasing. The general construction of locomotive boilers is such that these have foaming more easily than the stationary boilers have.

To a certain extent foaming can be controlled by the use of anti-foaming chemicals, some of which act by counteracting the reduction in surface tension, and some by mechanical action, like castor oil which spreads on the surface of the water. Castor oil may be used for low pressure boilers. The only satisfactory remedy for foaming is *blow down*, i.e., removal of concentrated boiler water and replacing it by fresh feed water.

Dissolved solids, suspended solids and alkalinity affect carry over. It is believed that, in general, the suspended solids are more responsible for carry over than the dissolved solids. Carry over is undesirable because it not only decreases the efficiency but also results in some practical difficulties in the successful operation of a boiler. Some of these are: 1. a considerable amount of heat is wasted by the removal of hot boiler water; 2. water reduces superheat of steam in the super heaters; 3. water takes along with it grit which impinges on the surface at high velocity, and damages the walls of the cylinders and valve seats; 4. water contains considerable amount of dissolved solids which gradually get deposited and the deposits restrict the passage in the pipelines, causing corrosion and insulation; 5. water entering the steam engine cylinder causes serious trouble because of its incompressibility and inability to escape quickly through the steam ports; 6. foaming of water makes it difficult to know the exact level of water in the gauge glass.

1.6 Boiler Corrosion

Corrosion is one of the most serious problems caused by the use of unsuitable water. Boiler tubes, drums, economizers, super heaters and condensers are the most affected parts. The corrosion problem extends even to parts which are not directly in contact with boiler water because gases like oxygen and carbon dioxide are released during heating of water. These gases have corrosive effects. Corrosion can be attributed to the following (i) dissolved oxygen; (ii) mineral acids; (iii) dissolved carbon dioxide and (iv) galvanic cell formation.

(i) *Dissolved Oxygen*

Oxygen dissolved in water is mainly responsible for corrosion in boilers. Dissolved oxygen in the presence of moisture and at the high temperature prevailing in the boiler readily attacks iron. The higher the pressure, the greater will be the dissolve oxygen content, and the higher the temperature the lesser the oxygen content. The amount of oxygen dissolved is also influenced by the other dissolved matter present. The solubility of oxygen in pure water is greater than in water containing other dissolved matter. In a boiler, oxygen is introduced through the fresh make up water supplied to it and also through infiltration of air into the condensate system. As the water is heated the dissolved oxygen is liberated and iron is corroded, (The mechanism of corrosion has been discussed in detail in the chapter on Corrosion.)

Dissolved oxygen can be removed by :

(a) ***Mechanical deaerator*** : Various types of deaerators have been designed. The general principle is that in a cylindrical chamber water falls through perforated plates, or is sprayed, so that a larger surface is exposed. The deaerator is connected to a vacuum line which sucks out the gases.

(b) ***Pre-heating*** : Advantage is taken of the fact that the solubility of a gas decreases with rise in temperature. Air begins to be expelled at a temperature of about 65° and is almost completely removed at the boiling point. For the removal of dissolved oxygen the sample of water is sprayed in cylindrical equipment. It meets steam moving in the opposite direction. Steam heats and carries away with it the gases which have been liberated.

(c) ***Chemical treatment*** : The above two methods remove most of the oxygen but for a complete removal, chemical treatment is required. This is done by adding a reducing agent, sodium sulphite being commonly used, which at boiler temperature reacts rapidly as follows :

$$2Na_2SO_3 + O_2 \rightarrow 2Na_2SO_4$$

Ferrous sulphate is also used. It gives precipitate of $Fe(OH)_2$ which is oxidized to $Fe(OH)_3$.

In high pressure boilers even traces of oxygen are undesirable. As it is necessary to keep dissolved salts also to a minimum, hydrazine is added in such amounts that after reacting as under, there is a residual hydrazine of about 0.02 mg/l:

$$N_2H_4 + O_2 \rightarrow 2H_2O + N_2$$

Thus no solid is added to the boiler. Hydrazine should be handled carefully, usually its solutions are used.

(ii) *Mineral acids*

Most of the natural waters are alkaline. Waters in the mining areas are often acidic. Sometimes, in industrial areas the water may become acidic because of the discharge of acidic industrial wastes into the surface water.

Some of the inorganic salts may hydrolyse to produce acidity, the most important of such salts being magnesium chloride. Its action may be represented thus:

$$MgCl_2 + 2H_2O \rightleftharpoons Mg(OH)_2 + 2HCl$$

(As $Mg(OH)_2$ is insoluble the forward reaction goes to completion.)

$$Fe + 2HCl \rightarrow FeCl_2 + H_2$$

$$FeCl_2 + 2H_2O \rightarrow Fe(OH)_2 + 2HCl$$

As a result of these reactions even a small amount of $MgCl_2$ will cause corrosion of a large amount of iron. If the amount of the acid formed is small, it may get neutralized by the alkalinity present in the water, otherwise alkali will have to be added. It has been suggested that except in high pressure boilers, boiler water should have a pH value of about 10.5 so that the acid may not attack iron. High pressure boiler water needs separate consideration.

(iii) *Carbon dioxide*

Water contains some dissolved carbon dioxide. In case some bicarbonates are present in the boiler water, more carbon dioxide will be produced on heating because of the decomposition of the bicarbonates. Carbon dioxide coming in contact with water produces carbonic acid, H_2CO_3, which has a corrosive effect on the equipment. If carbon dioxide is released inside the boiler, it will be carried along with steam and H_2CO_3 will be formed where the steam condenses, and corrosion will take place. Local corrosion caused by carbon dioxide is called *pitting.*

Carbon dioxide in water can be removed by the addition of lime. Another method is to filter water through limestone when it not only becomes clear but carbon dioxide is also removed:

$$CaCO_3 + CO_2 + H_2O = Ca(HCO_3)_2$$

Of course, this reaction adds to the hardness of water.

The problem of carbon dioxide present in the steam can be tackled by the use of ammonia. It may be noted that ammonia itself has a corrosive action on copper and copper alloys; therefore its use is being replaced by amines.

(iv) *Galvanic cells.*

Corrosion can also be due to simple galvanic cells which iron forms with some of the boiler fittings made of other materials or with impurities. This can be avoided by suspending zinc plates. Zinc is more electropositive than iron hence it is anodic to iron and gets dissolved first and iron is saved.

Galvanic couples set up cells which cause formation of corrosion products. These deposits in themselves lead to further corrosion. This type of damage can be reduced by the use of proper inhibitors. Care has to be taken to see that the surface is clean and the inhibitor film covers the surface completely. Use of organic as well as inorganic inhibitors for protection against corrosion has been discussed in the

Chapter on Corrosion. It may be added that in high pressure boilers often chromates are used as inhibitors. These have to be removed from the blow down before discharge otherwise fresh water may get poisoned. The pH is reduced to < 5 and then deionized and finally the pH is increased to 6 and then discharged.

Silicates, amines, polyphosphates, tannins and lignins have been used in recirculated water systems for protection against corrosion.

1.7 Scales and Sludges

From the standpoint of the chemical constituents present in water, a boiler may be considered to be a huge concentrator of impurities. Water containing impurities is fed into the boiler and pure water, in the form of steam, is removed, leaving behind the impurities. In a high pressure boiler, steam equivalent to the total water capacity may be produced within 15 minutes. Boiler tubes completely blocked by deposits and fragments of scales are not very uncommon. Even the pipes feeding water in a boiler are found to have sufficient deposits.

Some of the deposits stick to the metal surface and are known as ***scales***. If they are in the form of soft and muddy deposits, which can be flushed out easily, or in the form of suspensions, they are known as ***sludges***.

Scale formation takes place by two distinct mechanisms. A salt like calcium bicarbonate gives off carbon dioxide and insoluble calcium carbonate is precipitated on the boiler surface giving the scale. This is the principal factor in fire tube boilers. Another important cause is the lower solubility of certain salts, notably calcium sulphate, at elevated temperatures. The data available on the solubility of various salts does not conform to the high temperature and pressure conditions of the boiler. Another complicating factor is that a number of salts are present in water which affects the solubilities of each other. Therefore, it is difficult to calculate the quantity of a salt that will crystallize out in a boiler.

Water near the hot metal surface is at a higher temperature than the main body of water. A salt like calcium sulphate crystallizes out on the metal surface because the solution there becomes super-saturated. Also, as a film of water on the metal surface is converted into steam, it leaves behind a deposit of the salts in spite of the fact that the rest of the water in the boiler is not saturated with those salts. Gradually, more and more salts are deposited giving a layer on the surface which is known as scale. The scales so formed are mostly of calcium sulphate which stick tenaciously to the metal surface and interfere seriously with the flow of heat.

The behaviour of the silicates is not well understood. Even when there are only a few milligrams of silica per litre of water it is likely to deposit silicate of calcium or magnesium or other silicates. The silicate deposits are particularly troublesome as their thermal conductivity is very poor and they stick very firmly to the metal surface making the removal of the scales very difficult.

Broadly speaking, the scales may be considered to be of three types namely, carbonate, sulphate and silicate types, whose formation has been described above. Generally, the scales found in the boilers are of mixed composition and are composed of several salts. Colloidal and suspended matter may get deposited by coagulation or may be trapped in the precipitates of other salts.

1.7.1 Disadvantages of scale formation

Scale formation is objectionable for the following reasons:

(a) ***Waste of fuel.*** The scales are bad conductors of heat, therefore, their effect is like that of an insulating material spread over the metal surface. This results in reduction in the rate of heat transfer with the result that a larger quantity of fuel has to be used. It has been reported that 0.25 cm thick scale would increase fuel consumption by about 2 to 3%.

(b) ***Overheating of the boiler tubes.*** A more serious effect than the waste of fuel due to overheating of the tubes is explained with the help of Fig. 1.3. The figure on the left shows a section of the boiler tube metal which is completely free from scales. In this case, in order to heat the boiler water to T_1^0 the metal surface towards the source of heat is heated to T_2^0 because of the temperature drop across the metal. The figure to the right represents a similar situation except that there has been scale formation. As the scale is an insulator, there is a greater drop in temperature. Therefore, for heating water to T_1^0 the temperature on the inner surface of the metal is T_3^0 and on the outer surface the temperature is T_4^0 which is considerably higher than the scale-free metal surface temperature of T_2^0.

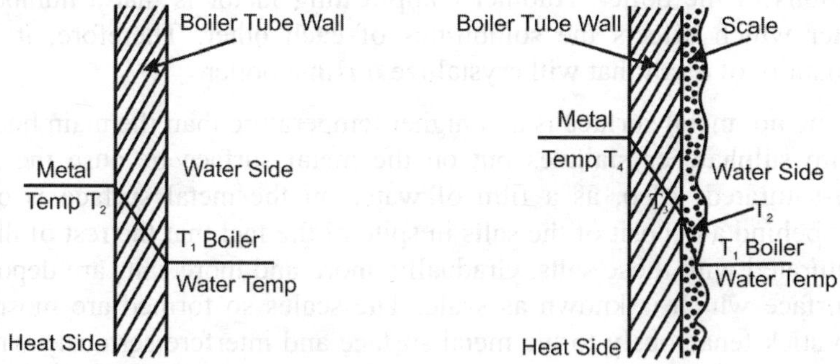

Fig. 1.3 Insulating effect of scales.

At higher temperatures the metal becomes weak and can withstand less pressure. Overheating also causes undue expansion of tubes leading to *'bagging'*. This becomes dangerous in modern high pressure boilers.

(b) ***Corrosion.*** At the higher temperature so maintained there is greater corrosion of the boiler tubes.

Thus it is seen that the disadvantages of scale formation are primarily due to the fact that scales are bad conductors of heat.

Scale formation can be prevented by the following means :

(i) ***External treatment.*** This treatment is given outside the boiler before the feed water enters it. Attempt is made to remove or reduce the amounts of that substance, mainly calcium and magnesium salts and silica, which form deposits. This requires an investment on external treatment equipment. In several cases this can be avoided, for example where the water requirement is not very exacting, the plant is small, the water available is of good quality or a combination of these reasons.

(ii) ***Internal treatment.*** In spite of the external treatment, some salts often remain in the water. Attempt is made, by adding chemicals to water entering the boiler, to convert the remaining salts into more soluble salts or such salts that the deposits will be in the form of nonstick sludge which can be easily removed. The treatment also aims to prevent corrosion not only inside the boiler but also in the pipe line and equipment handling steam in general and condensate in particular, due to the dissolved gases liberated in the boiler. In the cooling water system it prevents biological growths.

(iii) ***Blow down.*** In spite of giving external and internal treatments the concentration of impurities, especially the soluble impurities, goes on increasing as steam generation progresses and the ultimate practical solution available is to make a blow down, i.e., remove a portion of the concentrated water which contains a considerable amount of dissolved matter. Along with it some sludge is also removed. Fresh water is then fed in its place which contains a smaller amount of impurities.

The percentage of blow down is an indication of the efficiency of the water treatment. It should be less than 10%.

1.8 Caustic Embrittlement

It has been found that sometimes the boiler failure takes place because the metal plate has become brittle and inter crystalline cracks have been formed in the metal. This happens in the areas which are under stress.

The alkali metal carbonates and the bicarbonates present in the feed water give rise to caustic alkalinity at the high temperatures and pressures prevailing in the boiler. The concentration becomes greater near the rivets because of bad circulation of water and also because of leaks through which steam escapes leaving behind water with a higher concentration of caustic alkalinity.

It has been observed that those boilers which were fed with water containing sodium sulphate were free from such troubles and therefore the usual practice has been to keep the ratio of sodium sulphate to sodium hydroxide as 1:1, 2:1 and 3:1 for boilers working at pressures up to 10, 20 and above 20 kg/sq. cm respectively. Some later authorities do not seem to favour this scheme and recommend the use of tannins, lignins, sodium nitrate or neutralization of the excess alkali or control of the pH value. The use of phosphates (discussed later in this Chapter) is now being adopted in the modern practice.

Turbine Deposits

These pose one of the most serious problems in the working of the large turbines. It would be well to remember that the turbine deposits may be traced to the following factors and remedial measures be taken:

1. The condensate should be, and is taken as, of high quality but periodically it is not of the required quality.
2. The chemical analysis throughout the steam/water circuit especially the steam portion of the system is insufficient or not of the required high standard.
3. The chemicals are either not fed properly or are not mixed properly.
4. The demineralization plant is not operated properly.

It hardly needs to be impressed on the engineers that for minimising turbine deposits these points should always be kept in mind and no slackness, for however short duration it be, should be permitted.

1.9 Water Treatment

The objective of boiler feed water treatment is firstly to avoid trouble in the boiler plant and secondly to obtain steam of sufficient purity for the steam turbine. If the steam is not of sufficient purity, there is fouling of the blades of the steam turbine which reduces the capacity of the turbines.

The actual treatment that needs to be given to boiler feed water depends to a great extent on the working pressure. As the boiler pressure goes up, the specifications for feed water become more rigid. Accordingly, a treatment has to be carefully chosen. The common treatments are described below :

1.9.1 Preliminary treatments

Preliminary treatments may include simple treatments like sedimentation, coagulation and filtration. These will not be needed for clear water, like well water. An operation like filtration may be postponed for a later stage, say after lime-soda process. Details of these treatments are similar to those given under municipal water supply and need to be studied carefully.

In case a plant is getting some clear water, say, from a well, and some turbid water, say from a river, it is advisable not to mix them but to remove the turbidity of the river water first. Otherwise turbidity will get diluted and will be more difficult to remove. Similar principle applies for other kinds of impurities.

1.9.2 Pre-heating

Pre-heating is heating feed water before it enters the boiler. Very often blow down water and low pressure steam is available at the plant at negligible cost and can be used for heating the feed water. Hot flue gas leaving the boiler furnace is also used for pre-heating. Water may be heated in a heat exchanger. The advantages of pre-heating are:

(a) Instead of feeding the boiler with water at that atmospheric temperature it is fed with hot water, thus saving fuel.

(b) A portion of temporary hardness may be removed. The bicarbonates are decomposed precipitating the insoluble carbonates thus :

$$Ca\ (HCO_3)_2 \rightarrow CaCO_3 + CO_2 \downarrow + H_2O$$

It may be noted that for complete removal of temporary hardness water has to be boiled for 10 to 15 minutes. Usually this does not happen in pre-heaters.

(c) Dissolved oxygen and carbon dioxide are removed. Initially dissolved gases are removed from water on heating to about 65° and are almost completely removed on boiling. As these gases have a corrosive effect their removal is advantageous.

(d) Use of hot water while giving chemical treatments, such as lime-soda treatment, greatly helps the operation. The effect of temperature is discussed along with the lime-soda process.

1.9.3 Lime-Soda Process

This process is used for feed water treatment for low and also medium pressure boilers. It is based on converting the soluble impurities into insoluble salts, which get precipitated. The precipitate is allowed to settle and is then filtered off. A calculated amount of milk of lime, generally with about 10% excess, is added to the sample. Its effects are :

(i) temporary hardness is removed by the following reactions:

$$Ca(HCO_3)_2 + Ca(OH)_2 \rightarrow 2CaCO_3 \downarrow + 2H_2O$$

$$Mg(HCO_3)_2 + Ca(OH)_2 \rightarrow CaCO_3 \downarrow + Mg\ CO_3 \downarrow + 2H_2O$$

$$Mg(HCO_3)_2 + 2Ca(OH)_2 \rightarrow 2CaCO_3 \downarrow + Mg(OH)_2 \downarrow + 2H_2O$$

$$2Na(HCO_3)_2 + Ca(OH)_2 \rightarrow CaCO_3 \downarrow + Na_2CO_3 \downarrow + 2H_2O$$

$CaCO_3$ being comparatively insoluble is readily precipitated. $MgCO_3$ formed in a similar 1:1 reaction is not so insoluble and is not precipitated. Therefore for the removal of Mg $(HCO_3)_2$ twice the amount of lime is needed that produces Mg $(OH)_2$, an insoluble magnesium salt. Excess $Ca(OH)_2$ could be removed by addition of $NaHCO_3$.

The removal of temporary hardness by the use of lime has an important advantage, namely, that of reduction in total dissolved solids. Generally speaking, a large portion of the hardness in natural water is temporary. The removal of temporary hardness by the use of lime does not introduce any soluble salt in water, as is the case with soda treatment or even zeolite process. This is a distinct advantage especially where water with low total solid content is required. As lime is inexpensive, for obtaining high quality water it may be more economical to give lime treatment to water of high carbonate hardness followed by zeolite or demineralizing treatment.

(ii) Lime also removes all permanent magnesium-hardness but is not useful to remove permanent calcium hardness.

$$MgSO_4(\text{or } MgCl_2)+Ca(OH)_2 \rightarrow Mg(OH)_2+CaSO_4(\text{or } CaCl_2)$$

An equivalent amount of permanent calcium hardness is introduced, which has to be removed by soda treatment. Therefore, in spite of almost quantitative precipitation of magnesium, there is no softening.

(iii) Lime also removes free acids, carbon dioxide, hydrogen sulphide and precipitates dissolved iron and aluminium salts as insoluble hydroxides. A large excess of lime should be avoided.

$$2HX + Ca(OH)_2 \rightarrow CaX + H_2O$$

$$CO_2 + Ca(OH)_2 \rightarrow CaCO_3\downarrow + H_2O$$

$$H_2S + Ca(OH)_2 \rightarrow CaS\downarrow + H_2O$$

$$Fe^{2+} + Ca(OH)_2 \rightarrow Fe(OH)_2\downarrow + Ca^{2+}$$

$$2Al^{3+} + 3Ca(OH)_2 \rightarrow 2Al(OH)_3\downarrow + 3Ca^{2+}$$

The lime treatment is followed by soda treatment which removes the remaining calcium permanent hardness (including Ca-hardness introduced during the removal of Mg-permanent hardness)

$$CaSO_4,(\text{or } CaCl_2)+Na_2CO_3 \rightarrow CaCO_3+Na_2SO_4, (\text{or } 2NaCl)$$

$$\therefore \quad \text{Soda} = \frac{106}{100}$$

$$[\text{Permanent}(Ca^{2+} + Mg^{2+} + Fe^{2+/3+} + Al^{3+} + \text{etc}) + \text{free } H^+ - HCO_3^-]$$

The process may be carried on in batches or continuously. In the batch process the chemicals are added in batches in water taken in tanks, agitated for the necessary time and the precipitate allowed to settle and then filtered. Fresh water is again taken into the tank and the operation is repeated. Filtration through sand beds should be avoided as the water is alkaline and will easily dissolve silica. It may be filtered through filter press, coal, coke or even dense iron ore. Silica may also be introduced by the use of lime of inferior quality.

In the continuous process, water together with lime and soda solutions enters a smaller tank continuously at a predetermined rate, bringing the correct amount of materials together. The chemicals and the water to be treated enter a bigger tank or chamber (Fig. 1.4). The second chamber is a tall structure having a larger cross section at the top than at the bottom. As a solution moves upwards the precipitates formed as a result of the chemical reactions settle down, because towards the upper portion of the chamber the velocity is less due to the larger cross section. Clear water is removed continuously from the top and sludge from the bottom. A part of the sludge is recirculated which not only enables the use of excess chemicals left behind in the sludge but also helps in the precipitation by providing nuclei.

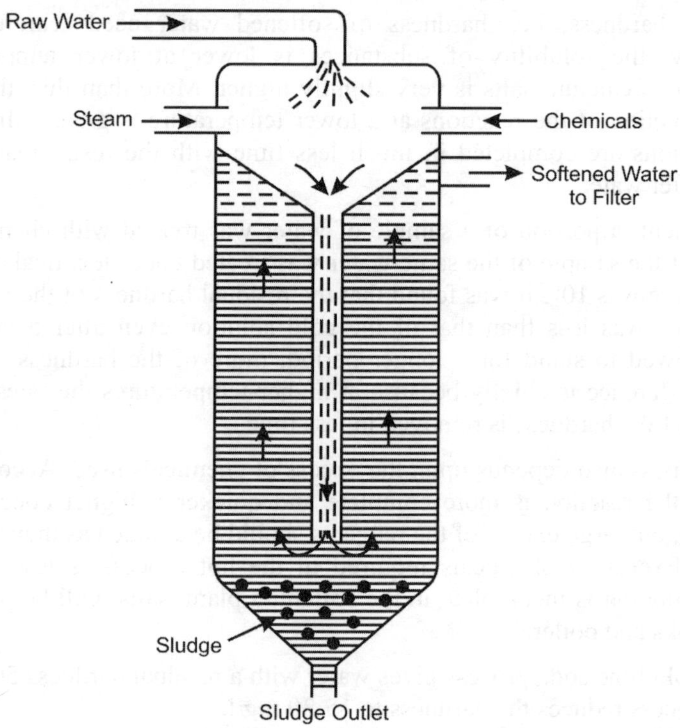

Fig. 1.4 Continuous hot lime-soda process.

When the chemicals are added to water at atmospheric temperature, the process is known *as cold lime soda process*. At this temperature the precipitates are finely divided and do not settle easily, nor can they be easily filtered. It is therefore, necessary to add coagulants like alum, aluminium sulphate, ferrous sulphate, sodium aluminate etc. These get hydrolysed and form flocculent and gelatinous precipitates which entrap the finely divided particles and settle down readily or are easily filtered off. Ferrous sulphate and sodium aluminate are costlier than alum. Ferrous sulphate helps to remove oxygen and sodium aluminate helps the removal of silica and also oil.

When this treatment is given at a higher temperature the process is known as *hot lime soda process*. Water is heated by waste steam or hot gases and a calculated amount of chemicals are added as in the cold process. The distinguishing features are the following :

(i) As the reactions proceed quickly at higher temperature, the softening reactions are completed in about 15 minutes, whereas in the cold process several hours are needed. This greatly increases the capacity of the plant.

(ii) Precipitates and sludges formed settle rapidly at higher temperatures, hence coagulants are not needed.

(iii) Filtration becomes easier because of the bigger size of the particles and the reduced viscosity of solution at the higher temperature; hence the capacity of the filter is also increased.

(iv) Residual hardness, i.e., hardness of softened water needs careful consideration. Generally, the solubility of substances is lower at lower temperatures but the solubility of calcium salts is very slightly higher. More than this, the time taken for the completion of the reactions at a lower temperature is greater. In the hot process the reactions are completed in much less time with the result that the hot process gives softer water.

In an experiment, a portion of a sample of water was treated with chemicals at 95° and another portion of the sample of the same water was treated under identical conditions except that the temperature was 10°. It was found that the residual hardness of the hot solution at the end of 10 minutes was less than that of the cold solution even after 5 hours. If the cold solution was allowed to stand for a longer period, more of the hardness would have been removed. This difference is chiefly because at higher temperatures the rates of reactions are greater, so more of the hardness is removed in less time.

Residual hardness also depends upon the excess of chemicals used. According to the Law of Mass Action the reaction is more complete and quicker if higher concentrations of the reagents are used, but large excess of the reagents should be avoided as they will appear in the softened water. Excess of chemicals required in the hot process is less than in the cold process. If precipitation is incomplete in the softening plant, salts, will be precipitated in the pipes, storage tanks and boiler.

In practice, cold lime soda process gives water with a residual hardness 50-60 mg/l and the hot lime soda process reduces the hardness to 15-30 mg/l.

In some places, water is highly alkaline and contains a considerable amount of sodium bicarbonate. The total alkalinity of this type of water is greater than the total hardness. On adding lime, calcium carbonate precipitates forming sodium carbonate.

$$2NaHCO_3 + Ca(OH)_2 \rightarrow CaCO_3 + Na_2CO_3 + 2H_2O$$

In such a case, soda may not be needed for the removal of permanent hardness. As a matter of fact, cases are known where gypsum has to be added to remove soda so produced.

1.9.4 Permutit or Zeolite Process

Permutits or zeolites are natural or synthetic complex compounds, consisting of hydrated sodium aluminium silicates approximately corresponding to the chemical formula $Na_2O.Al_2O_3.2SiO_2.6H_2O$. The artificial zeolites have greater capacities per unit weight for softening water than natural zeolites but are less durable. Zeolites are insoluble. They introduce Na^+ in place of Ca^{++} and Mg^{++}. As these exchange bases, they are also known as *base exchangers.* If Na_2Z represents permutit or zeolite, the reactions taking place may be represented thus :

$$Ca(HCO_3)_2 + Na_2Z \rightarrow 2NaHCO_3 + CaZ$$

$$MgSO_4, + Na_2Z \rightarrow Na_2SO_4 + MgZ$$

Permutits or zeolites are kept in cylinders and used as filter beds. Hard water is percolated through them at a certain specified rate. The hardness causing substances (or ions) are retained as insoluble compounds according to the reactions given above. Thus the outgoing water contains sodium salts corresponding to the calcium and magnesium salts present in the incoming hard water. After use for some time the zeolite is converted into calcium or magnesium zeolite and then it ceases to soften water. At this stage it has to be regenerated by treatment with a concentrated sodium chloride solution when the original sodium zeolite is recovered :

$$CaZ + 2NaCl \rightarrow Na_2Z + CaCl_2$$

The zeolite bed is then washed with water and is again ready for use. See Fig. 1.5.

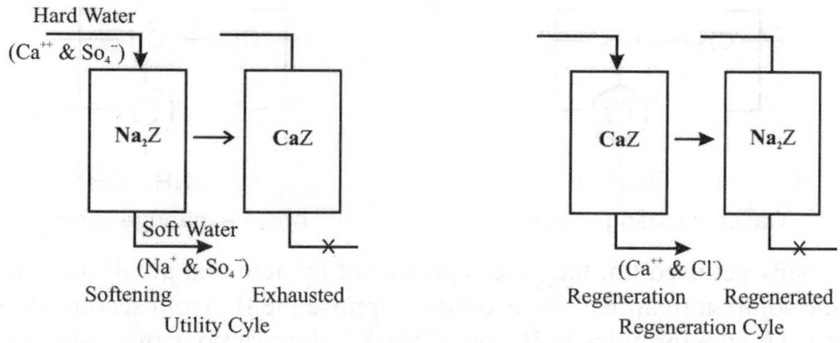

Fig. 1.5 Water softening by zeolite process.

Water having turbidity or excess of acidity or alkalinity or containing dissolved iron or manganese salts should not be fed into a zeolite softener. Turbidity will clog the pores and restrict the flow; excess of acidity or alkalinity will attack the zeolite, there fore the pH value of water to be passed through a zeolite bed should be around 7; iron or manganese salts will produce iron or manganese zeolites from which it cannot be easily regenerated. Therefore, these should be removed before passing water through the zeolite bed. Generally, hot water is not passed through a zeolite bed because zeolite tends to dissolve in hot water.

It is to be further noted that the acid radicals are not removed. The temporary hardness will give water containing $NaHCO_3$ which dissociates inside the boiler giving high alkalinity:

$$NaHCO_3 \xrightarrow{\Delta} NaOH + CO_2$$

Alkalinity causes "caustic embrittlement". Carbon dioxide goes along with steam and makes the condensed water acidic, hence corrosive. Thus waters containing a large amount of temporary hardness should not be directly treated by the zeolite process. Such water may be boiled or treated by lime to remove temporary hardness and then softened by this process.

The *advantages* of the zeolite process are several. It removes the hardness completely; the equipment is a compact unit occupying a small space; it automatically adjusts itself to different hardness and can be worked under pressure, therefore the plant can be installed in the water supply line and double pumping is avoided. Moreover, as the impurities are not precipitated, there is no danger of sludge formation at a later stage.

A combination of lime soda process and zeolite process gives better results. Zeolite process has now been superceded by deiozination process.

1.9.5 Ion-Exchange Process (Deionizers or Demineralizers)

Several organic polymeric–resins, such as polystyrene with functional groups ($-^{+}NH_3OH^-$, $-SO_3^-H^+$) on benzene rings, are now available which behave in a manner similar to the zeolites. This process differs from all other processes in the fact that by it all the ions are removed. The cations are replaced by H^+ and anions by OH^-. Water to be treated is generally passed through two cylinders containing the resins, one of which removes cations and the other anions. See Fig. 1.6.

Polystyrene resins :

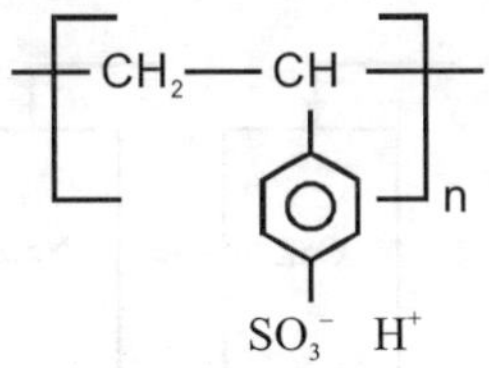

Cation exchange resin

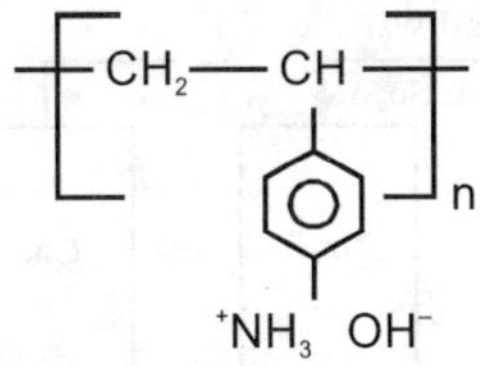

Anion exchange resin

When the resins get used up, they are regenerated by acid and alkali treatments similar to sodium chloride solution treatment for zeolites. Suppose the first and second cylinders contain resins represented by the formulas H_2R' and $(OH)_2 R''$, the reactions may be represented thus :

$$CaSO_4 \leftrightarrow Ca^{++} + SO_4^{--}$$

(Any other hardness causing salt or ionisable substance will behave in a similar manner).

Cylinder I : $H_2R' + Ca^{++} \rightarrow 2H^+ + CaR'$

It is easy to see that the water coming out of the cation exchanger is highly acidic. This has to be kept in mind in selecting the materials handling it. Water from natural sources usually contains sufficiently large quantities of carbonates and bicarbonates. These ions like other anions can be removed by passing through the anion exchanger. CO_3^{2-} and HCO_3^- in the presence of H^+ form H_2CO_3 (or H_2O+CO_2). The load on anion exchanger is reduce by having a degasifier (which removes gases by mechanical means like blowing air or suction, as shown in Fig. 1.6, which will remove CO_2. Thus only mineral acidity will have to be removed by the anion exchanger.

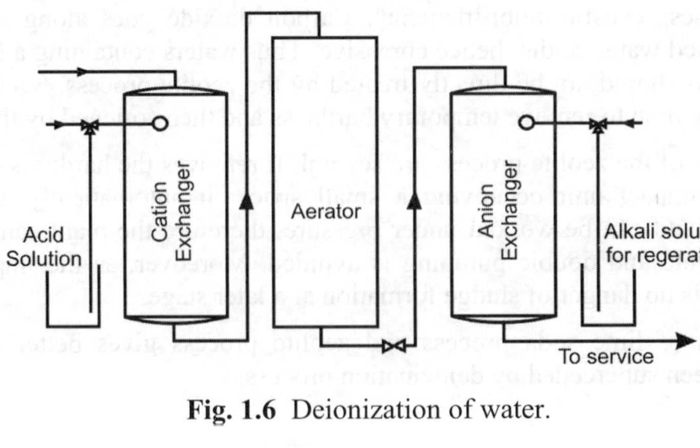

Fig. 1.6 Deionization of water.

Cylinder II : $(OH)_2R'' + SO_4^{--} \rightarrow 2OH^- + SO_4R'$

H^+ and OH^- unite to form H_2O, thus no solid is added to water. When the plant does not soften any more, the first cylinder is regenerated by passing a dilute acid solution through it and the second cylinder by passing a dilute alkali solution. The reactions during regeneration may be represented thus:

$$\text{Cylinder I : } CaR' + 2HCl \rightarrow H_2R' + CaCl_2$$

$$\text{Cylinder II : } SO_4R'' + NaOH \rightarrow (OH)_2R'' + Na_2SO_4$$

Mixed bed deionizers are also now available. In these the sample of water is passed through only one cylinder and all the ions get removed.

This process can also be used to purify highly acidic or alkaline water. As a matter of fact, it removes all the ionizable substances and the water obtained is like distilled water. Many of the commercial plants give water with total dissolved solids as low as 3 mg/l and silica 0.1 mg/l.

The main disadvantage is that the equipment is costly and more expensive chemicals, as compared with the zeolite process, are needed for regeneration. Another important factor to be considered is the quality of water that is being softened. If it contains turbidity or a large amount of dissolved salts then the output is reduced. Turbidity must be below 10 ppm and has to be removed first by coagulation and filtration. If the water contains sufficient temporary hardness it is advisable to treat it first by lime, which removes hardness without adding dissolved salts, and then remove the remaining hardness by this process.

This process is very good for water for very high pressure boilers. For low pressure boilers, working say below 50 kg/cm^2, the equipment is rather costly and water may be treated more economically by other processes.

1.10 Water Conditioning

The term 'conditioning' is a general term. It refers to a treatment given to correct undesirable tendencies or to complete softening process or to remove a difficulty created by the main process. The following are a few of such treatments.

(i) Phosphate Treatment : In the lime-soda process calcium is precipitated as $CaCO_3$ which is quite insoluble. However, a small amount remains in solution. This can be further removed by the addition of sodium phosphate (Na_3PO_4), which precipitates the less soluble $Ca_3(PO_4)_2$; thus the hardness is further reduced. For best results, the pH value of the water should not be less than 9.5. For this purpose caustic soda is added.

1. $MgCl_2 + 2NaOH \rightarrow Mg(OH)_2 + 2NaCl$
2. $Ca(HCO_3)_2 + 2NaOH \rightarrow CaCO_3 + Na_2CO_3, + 2H_2O$
3. $3CaCO_3 + 2Na_3PO_4 \rightarrow Ca_3(PO_4)_2 + 3Na_2CO_3$

On a small scale, water almost free from hardness can be produced by adding a calculated amount of sodium hydroxide and sodium phosphate. These precipitate magnesium as hydroxide and calcium as phosphate. These reagents are costlier than lime and soda, hence for larger requirements the lime-soda treatment is preferred. To get the maximum benefit, the phosphates are added to hot water. In boilers operating at pressures up to 50 kg/cm^2 pressure about 40 mg/1 of the phosphates in excess of the theoretical amount are used to complete the reactions. Excess phosphate may be 5 mg/1 or less when the pressure is 70 kg/cm^2 or higher.

Phosphate treatment also helps to control the pH value (i.e., the degree of acidity or alkalinity) of water. Trisodium phosphate, Na_3PO_4, gives alkaline reaction and can be used to increase the pH value. Sodium dihydrogen phosphate, NaH_2PO_4, is acidic and can be used to reduce the pH value.

This treatment is often given after lime-soda treatment. It has the disadvantage of causing "after precipitation" that is, gradual precipitation in the pipes, equipment, etc., (Also see internal treatment, described later in this Chapter).

(ii) Sodium Aluminate Treatment : Sodium aluminate precipitates calcium and magnesium as insoluble aluminates. Sodium aluminate hydrolyses also, giving NaOH and $Al(OH)_3$. NaOH so produced precipitates some of the magnesium as $Mg(OH)_2$. Thus on the addition of $NaAlO_2$ flocculent precipitates of $Al(OH)_3$ and $Mg(OH)_2$ are produced which enmesh finely suspended and colloidal impurities, including silica and oil droplets, which are ordinarily difficult to remove. If sodium aluminate treatment is given after lime-soda treatment no other coagulant will be necessary.

Sodium aluminate is cheaply available from bauxite refineries in the form of a thick solution and may be used as such for water softening.

(iii) Treatment for Microbiological Growth: The growth of greenish matter may be commonly observed in stagnant waters. Similar growth may take place in heat exchangers, condensers and cooling systems. They restrict the flow of water and also retard the heat transfer. Sometimes they may come out suddenly in large masses and clog the pipe or the pump. Small plant growth, which develops rapidly, is known as algae. Gelatinous matter, which sticks to surfaces and is due to micro-organism is called slime.

Micro organisms present in air and water can accumulate on heat-exchanger surface and isolate the surface from the inhibitor used to protect the surface. Bacteria produce gas which can attack stainless steel and cause pitting. A good biological control programme, which is often neglected, is needed.

The simplest way to remove them is to clean mechanically. Chlorine is very effective for the removal of both types of growth. The residual chlorine should be about 0.5 mg/l. Excess chlorine is removed by adding SO_2 or Na_2SO_3. Copper sulphate is more effective for algae than for slime. It cannot be used if pH value is above 8.5 as copper hydroxide will be precipitated. Chromates used for corrosion control help in the control of algae and slime.

Chloramine is also effective in checking these growths. Bleaching powder mixed with ammonia solution is fed in when the ammonia and chlorine combine to form chloramine and destroy the microbiological growth.

(iv) Iron and Manganese Removal : These are generally present in small amounts in water in the absence of dissolved oxygen. Iron and manganese give red and black precipitates, respectively, when the water containing these as dissolved impurities come in contact with air. They are highly undesirable in water for dyeing and tanning industries, and for laundries. They also give a metallic taste to water and are conveniently removed in the form of insoluble $Fe(OH)_3$ and $Mn(OH)_2$ by oxidation followed by coagulation and filtration. The oxidation may be done either by aeration i.e., bringing in contact with air (e.g., by spraying the water in the air or blowing air) or by chlorination.

(v) Silica Removal : About 5 to 8 mg/l of silica is present in natural waters. In some cases the amount is much more. Inside the boiler its concentration increases. Its deposits form a silicate scale which is very hard and very bad conductor and difficult to remove. One important source of silica is the sand filter. If water treated by lime-soda process is filtered though a sand filter a large amount of silica dissolves in water because of its alkalinity. Calcium and magnesium silicates act as cements and bind together other particles in to a hard scale.

In high pressure boilers the amount of silica has to be kept very low. At a pressure of about 40 kg/cm^2, 5 ppm of silica dissolves in steam and goes along with it. Its amount in steam increases to 40 ppm at 100 kg/cm^2 and is still higher as the pressure increases. Several theories of silica carry-over have been proposed. It is believed that in most cases droplets of water are carried along with steam. Water dries up in the super heaters and silica is carried forward by dry steam. It deposits on and distorts the shape of turbine nozzles and blades. Sometimes the surface becomes uneven and resistance to the flow of steam increases. In some turbines loss in efficiency of about ten per cent in five days has been reported. Catastrophic failure of the rotating parts has also been reported. Soluble silica is removed by the use of magnesium compounds during the hot lime-soda process, the sludge being recirculated. It is the sludge which absorbs silica. Magnesium oxide has been found to be very efficacious. It does not increase the dissolved solid content. Removal of silica is not a stoichiometric reaction (i.e., there are no definite combining weights). The removal proceeds by absorption from the solution.

For high pressure boilers silica is removed by strongly basic anion exchangers.

(vi) Oil Removal : Generally, natural waters are free from oil but small amounts are likely to come from the oils used for lubrication of the pumps. A large amount of water is obtained by condensing the exhaust steam, which may get mixed with the lubricating oil in the steam cylinder. In many cases the steam generated is used for driving the engines and exhaust steam is always available. The steam on being condensed gives water which is as good as distilled water except that it contains droplets of oil brought from the engine. In efficient plants 90 to 95% of the water requirement of a boiler is met from this source.

For the removal of oil the exhaust steam is passed through mechanical oil separators which consist of a chamber with a number of metal plates. As the steam with droplets of oil passes over the plates, droplets of oil are retained on the metal surface. Thus most of the oil is removed. The steam is then condensed for boiler use. If very high quality water is needed, sodium aluminate is added to the condensate which removes the remaining oil, as explained under sodium aluminate treatment.

According to Stromeyer, a 3 mm thick scale will raise the temperature of a plate by 150° whereas less than 0.5 mm of grease will produce a worse effect. If the oil is a vegetable or animal oil, it may get hydrolysed producing fatty acids which may have corrosive action. There is a greater possibility of the fatty acids combining with the alkalinity, producing soaps which will form sticky coatings on the metal surface. These will attract other impurities and form thicker deposits.

(vii) **Oxygen Removal :** Oxygen removal assumes greater and greater importance as the boiler pressure increases. For methods of removal see the section on "Corrosion" in this chapter.

(viii) **Boiler Compounds :** In order to check the bad effects of raw feed water, a number of substances have been used in the past, mostly without any scientific basis. A number of fraudulent compounds have been well advertised and widely used. Some of these have been found to be harmful. Therefore, one has to be cautious while using them.

The practice of using chemicals for feed water is now established on a scientific basis. Some reputed companies marketing the boiler compounds also give their compositions. If these are used by well-informed operators after taking into account the analysis of water, the water treatment becomes simpler. But such compounds are more expensive than the ingredients bought separately.

The inorganic constituents of the boiler compounds precipitate the scale forming salts and also help to control corrosion. The organic constituents are characterised by high molecular weights. These form large particles which are adsorbed on the metal or crystal surface and thus prevent formation of hard scales. The materials deposited get removed in blow down. A boiler compound used by the U. S. Navy is reported to consist of 47% anhydrous disodium phosphate, 44% soda ash and 9% starch.

It is desirable to study the analysis of water and make suitable formulations of the chemicals to be added. If such facilities are not available, boiler compounds of repute may be used.

1.11 Internal Treatment of Boiler Water

Internal treatment consists of adding chemicals to the water inside the boiler. It is mainly used as a corrective treatment, viz., to prevent scale formation, also to inhibit corrosion, steam contamination and caustic embrittlement.

After the external treatment the feed water contains only a small amount of calcium and magnesium salts. As steam generation proceeds, the concentrations of these salts increases and scales begin to form. The formation of the scales is prevented by injecting sodium phosphate directly in the boiler. This causes the residual hardness to be precipitated as calcium phosphate and magnesium hydroxide, which do not stick to the metal surface and are removed by blow down.

Phosphate – calgon: An ideal condition would be to convert the salts into such highly soluble ones that they are not precipitated. It is achieved to some extent by the use of sodium hexameta phosphate $(NaPO_3)_6$, which is sold under the name 'calgon'. It forms complex ions with calcium and prevent its precipitation. As it is difficult to make all the salts indefinitely soluble, it can be said that the objective of internal treatment is to precipitate them in such a physical state that scales are not formed. In such cases the physical state of the precipitate is more important than the chemical nature. Consequently the temperature, concentration pH value, reagents, etc., affect the success to a great extent. Calcium salts, if precipitated as calcium silicate, will give hard scales as they adhere to the metal surface, whereas the physical state of calcium phosphate is such that no scales are formed.

There are some cases in which the external treatment can be completely avoided and only internal treatment given. This is recommended when (i) the size of the plant is so small that installation of external treatment facilities prove expensive, (ii) the boiler is working at low or medium pressure, (iii) raw water of good quality is available and (iv) only a small amount of make up water is needed (i.e., most of the water requirement is met from water obtained by condensing exhaust steam and only a small amount of fresh water is required).

At present phosphates are widely used for internal treatment (for details see phosphate treatment). The phosphates get precipitated in the form of sludge; thus drying of the calcium salts on the metal surface and scale formation is prevented. After precipitation it is necessary to keep the precipitate as a fluid sludge which can be removed by blow-down. In low pressure boilers organic compounds like starch, lignin, tannins, etc., are added for preventing the inorganic precipitates from sticking to the metal surface. These prevent the coming together of the precipitated particles, and hence avoid their forming scales.

The above approach, of adding Na_3PO_4, which introduces caustic and phosphate in the ratio of 1:3, is known as "coordinate pH-Phosphate" method. The pH and phosphate concentrations are kept along the curve A of Fig. 1.5. Above this curve there will be free NaOH which can cause caustic embrittlement. As some PO_4^{3-} get precipitated due to Ca^{2+} and Mg^{2+} escaping prior to removal, some caustic becomes "free". To avoid this danger *"Congruent phosphate-pH control"* has been now recommended for the high pressure boilers. The ratio of caustic-to-phosphate is kept between 2.6 : 1 and 2.85 : 1. Curve B of Fig. 1.7 shows caustic to phosphate ratio of 2.6 : 1. This range is maintained by adding sodium phosphates, caustic soda, and phosphoric acid, as needed.

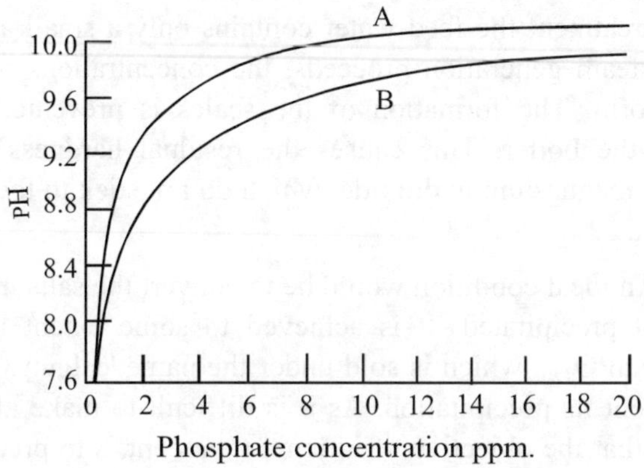

Fig. 1.7 Congruent phosphate – pH control. Exact ratio between PO_4^{3-} and OH^- is to be decided by the operating factors. For high pressure boilers, values on curves B are preferred. Above curve A there is the danger of free NaOH.

Sodium sulphite or preferably hydrazine is added to boiler water so as to remove dissolved oxygen which causes corrosion.

Chelating Agents like sodium salts of EDTA (ethylene-diamine tetra acetic acid) and NTA (nitrilo triacetic acid) are also being used, especially for treating iron. Excess of these should be avoided as they attack the metal. These produce heat-stable molecular complexes which remain in solution.

1.11.1 Blow Down

In spite of giving the external and internal treatments concentration of dissolved solids in the boiler drum and tubes increases. The low pressure boilers can have 2,500 to 5,000 ppm of solids but for the high pressure boilers it has to be 500 ppm or less so as to produce steam of acceptable purity. It therefore becomes necessary to remove a portion of water from the boiler and replace it by fresh feed water. This is a wasteful process as far as the heat aspect is concerned but is the last resort taken for the safety of the boiler. Blow down is made every few hours or made continuously.

While removing impurities that have concentrated in the boiler during steam generation blow down discharges heat that could otherwise be useful. In intermittent blow down the heat cannot be recovered. In large boilers blow down invariably is continuous. Heat recovery in such a case depends on the discharge rate and also the impurities present.

Simple calculations can be very helpful in judging the performance of the boiler and thereby showing in what direction to make efforts to improve the performance. Suppose, in a boiler the maximum total solids allowed are 1000 mg/l and it produces 50,000 kg/hour (S) of steam. The feed water has 10 mg/l of solids. To keep the concentration of solids at a constant level in the boiler, the amount removed with blow down (B) must equal the amount entering with feed water (F).

Therefore :

$$10 \text{ mg/l} \times F = 1000 \text{ mg/l} \times B$$

$$B = 10/1000 \times F = F/100$$

or $B = 1\% \times F$ i.e., 1% of the feed water is removed by making a blow down.

In case silica, suspended solids, etc. determine the blow down, calculations can be made in a similar manner.

From the above it is easy to calculate the amount of water fed in the boiler per hour.

$$F = S + B = 50{,}000 + 0.01 \times F$$

$$0.99\ F = 50{,}000$$

$$F = 50{,}000/0.99 = 50{,}505 \text{ kg/hour}.$$

These can be counterchecked from the material balance thus

$$B = F - S - 50{,}505 - 50{,}000 = 505 \text{ kg/hour}$$

As the blow down is 1% of the feed water, we have $50{,}505 \times 0.01 = 505$ kg/hour. If in actual working discrepancy is noticed, a thorough check up is required. There may be leakages of steam, water and more commonly and dangerous–leakage of cooling water in the heat exchanger, which results in the contamination of feed water.

The blow down water from a high pressure boiler is discharged into a flash tank where low pressure steam is obtained and hot water passes through heat exchanger to heat the feed water. Thus maximum heat economy is obtained (These calculations are also simple but are beyond the scope of this book).

1.12 Treating Brackish and Saline Water

The treatments discussed so far are applied to water with relatively low dissolved salt content. Due to vast regions having poor water supplies attention has been turning to water with a large amount of salt content, say, from 1,000 to 35,000 ppm. Some deep well waters, especially in the deserts, have several thousand ppm of total dissolved solids and usually water has 10,000 ppm or more of dissolved solids. Sea water has around 35,000 ppm dissolved solids. Treatment of such water directly by demineralization is very expensive.

1.12.1 Distillation

Distillation has been the common technique for getting good water from the sea. Several kinds of distillation equipment are in use. Vacuum freezing has also been adopted. Ice crystals coated with salt water are separated, washed with cold water and melted to give good water. Both these processes require a large amount of energy and the escalating cost of energy are turning attention in other directions. It may, however, be mentioned that distillation is still economically viable in the plants where a large amount of waste heat is available.

1.12.2 Membrane Processes (Electro Dialysis, Reverse Osmosis)

These processes have now been developed to such an extent that some large plants are using them for treating high dissolved salt containing water prior to the common demineralization. Two of the methods are electro-dialysis and reverse osmosis techniques.

In the electro dialysis or E. D. technique the dissolved mineral is driven away from the water. The basic principle is that the water is passed between paired sets of parallel membranes. An electric field is applied perpendicular to the direction of flow of water. The ions move towards the electrodes of the opposite charge. Selectively permeable membranes are chosen. The membrane near the negative electrode allows the cations to pass and the one near the positive electrode allows the anions to pass (Fig. 1.8). A large number of pairs of membranes are assembled. Plastic spacers not only separate the membranes but also direct the flow of water through a winding path across the exposed membrane surfaces. Water is sent at a pressure of about 6 kg/cm^2. This technique is favoured for water with very high dissolved salt contents, even for sea water.

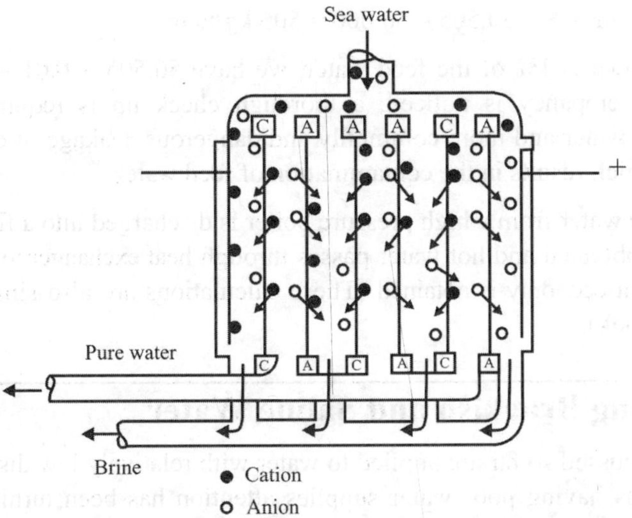

Fig. 1.8 Electro dialysis cell. In brackish water cations are removed through cation permeable membranes (C) and anions are removed through anion permeable membranes (A).

Reverse Osmosis

Diffusion is the movement of molecules from a region of higher concentration to a region of lower concentration. Osmosis is a special case of diffusion in which the molecules are water and the concentration gradient occurs across a semipermeable membrane. The semipermeable membrane allows the passage of water, but not ions (e.g., Na^+, Ca^{2+}, Cl^-) or larger molecules (e.g., glucose, urea, bacteria). Diffusion and osmosis are thermodynamically favorable and will continue until equilibrium is reached. Osmosis can be slowed, stopped, or even reversed if sufficient pressure is applied to the membrane from the 'concentrated' side of the membrane. Since reverse osmosis is thermodynamically unfavourable process, it requires a driving force to push the fluid through the membrane, and the most common force is pressure from a pump. The higher the pressure, the larger the driving force. As the concentration of the fluid being rejected increases, the driving force required to continue concentrating the fluid increases.

The reverse osmosis process is used in a wide range of applications. This includes waste water treatment, production of pure water for industries [e.g., wafer fabrication], and removal of salts from salty or sea water [desalination].

How does it work?

Let us start with the normal osmosis process. Consider a semi-porous membrane that separates equal volumes of liquid into two compartments. In one compartment is water, and the other dilute salt solution [see Fig. 1.9(a)]. Due to the effect of osmosis, water flows from side containing water to the side with the salt solution. Water flows over because it is trying to dilute the salt solution. This continues until an equilibrium state is reached.

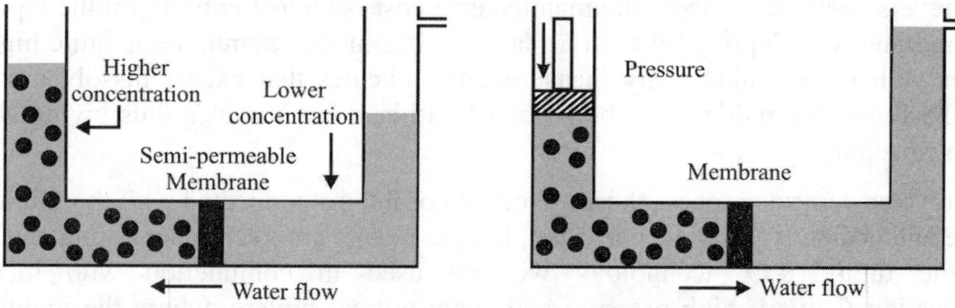

(a) Osmosis (b) Reverse Osmosis

Fig. 1.9 Reverse Osmosis Cell.

For reverse osmosis, the aim is not to dilute a salt solution, but to purify water by removing salts and other contaminants. This is done by applying a pressure to the side containing the salt solution. It causes water to flow to the other side of the membrane [see Fig. 1.7(b)]. This purifies the water, since the membrane prevents salts and other particles from passing through.

Semipermeable membrane : Many synthetic membranes that can withstand high pressures have been developed. A typical reverse osmosis membrane has 2 layers :

- a dense top layer of less than 1μm think
- a porous sublayer of 50-150μm thick, for supporting the top layer

Reverse osmosis is also known as hyperfiltration and is the finest filtration known. Reverse osmosis is not only used to purify water and remove salts but to remove other impurities in order to improve the color, taste or properties of the fluid. Reverse osmosis is capable of rejecting bacteria, salts, sugars, proteins, particles, dyes, and other constituents that have a molecular weight of greater than 150-250 daltons. The larger the charge and the larger the particle, the more likely it will be rejected.

Further this process can be used to purify fluids such as ethanol and glycol, which will pass through the reverse osmosis membrane, while rejecting other ions and contaminants present in them.

Reverse osmosis units remove ionizable as well as non-ionizable matter, both colloidal and high molecular weight organic matter. This is a distinct advantage and sounds very attractive in dealing with the problem of colloidal silica, which is not removed by demineralization, for water for supercritical boilers. Electro dialysis units require more skilled labour who can handle the electrical and mechanical auxiliary units. Their initial cost for low dissolved solid range is slightly lower than the R.O. units. Cellulose acetate affixed to either side of perforated tube is a popular membrane. Plants giving several million litres of water are available. At the present R.O. units are being preferred. Suppliers claim efficiencies from 85 to 99%.

The E.D. and R.O. units have attained prominence during the last few years and are particularly useful when feed water has a large amount of dissolved solids. When used before demineralization the chemical requirement for demineralization is reduced and the life of the resin is increased.

In the reverse osmosis process the maintenance cost is almost entirely on the replacement of the membrane. It is reported that in a plant the life of the membrane is quite high, about three years. When the conductivity measurement indicates that excess dissolved solids are going in the water, the module can be replaced within a few minutes thus giving an almost uninterrupted supply.

In the reverse osmosis process there are certain problems connected with the fouling of the membrane and design is yet to be perfected, but the results are very encouraging and it seems that in the future R.O. technology will be used in conjunction with mixed bed demineralization for very high pressure and super-critical boilers, where the impurities are measured in parts per billion instead of parts per million.

Just to give a rough idea it may be mentioned that in one case the dissolved salt decreased in the feed water from about 4500 ppm to about 350 ppm, conductivity from about 6000 μmho to about 500 μmho and silica from 7 ppm to 2 ppm after passing through a spirally wound polyamide reverse osmosis unit. The water now becomes suitable for treatment by demineralization.

An interesting case is that of a two-unit 800 MW power station at Cadafe, Venezuela, S. America. The plant was nearing completion in 1973 but public water supply was not ready. Alternative sources considered were well and sea water. Treatment methods for water from the sea–which provides uninterrupted water supply–considered were waste–heat desalination, distillation and reverse osmosis technique. The deciding factors were capital cost, operating cost, reliability, simplicity of operation, ease with which capacity could be expanded in the future and delivery time. In that area well water was not considered a reliable source for the large quantities of water needed.

In 1978 the capital investment required for a 12 million litre per day plant was estimated to be 1.5 crore rupees for reverse osmosis technique and 3.8 crore for multistage distillation; operating costs were almost similar, the distillation being slightly costlier. In spite of higher investment costs the reverse osmosis was preferred because the energy costs are rapidly increasing and it was felt that distillation cost would increase considerably.

In the Cadafe plant the sea water is treated with chlorine and ferrous sulphate, to destroy biological growth and to coagulate, filtered twice to remove fine particles, excess chlorine removed, pH adjusted and then pumped at 60 kg/cm^2 pressure through the reverse osmosis unit which has a number of modules of the semi-permeable membranes. The primary unit reduces the total dissolved solids from 38,000 ppm of the sea water to 1,000 ppm which is treated in the secondary unit that gives water with 100 ppm of the total dissolved solids. Necessary quantity of water is taken for supply as potable water and the remaining passed through demineralizers giving high quality water for the high pressure boilers.

1.12.3 Use of Condensate

After performing the intended mechanical work the steam is condensed to give water of high purity. In the process, water picks up corrosion and erosion products which tend to deposit in boilers and super heaters and on turbine blades. Deposition of corrosion products promotes further corrosion. Leakages of condenser tubes are not uncommon. This is an important source from which considerable amount of dissolved salts are introduced. It is, therefore, absolutely necessary that chemical analysis of the condensate is made frequently and accurately (Fig. 1.10). In very high pressure and supercritical boilers it is a routine practice to filter the condensate and pass through demineralizer.

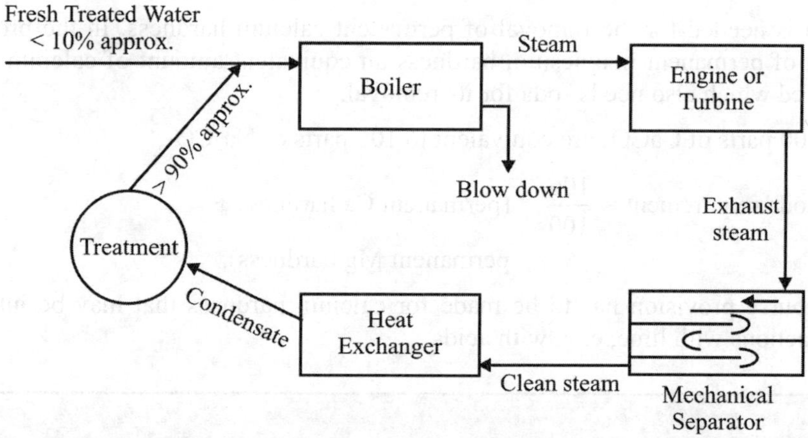

Fig. 1.10 Use of condensate.

1.13 Calculation of Water Treatment Problems

I. *Calculation of Hardness*. It has already been explained, with examples, under the Degree of Hardness that the hardness is expressed in terms of $CaCO_3$. The weights of the substances are converted in terms of weight $CaCO_3$ and then the results are expressed in units of milligrams per litre of water (or parts per million or degree Clark, etc.).

II. *Lime-Soda Requirement*. In Water Chemistry, by lime is meant $Ca(OH)_2$ and not CaO, and by soda is meant Na_2CO_3. 'The weights of the substances should be converted into weight of $CaCO_3$.

The following should be noted for lime requirement:

(i) Lime is required for $Ca(HCO_3)_2$

(ii) Lime requirement for temporary magnesium hardness is *double* that required for temporary calcium hardness.

(iii) Lime removes permanent magnesium hardness, *but it* introduces an equivalent amount of permanent calcium hardness.

As 100 parts of $CaCO_3$ are equivalent to 74 parts of $Ca(OH)_2$, the lime needed for softening = $\frac{74}{100}$ × (temporary Ca hardness + 2 × temporary Mg hardness + permanent Mg hardness).

(*Note*: the hardness is in terms of $CaCO_3$).

In case the water analysis shows the presence of acidic substances, e.g. H_2SO_4, etc., lime has to be calculated for these also. It should be noted that the result of this would be the introduction of the corresponding calcium salts, e.g. $CaSO_4$, which will cause permanent hardness and soda requirement for this will also have to be calculated.

Soda is needed for the removal of permanent calcium hardness. In the process of the removal of permanent magnesium hardness an equivalent amount of calcium hardness is introduced which also needs soda for its removal.

As 100 parts of $CaCO_3$ are equivalent to 106 parts of Na_2CO_3,

the soda requirement = $\frac{106}{100}$ × (permanent Ca hardness + permanent Mg hardness).

Of course provision has to be made for calcium hardness that may be introduced by other reactions with lime, e. g. with acids.

Example 1.

Calculate the amount of lime-soda needed for softening water containing the following per litre:

$Ca(HCO_3)_2$ ... 162 mg; $Mg(HCO_3)_2$ ... 73 mg;

$MgCl_2$ ... 95 mg; $CaSO_4$ ... 136 mg;

NaCl ... 585 mg;

What is the temporary and total hardness of the sample?

Solution: NaCl neither causes hardness nor has any reaction with lime-soda. The other substances on conversion in terms of $CaCO_3$ give the following results :

$Ca(HCO_3)_2$	...	$162 \times \frac{100}{162} = 100$ mg.
$Mg(HCO_3)_2$	...	$73 \times \frac{100}{146} = 50$ mg.
$MgCl_2$	...	$95 \times \frac{100}{95} = 100$ mg.
$CaSO_4$	...	$136 \times \frac{100}{136} = 100$ mg.

(i) Temporary hardness is due to the bicarbonates which are 100 + 50 = **150 mg/l.**

(ii) Total hardness is the sum of all the hardness causing constituents which are 100 + 50 + 100 + 100 = **350 mg/1.**

(iii) Lime requirement is (for temporary Ca hardness + 2 × temporary magnesium hardness + permanent Mg hardness).

$$\therefore \text{ Wt. of lime needed } = \frac{74}{100}(100 + 2 \times 50 + 100$$

$$= \frac{74}{100} \times 300 = \mathbf{222\ mg/l}$$

(iv) Soda is needed for permanent Ca hardness + permanent Mg hardness,

(***Note :*** Permanent magnesium hardness appears under soda requirement also, because in the lime treatment permanent calcium hardness is introduced in place of magnesium hardness. Since the amount of calcium hardness introduced is equivalent to the hardness originally present, the weights in terms of $CaCO_3$ remain the same for calculations:

$$MgCl_2 + Ca(OH)_2 = Mg(OH)_2 + CaCl_2 \text{ (this } CaCl_2 \text{ causes permanent hardness).}$$

$$\therefore \text{ Wt, of soda needed} = \frac{106}{100} \times (100 + 100) = \frac{106}{100} \times 200$$

$$= 212 \text{ mg/l.}$$

Example 2.

In example I calculate the amounts of lime soda needed per 10,000 litres of water if the sample of lime is 85% pure and soda is 98% pure and 10% excess of the chemicals are to be added in order to complete the reactions quickly.

Solution : The quantities of pure lime and soda needed will remain the same but as 100 parts of the sample of lime contain only 85 parts of the pure substance the weight of the sample of lime $= 222 \times \frac{100}{85}$ mg/1.

Further, 10% excess chemical is added, i.e., 110 parts are added for 100 parts are added for 100 parts of the calculated.

$$\therefore \text{Actual requirement} = 222 \times \frac{100}{85} \times \frac{110}{100} = 287.3 \text{ mg/1.}$$

$$\therefore 10{,}000 \text{ litres of water needs } 287.3 \times \frac{10{,}000}{1{,}000}$$

$$= 2873 \text{ g or } 2.87 \text{ kg of lime.}$$

Similarly soda requirement per 10,000 litres

$$= 212 \times \frac{100}{98} \times \frac{110}{100} \times \frac{10{,}000}{1{,}000} = 2380 \text{ g or } 2.38 \text{ kg.}$$

Example 3.

Calculate the amounts of lime-soda needed for the treatment of 1,000 litres of water containing the following:

$CaCO_3$ - 120 mg/l; $MgCO_3$ - 42 mg/l; $Mg(NO_3)_2$ - 148 mg/l;

$CaCl_2$ - 111mg/l and KCl - 50 mg/l.

Solution : $CaCO_3$ and $MgCO_3$ should be regarded as being present in the form of their bicarbonates and that their weights have been expressed in terms of $CaCO_3$ and $MgCO_3$. Permanent hardness will be caused by $Mg(NO_3)_2$. KCl will not react with lime and soda. On converting the weights in terms of the weight of $CaCO_3$ the following results are obtained :

$$CaCO_3 \ldots 120 \text{ mg/l}; \quad Mg(NO_3)_2 \ldots 148 \times \frac{100}{148} = 100 \text{ mg/l}$$

$$MgCO_3 \ldots 42 \times \frac{100}{84} = 50 \text{ mg/l}; \quad CaCl_2 \ldots 111 \times \frac{100}{111} = 100 \text{ mg/l.}$$

Lime is required for $CaCO_3$ and $MgCO_3$ (present as bicarbonates) and for $Mg(NO_3)_2$.

$$\text{Lime requirement} = \frac{74}{100} \times (120 + 2 \times 50 + 100) = 236.8 \text{ mg/1.}$$

$$1000 \text{ litres of the sample needs } 236.8 \times \frac{1000}{1000} = \mathbf{236.8 \text{ g of lime}}$$

Soda is required for $CaCl_2$ and for Mg $(NO_3)_2$ (Now present as $Ca(NO_3)_2$ as a result of reaction with $Ca(OH_2)$.

Soda required for 1000 litres of the sample

$$= \frac{106}{100} \times (100 + 100) \times \frac{1000}{1000} = \mathbf{212 \text{ g of soda.}}$$

Example 4.

A sample of polluted water on analysis gave the following results:-

Suspended matter –200 mg/l; $CaSO_4$ –250 mg/l; $MgSO_4$ –240 mg/l; $MgCl_2$ –190 mg/l; H_2SO_4 –98 mg/l and organic matter –750 mg/l. Calculate the amounts of lime soda needed per litre for its treatment if the purity of lime is 90% and that of soda is 98%. What is the hardness of the sample?

Solution : The suspended and organic matter does not need lime and soda. Sulphuric acid does not cause hardness but reacts with the alkali added for softening. Let it be assumed that lime is first added followed by soda treatment. In such a case lime will react with H_2SO_4 producing $CaSO_4$ which will cause permanent hardness for which soda will be needed.

Expressing the weights, in terms of equivalents of $CaCO_3$ the following results are obtained ;-

$$CaSO_4 \quad \ldots \quad 250 \times \frac{100}{136} = 184 \text{ mg/l}$$

$$MgSO_4 \quad \ldots \quad 240 \times \frac{100}{120} = 200 \text{ mg/l}$$

$$MgCl_2 \quad \ldots \quad 190 \times \frac{100}{95} = 200 \text{ mg/l}$$

$$H_2SO_4 \quad \ldots \quad 98 \times \frac{100}{98} = 100 \text{ mg/l}$$

Lime is needed for $MgSO_4$, $MgCl_2$ and H_2SO_4.

$$\therefore \text{ Weight of pure lime needed} = \frac{74}{100} \times (200+200+100) = 370 \text{ mg/1.}$$

As the sample is 90% pure, the weight of the sample of lime needed $370 \times \frac{100}{90}$ = 411mg/1.

Soda is needed for $CaSO_4$ and for the calcium salts produced during the removal of $MgSO_4$, $MgCl_2$ and H_2SO_4 by lime treatment.

∴ Wt. of 98% pure soda

$$= \frac{106}{100} \times (184 + 200 + 200 + 100) \times \frac{100}{98} = 740 \text{ mg/l}$$

Hardness is produced by $CaSO_4$, $MgSO_4$ and $MgCl_2$ only

∴ Hardness = 184 +200 + 200 = 584 mg/l.

Example 5.

In an experiment to determine the hardness of a sample of water 25 ml of N/50 Na_2CO_3 solution was added to 100 ml of the sample of water. After the completion of the precipitation of the insoluble carbonates, the unreacted Na_2CO_3 was titrated against N/50 H_2SO_4 solution when 10 ml of the acid was required. Calculate the degree of hardness and comment on the nature of the hardness so determined.

Solution: Na_2CO_3 will precipitate the calcium hardness as $CaCO_3$. Hardness due to magnesium salts will not be precipitated because $MgCO_3$ is comparatively soluble. Therefore this method does not report the magnesium hardness.

Volume of N/50 Na_2CO_3 solution left unreacted

= volume of N/50 acid solution used for titration = 10 ml.

Volume of N/50 Na_2CO_3 solution used in precipitation

= 25 – 10 = 15 ml

1 litre of N solution = 50 g of $CaCO_3$

∴ 15 ml of N/50 Na_2CO_3 solution

$$= \frac{50}{50} \times \frac{15}{1000} = 0.015 \text{ g of } CaCO_3$$

∴ Hardness equivalent to 0.015 g $CaCO_3$ was present in 100 ml sample.

Hardness of the sample is $0.015 \times \frac{1000}{100}$ g/l

OR $$0.015 \times \frac{1000}{100} \times 1000 = 150 \text{ mg/l.}$$

Example 6.

The hardness of 10,000 litres of a sample of water was completely removed by passing it through a zeolite softener. The zeolite softener required 200 litres of sodium chloride solution containing 20,000 mg/l of NaCl for regeneration. Calculate the hardness of the sample of water.

Solution : As 200 litres of sodium chloride solution containing 20,000 mg/l of NaCl was needed to regenerate the zeolite through which 10,000 litres of hard water had been passed, the hardness of 10,000 litres of the sample of hard water is equivalent to 200 × 20,000 mg of NaCl.

Since 58.5 parts of NaCl ≡ 50 parts of $CaCO_3$,

The hardness of 10,000 litres of the sample

$$= 200 \times 20{,}000 \times \frac{50}{58.5} \text{ mg of } CaCO_3$$

or the hardness of 1 litre of the sample

$$= 200 \times 20{,}000 \times \frac{50}{58.5} \times \frac{1}{10{,}000} \text{ mg of } CaCO_3$$

$$= 341.9 \text{ mg of } CaCO_3$$

Hardness of the sample of water is 341.9 mg/1.

Examples

1. A sample of water contains the following $Ca(HCO_3)_2$ – 162 mg/l; $Mg(HCO_3)_2$ – 146 mg/l; $CaSO_4$ – 68 mg/l; $MgCl_2$ – 95 mg/l, and KCl-745 mg/l.

   Calculate (a) temporary hardness, (b) permanent hardness and (c) amounts of lime and soda needed for the softening of 1 litre of the sample of water.

   (**Answer :** (a) 200 mg/l; (b) 150 mg/l; (c) 296 mg and 159 mg.)

2. A sample of water contains the following impurities per litre :

   $Mg(HCO_3)_2$ – 73 mg : $CaCl_2$ – 222 mg;

   $MgSO_4$ – 120 mg and $Ca(NO_3)_2$ – 162 mg.

   The sample of lime available is 74% pure and soda is 90% pure. 10% excess chemicals have to be added. Calcuiate the amounts of lime and soda required for the treatment of 5,000 litres of water.

   (**Answer :** 1100 g; 2591 g).

3. Calculate the amounts of lime and soda needed for the treatment of 10,000 litres of water containing the following:

   $Al_2(SO_4)_3$ – 34.2 mg/l; HCl – 7.3 mg/l;

   $MgCl_2$ – 9.5 mg/l; and NaCl – 58.5 mg/l.

   (Lime treatment is given first, followed by soda treatment.)

   [Hint. Al = 27. Try to solve this question by writing chemical equations also.]

   (**Answer :** 37 mg/1, 53 mg/l).

1.14 Questions

1. Among chloramine and Bleaching powder, which is the powerful disinfectant?
2. Name the indicator used in the EDTA method.
3. Why demineralization process preferred over zeolite process.
4. Write the general formula of zeolite.
5. Differentiate between sludge and scale

6. What is reverse osmosis?
7. Convert 219 ppm of Mg $(HCO_3)_2$ to $CaCO_3$ equivalent.
8. Treating boiling water with soda will remove what type of hardness.
9. What is meant by caustic embrittlement.
10. Explain Break point chlorination.
11. (a) Explain the scale and sludge formation in boilers.
    (b) What problems are caused by these?
    (c) Calculate the total hardness in ^{0}Fr and ^{0}Clarke for a sample of water obtained from a well in Khammam which showed the followed analysis:

    $MgCl_2$ = 9.5 mg/litre; $CaCl_2$ = 11.11 mg/litre
    Mg $(HCO_3)_2$ = 14.6 mg/litre; NaCl = 5.84 mg/litre
    Ca $(HCO_3)2$ = 16.2 mg/litre; Organic impurities = 20 mg/litre

12. (a) Outline the lime-soda process for the softening of hard water.
    (b) A sample of water from Nalgonda showed the following

    NaOH = 4.0 ppm; $MgSO_4$ = 24 ppm
    $MgCl_2$ = 94 ppm; $Ca(HCO_3)_2$ = 16.2 ppm
    $Ca(NO_3)_2$ = 16.4 ppm; KCl = 7.45 ppm

    Calculate the amount of lime and soda required (in kgs) to soften 10,000 litres of this water.

13. (a) What are boiler troubles and how they can be avoided?
    (b) Describe the ion-exchange process for softening of hard water.
14. (a) How is the softening of water carried out using the Zeolite process?
    (b) A sample of water from a well in Adilabad gave the following analysis:-

    NaCl = 5.85 mg/litre ; $MgSO_4$ = 12.0 mg/litre
    $Ca(HCO_3)_2$ = 8.1 mg/litre ; $CaSO_4$ = 13.6 mg/litre
    $Mg(HCO_3)_2$ = 14.6 mg/litre ; Organic matter = 4.0 mg/litre

    Calculate the quantities (in kgs) of lime and soda required to soften 12,000 litres of this water.

15. (a) Discuss the permutit process of softening of hard water
    (b) Calculate in kgs the amount of lime and soda required to soften 10,000 litres of water from a bore well in Medak which showed the following analysis:-

    KOH = 0.0056 gm/litre ; $MgSO_4$ = 0.024 gm/litre
    $Ca(NO_3)_2$ = 0.0164 gm/litre ; Mg Cl_2 = 0.094 gm/litre
    $Ca(HCO_3)_2$ = 0.0162 gm/litre ; KCl = 0.0745 gm/litre

16. Write a note on Boiler troubles

17. (a) Explain the processes of scale and sludge formation in boilers.

(b) What is caustic embrittlement and how it can be prevented?

(c) Determine the temporary, permanent and total hardness in ppm units for a water sample from Kurnool which showed the following analysis:-

$Ca(CO_3)_2$ = 32.4 mg/litre ; $Mg(HCO_3)_2$ = 29.2 mg/litre

NaCl = 5.85 mg/litre ; $CaCl_2$ = 22.2 mg/litre

Mg SO_4 = 1.2 mg/litre ; Organic matter = 15.5 mg/litre

18. Discuss the hot and cold lime-soda processes for the softening of hard water.

19. (a) How is the softening of hard water carried out using Zeolite process?

(b) A sample of water from a tank in Jagtial showed the following analysis :

$CaCl_2$ = 11.1 mg/litre ; KCl = 40 mg/litre

Organic matter = 30 mg/litre ; $MgCl_2$ = 9.5 mg/litre

$Ca(HCO_3)_2$ = 16.2 mg/litre ; Mg $(HCO_3)_2$ = 14.6 mg/litre

Calculate the amount of lime and soda required in kg to soften 50,000 litres of this water.

20. Write a short note on municipal water

21. (a) Describe the ion-exchange process for softening of hard water.

(b) One litre of water from a well in Nirmal Town, near Adilabad, gave the following analysis:-

Suspended matter = 100 mg ; $Ca(HCO_3)_2$ = 81 mg;

Mg $(HCO_3)_2$ = 146 mg ; $MgSO_4$ = 120 mg

$Ca(SO_4)$ = 136 mg; ; NaCl = 58.5 mg.

Calculate the total hardness in ^{0}Fr and ^{0}Clarke of this sample of water.

22. Write a note on Sterilization of drinking water.

23. (a) Describe the lime-soda process for softening of hard water.

(b) Sample of water from an underground reservoir in Anatapur town gave the following analyses:-

$Ca(HCO_3)$ = 81 mg/litre ; NaCl = 5.85 mg/litre.

Organic matter = 25 mg/litre ; $MgSO_4$ = 120 mg/litre.

$Mg(HCO_3)_2$ = 146 mg/litre ; $CaSO_4$ = 136 mg/litre.

Calculate the quantities (in kgs) of lime and soda required to soften 20, 000 litres of this water.

24. A sample of water contains the following $Ca(HCO_3)_2$ – 162 mg/l; $Mg(HCO_3)_2$ – 146 mg/l; $CaSO_4$ – 68 mg/l; $MgCl_2$ – 95 mg/l, and KCl-745 mg/l.

Calculate (a) temporary hardness, (b) permanent hardness and (c) amounts of lime and soda needed for the softening of 1 litre of the sample of water.

(**Answer :** (a) 200 mg/l; (b) 150 mg/l; (c) 296 mg and 159 mg.)

25. A sample of water contains the following impurities per litre :

$Mg(HCO_3)_2$ – 73 mg : $CaCl_2$ – 222 mg;

$MgSO_4$ – 120 mg and $Ca(NO_3)_2$ – 162 mg.

The sample of lime available is 74% pure and soda is 90% pure. 10% excess chemicals have to be added. Calcuiate the amounts of lime and soda required for the treatment of 5,000 litres of water.

(**Answer :** 1100 g; 2591 g).

26. Calculate the amounts of lime and soda needed for the treatment of 10,000 litres of water containing the following:

$Al_2(SO_4)_3$ – 34.2 mg/l; HCl – 7.3 mg/l;

$MgCl_2$ – 9.5 mg/l; and NaCl – 58.5 mg/l.

(Lime treatment is given first, followed by soda treatment.)

[Hint. Al = 27. Try to solve this question by writing chemical equations also.]

(**Answer :** 37 mg/1, 53 mg/l).

27. In an experiment to determine the hardness of a sample of water the following results were obtained;

Volume of the sample of water to be tested = 50 ml.

N/50 Na_2CO_3 solution added to the sample = 12 ml.

N/50 H_2SO solution used to titrate = 7 ml.

Calculate the hardness of water in grains per gallon and mg/l

(**Answer :** 7 gr/gal.; 100 mg/l).

28. 100 ml of a sample of water was titrated against a N/10 acid solution using phenolphthalein as indicator when 6 ml of the acid solution was required for discharging the colour. Then on the addition of methyl orange 4 ml of the acid solution was further required for neutralization. Calculate the hydroxyl and carbonate alkalinities in terms of mg/l of $CaCO_3$.

(**Answer :** 100; 400).

29. 100 ml of a sample of water required 5 ml of N/10 acid solution for titration while using phenolphthalein as indicator. 100 ml of the sample of water was again taken and methyl orange was used as indicator when 15 ml of the same acid solution was required for neutralization. Interpret the results of alkalinities in mg/l.

(**Answer :** Carbonate alkalinity 500 mg/l; bicarbonate alkalinity 250 mg/l.)

30. A zeolite softener was completely exhausted and was regenerated by passing 100 litres of a sodium chloride solution containing 120 g/l of sodium chloride. How many litres of a sample of water of hardness of 500 mg/l can be softened by the softener before regenerating it again?

(**Answer :** 20,513 litres).

1.15 Fill in the Blanks

1. NaCl causes_____ hardness
2. _____ water contains highest organic impurity.
3. Temporary hardness is caused by _____.
4. Formation of _____ causes caustic embrittlement.
5. Ca $(HCO_3)_2$ causes_____ hardness.
6. Equivalent weight of $CaCO_3$ is_____.
7. Hardness of water is measured in _____.
8. One ppm of hardness is equal to _____ °Cl.
9. Dissolved oxygen in water is removed by _____.
10. Dissolved carbon dioxide in water is removed by _____.
11. Caustic embrittlement is avoided by adding _____softening agent
12. Phosphate conditioning of water is done by using _____ salt.
13. Radio active conditioning prevents formation of_____.
14. $MgCO_3$ forms_____ (scale/sludge)
15. $CaCO_3$ forms_____ (scale/sludge)
16. _____ forms hardest scale.
17. In Hot-Lime soda process the function of lime is to remove _____ type of hardness.
18. Alkalinity in water cannot be due to the simultaneous presence of_____
19. The process of removing extra common salt from water is called as _____
20. When P = M the alkalinity is due to the presence of_____ ions.

21. A water sample contains 136 mg of calcium sulphate per liter. The hardness of water in terms of calcium carbonate ______
22. The hardest scale is formed from _____ ions.
23. _____ act as coagulant simultaneously as softening agent.
24. Chlorine is used in purification of drinking water for_____.
25. Removal of micro organisms in water is called as _____.
26. _____ is the indicator used in EDTA method.
27. EDTA is_____ acetic acid.
28. Sterilization of water is done to remove_____.
29. The best sterilization agent is _____.
30. Powerful disinfectant is _____.
31. Removal of microorganisms in water is called as _____.
32. Brackish water is treated by _____method.
33. O_3 can act as_____.

2 Science of Corrosion

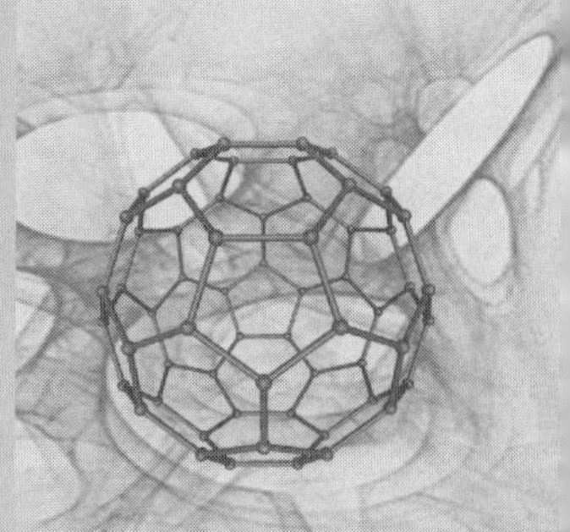

2 Science of Corrosion

2.1 Definition

One of the most widely accepted definitions of corrosion is that corrosion is the 'destruction of a metal (or alloy) by chemical or electrochemical reaction with its environment'. Thus corrosion involves chemical or electrochemical reaction and the use of the term is restricted to the destruction of metals or alloys. Destruction of wood, for instance, by rotting is not referred to as corrosion of wood. Physical or mechanical wearing away of a metal is not called corrosion but is called erosion. If corrosion and erosion take place together the destruction of the metal takes place very much faster.

Corrosion has also been defined as the deterioration of solids by liquid electrolyte. This definition covers the deterioration of refractory lining by molten glass, deterioration of concrete, etc. A still more comprehensive definition is that corrosion is an unintentional attack on a material by reaction with a surrounding medium. In this chapter corrosion will be considered mainly from the electro-chemical point of view.

2.1.1 Magnitude of the Problem

It is difficult to correctly assess the loss caused by corrosion. According to certain estimates, the direct loss due to corrosion in the U S A amounts to about Rs. 250 crores per annum. This does not include effects such as loss of lives, injuries to workmen, shut down of plants for repairs, etc. According to another estimate, the money spent in controlling corrosion is of the order of Rs. 350 crores in India, 2,500 crores in U K, Rs. 4000 crores in Australia and Rs. 10,000 crores in the U S A each year. In India due to its tropical climate the corrosion problem is more serious than in the colder countries.

The corrosion problem should be very seriously considered in all those cases where the structure or the equipment is meant to last a number of years, say 20 years or more. In some cases, like in some chemical equipment, the equipment is meant to last a few years only because of the changes in the process or in the demand for the product. In such cases it may not be economical to spend too much money on corrosion control or use expensive material for construction.

2.2 Types & Theories of Corrosion

In atmospheric corrosion the metals tend to revert to the states in which they occur in nature, i.e., the ore, because that is the most stable state. Such compounds always have less free energy than the pure metal. Fundamentally this is the cause of corrosion. Several theories of corrosion have been proposed from time to time.

2.2.1 Dry Corrosion (Chemical Corrosion)

Direct attack of dry gases on metals does not take place to any appreciable extent at atmospheric temperatures. When it does take place, a thick film of the solid corrosion product formed on the surface protects the remaining metal from further attack. If a volatile product is formed the attack proceeds rapidly, as happens when dry chlorine attacks tin sheets, cans, etc., (which are made of low carbon steel sheet coated with tin) above say 120° causing escape of volatile stannic chloride. Ferric chloride being solid protects the iron surface.

$$M \rightarrow M^{2+} + 2e^-, M^{2+} + O^{2-} \rightarrow MO$$

$$Sn \rightarrow Sn^{2+} + 2e^-, Sn^{2+} + 2Cl_2 \rightarrow SnCl_4$$

Direct Chemical action of gases from environment such as O_2, halogens, H_2S, SO_2, N_2 and inorganic material is called dry corrosion. There are mainly three types of chemical corrosion.

(i) *Oxidation corrosion*: Brought about by direct action of oxygen, usually in the absence of moisture. At ordinary temperature the alkali and alkaline earth metals are attached while metals like iron require temperature above 800 °C.

Mechanism: The metal acts as an electron donor

$$M \rightarrow M^{n+} + n\bar{e}$$

These electrons are absorbed by oxygen giving oxide anion

$$\frac{n}{4}O_2 + n\bar{e} \rightarrow \frac{n}{2}O^{2-}$$

$$\text{Net reaction } M + \frac{n}{4}O_2 \rightarrow \frac{n}{2}M_2O$$

$$\text{Metal} + \text{oxygen} \rightarrow \text{Metal oxide}$$

"The metal oxide is protective or non porous if the volume of the oxide is atleast as great as the volume of the metal from which it is formed. This is called ***Pilling-Bedworth rule***.

(ii) *Corrosion by gases like SO_2, Cl_2, H_2S, F_2, CO_2 etc.*: The extent of corrosion depends on the chemical affinity between the metal and gas involved and the porosity of the film formed.

If the film formed is non porous or protective, (*e.g.*, AgCl film formed by attack of Cl_2 on Ag metal) then corrosion is limited to only external surface.

If the film formed is *porous or volatile*, *e.g.*: $SnCl_4$ formed by attack of Cl_2 on Sn metal, the corrosion is deep.

(iii) *Action of Liquid Metals:* Liquid metals like Hg form alloys with the metal. This dissolution is either localized or lead to inter penetration thereby weakening the metal-metal bands and lead to corrosion.

The important industrial example of corrosion by direct chemical attack is the formation of scales during hot working of steel, as in the rolling of steel ingots into various sections. Attack of dry oxygen on iron does not take place at any appreciable rate below 800°, hence is not a major cause of destruction of iron in common usage.

It has been found and stated by pilling and Bedworth as a rule that those metals whose oxides have lesser volume than the original metals get corrode a faster than those whose oxides are more voluminous. Alkali and alkaline earth metals belong to the former group. As the oxide has less volume, the deposit is porous and open space is left through which oxygen diffuses to the metal surface. Thus corrosion proceeds. The oxides of some other metal have larger volumes, therefore they cover the metal surface and prevent diffusion of oxygen. If the metal is heated then the protective oxide layer breaks, more especially if there is temperature fluctuation, and corrosion proceeds.

Addition of resistant metals to other metals or alloys makes them more corrosion-resistant for the same reason. Addition of chromium to iron causes, on exposure to oxygen, the formation of a film which is richer in chromium oxide. It is this continuous resistant film which is responsible for the heat-resisting properties of chromium steels.

2.2.2 Wet Corrosion (Electrochemical Corrosion)

It is now accepted that most of the common cases of corrosion can be explained on the basis of the electrochemical theory. At ordinary temperatures the corrosion process involves oxidation. Very few cases of non-oxidative corrosion are met with under common conditions. Oxidation of a metal involves its loosing of electrons, thereby acquiring a positive charge.

If a piece of metal is dipped in water or a solution, it has a tendency to go into solution. Some of its atoms may loose electrons and go into solution as positively charged ions. These ions are held near the metal piece because of electrostatic attraction and the process of dissolution stops unless this balance is disturbed. Therefore, for the progress of corrosion, anodic and cathodic areas must exist in electrical contact with each other. Thus wherever corrosion is in progress, a number of tiny galvanic cells are active. In a sense metallic corrosion is opposite to electroplating.

1. ***Galvanic Corrosion:*** A galvanic cell is formed when two dissimilar metals, electrically connected, dip in an electrolyte. The principle of their working can be explained by considering a Daniel cell, Fig. 2.1 which consists of zinc and copper plates dipping in solutions having zinc and copper ions, the solutions being separated by a porous

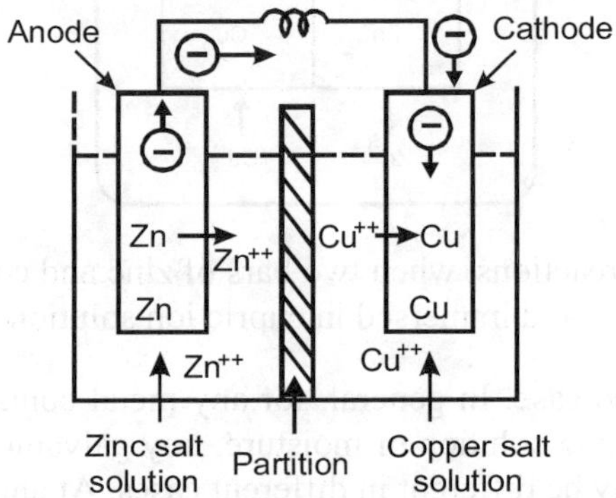

Fig. 2.1 Daniel cell.

partition. The plates are connected by a metallic conductor. The emf developed is due to two separate reactions taking place at the electrodes. Zinc atoms have a tendency to go into solution as Zn^{2+} making two electrons available. This is an oxidation reaction:

$$Zn\ (solid) \rightarrow Zn^{2+}\ (aq) + 2e$$

The electrons move along the conductor and are responsible for the reduction reaction at the copper plate.

$$Cu^{2+}\ (aq) + 2e^- \rightarrow Cu\ (solid)$$

Thus zinc dissolves and copper is deposited at its cost. Exactly similar reactions occur in corrosion. The interest of the corrosion engineer is in preventing this kind of reaction.

In Daniel cell the zinc plate is called an *anode* because oxidation reaction, i.e., liberation of electrons, takes place at the zinc plate. Similarly, because of reduction reaction, i.e., consumption of electrons, takes place at the copper plate, the plate is called a *cathode*.

The concept of corrosion reaction can be developed on the basis of reaction occurring in a Daniel cell. If the two electrodes of the cell, instead of being kept in different solutions and connected together through an external circuit, are short circuited (connected internally) and kept in a solution containing cupric ions, there will be no change in the nature of chemical reactions. Since, in such a system the resistance is made very small, the reactions take place with maximum current and the rate of dissolution of Zn becomes much faster than in a simple Daniel cell. This is illustrated in the Fig. 2.2. The current in the system can be increased to a still higher value if an arrangement of alternate strips of zinc and copper is considered to be dipped in the Cu^{2+} solution. Such galvanic cells can be formed even if one considers a zinc sheet containing microscopic inclusions of copper kept in contact with a solution containing the same ions.

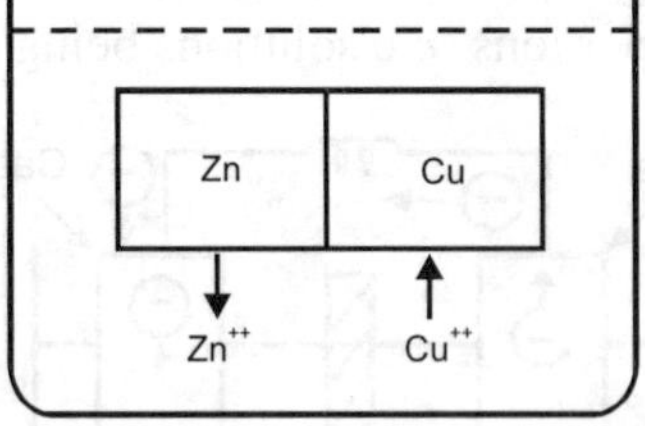

Fig. 2.2 Electrochemical reactions, when two bars of zinc and copper are welded together and immersed in cupric ion solution.

This is the most idealized case. In general, for any metal containing impurities and being in contact with a conducting solution or moisture, tiny galvanic cells are formed. The type of chemical reactions may be different in different cases. At anode there will always be the dissolution of a metal. The cathodic reaction may be the reduction of hydronium ion or oxygen.

The greater the potential difference between the two metals in actual contact, the faster will be the corrosion. Therefore, it is helpful to be able to determine the emf. The emf developed is the difference between the electrode potentials of cathode and anode. The electrode potential is a measure of the tendency of a metal to go into solution and depends upon several factors, like the metal, the concentration of the solution and the temperature. It is measured with reference to a hydrogen electrode whose potential has been taken as zero (A hydrogen electrode is prepared by bubbling hydrogen at 760 mm over platinum black plate immersed in a solution having 1 g of hydrogen ion per litre). The *normal electrode potential* of a metal is the potential between a hydrogen electrode and the metal immersed in a normal solution of its own ions (usually measured at 25°). When the elements are arranged in the order of their electrode potentials the well-known *electrochemical series* is obtained. A metal higher in the series, i.e., a more electro-positive metal, will displace the metal lower in the series, i.e., a less electro-positive metal. For example, iron will precipitate nickel and copper from their salt solutions.

The electrochemical series, though useful, does not give a basis for predicting, in a particular environment, whether a metal will or will not corrode. In actual circumstances some side reactions may take place which affect corrosion reactions. For example, copper will not corrode in acetic acid solution, but if dissolved oxygen is also present, corrosion will take place. There may be factors opposing corrosion reactions. Considering the conditions met with under many conditions of exposure, another series, known as the *galvanic series*, has been prepared. This series is of more direct help in corrosion processes than the electrochemical series. The galvanic series listing the more common metals and alloys is shown in the Table 2.1, from which the relative corrosion tendencies can be judged.

In the galvanic series chemically more active metals are given higher place. These metals easily react with oxygen, sulphur etc., hence, corrode readily. The metals at the lower end do not easily react, and therefore do not corrode so easily under ordinary atmospheric condition. When two metals are in electrical contact and an electrolyte is present, the one higher up in galvanic series corrodes and the other is protected. The greater the difference in their positions in the galvanic series the faster will be the corrosion. Metals close to each other in the galvanic series show less tendency to corrode when in contact with each other.

If to the Daniel cell just described a small electromotive force from an external source is applied in the reverse direction, the rates of copper deposition and zinc going into solution are reduced. As the applied emf is increased the reaction rates continue to decrease and when it is equal to the original emf of the cell neither current flow nor any reaction takes place, i.e., there is no corrosion. This principle is important as it provides a method for preventing corrosion. If a higher emf is applied the reverse reaction, i.e., deposition of zinc and corrosion of copper, can take place.

Table 2.1 Comparison of Electrochemical series (E.C) and Galvanic series

E.C series	Galvanic series (Anodic or corroded end)
Magnesium (impure)	Magnesium
Magnesium	Magnesium alloys
Aluminum	Zinc
Zinc	Aluminum
Cromium	Steel
Ferrous iron	Cast iron
Nickel	Stainless steels (active)
Tin	Lead-tin solder
Lead	Lead
Ferric iron	Tin
Copper	Muntz metal
Silver	Brasses
Pt	Copper
Au	Bronzes
	Copper nickel alloys
	Stainless steel (passive)
	Silver
	Graphite
	Platinum

There is another mechanism by which the reactions slow down. In the cell the concentration of zinc ions increases and that of cupric ions decreases as the current is withdrawn. This reduces the tendencies of zinc to ionize and copper ions to deposit. Thus gradually the cell is run out or *polarised* and the emf generated decreases.

2. ***Concentration Cell Corrosion:*** It appears, at the first instance, from the above discussions that for a metal to corrode it must be in contact to some other metal which is cathodic to it and that a pure metal should not corrode, at all. But, the electrochemical theory also provides an explanation for the corrosion of a single metal through the formation of electrolyte *concentration cell.* When two plates of the same metal dip in a salt solution of the same strength and are connected externally, no current flows, because both the plates have the same potential. If the concentrations of the solution around the plates are different there will be: the tendency for the metal around which concentration is lower, to go into solution, and metal will be deposited at the other plate. The surface exposed to a solution of lower concentration looses electrons hence becomes anode and dissolves, and the surface exposed to a solution of higher concentration becomes cathode as it receives electrons. This is clear from Fig. 2.3. Thus an emf will be produced. This shows that if a metal piece is in contact with a solution whose concentration is different at different ends of the piece, a concentration cell will be set up and corrosion will take place where the concentration of the electrolyte is lower. Such concentration difference of an electrolyte

can be caused by differential aeration at the electrodes. One is well aerated and another is less aerated so that less aerated has less concentration of O_2, hence it acts as anode. Thus, according to the electrochemical theory, corrosion of a metal occurs either through the formation of a galvanic cell or concentration cell. The galvanic cells are formed due to inhomogeniety at the surface of metal whereas the heterogeneity of the environment at different parts of the structure causes the formation of concentration cell. When pure and impure electrode metals are dipped in the same solution, impure metal becomes anode and gets corroded. Various types of concentration cells are explained in 1.11.

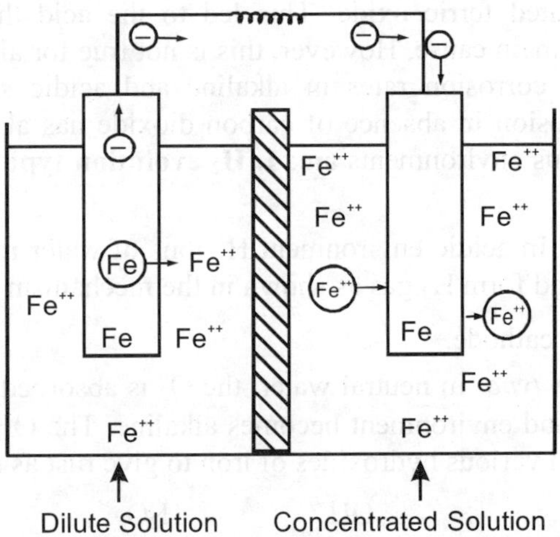

Fig. 2.3 Formation of a concentration cell.

From the foregoing, it is clear that for protection against corrosion all the factors that increase the resistance of the galvanic cell need to be promoted. The effect of such factors is called *polarization.* Polarization provides another approach to corrosion control. Polarization is due to several factors. The circuit resistance is one of these. As the current is withdrawn the ionic concentration around the anode increases and around the cathode it decreases. This causes concentration polarization. Yet another common cause of polarization is the formation of insoluble products on the cathode and anode surfaces. If the product forms a continuous, non-porous and adherent film the corrosion reactions may be completely stopped.

Among the metals that are used on a large scale the case of aluminium may be mentioned as a good example of protection against atmospheric corrosion. It readily reacts with oxygen of the air but a thin, non-porous film which adheres to the surface is formed. This film makes aluminium passive. Similar is the reason for the chromium and nickel plated surfaces being corrosion resistant. Passivity may be defined as a situation in which a thin film prevents a metal from corroding even in an environment in which it is unstable (i.e., it would have corroded if the protective film was not present).

2.2.3 Corrosion Reactions – H_2 Evolution & Oxygen Absorption Type

The corrosion occurs when a medium like water (moisture) is there in the environment which acts as electrolyte to carry electrons from anode to cathode. The environment may become acidic/alkaline as explained in the reactions below.

In the early nineteenth century it was demonstrated that for rusting of iron in the presence of both oxygen and water is necessary. It was also noticed that rusting takes place faster in acid solutions than in neutral or alkaline solutions. This was further confirmed by the detection of ferrous and ferric carbonates (due to carbonic acid in water) in iron rust which consists mostly of hydrated ferric oxide. This led to the acid theory of corrosion which considers acids to be the main cause. However, this is not true for all the metals. For instance, in the case of zinc the corrosion rates in alkaline and acidic solutions are comparable. Appreciable rate of corrosion in absence of carbon dioxide has also been reported in many cases. Corrosion in various environments are: **1. H_2 evolution type 2. O_2 absorption** type as indicated below.

1. *H_2 evolution type:* In acidic environment H^+ ions of water move towards cathode and absorb electrons and form H_2 gas as shown in the mechanism

   $2H^+ + 2e^- \rightarrow H_2$ at cathode.

2. *Oxygen absorption type:* In neutral water, the O_2 is absorbed to give OH^- ions as $H_2O + ½\, O_2 \rightarrow 2OH^-$ and environment becomes alkaline. The OH^- ions attack Fe^{3+} ions at the anode and form various hydroxides of iron to give rust as shown below.

$$Fe^{3+} + 2OH^- \longrightarrow Fe(OH)_2 \xrightarrow{OH^-} Fe(OH)_3 \xrightarrow[\Delta]{H_2O \uparrow} Fe_2O_3 \text{ or } Fe_3O_4$$

Under most of the common conditions of corrosion, according to the electrochemical theory, a number of small galvanic cells are formed. It has just been explained that in a galvanic cell the anode dissolves and some material is liberated at the cathode. Often the formation of anodic and cathodic areas takes place in quite an unexpected manner, hence the equipment fails at unsuspected points. It is, therefore, necessary to understand the corrosion reactions somewhat clearly.

On a corroding surface, at the anode, the metal atoms give up their valence electrons to the environment and are oxidized to positive ions. The reaction taking place in iron in the presence of moisture is typical and can be represented thus :

$$Fe \rightleftharpoons Fe^{2+} + 2 \text{ electrons} \quad \text{.....(2.1)}$$

If the ions and the electrons are not removed from the environment the forward reaction will stop and corrosion will not proceed. When corrosion is taking place the iron ions move away and come across the hydroxyl ions of the solution and form less soluble ferrous hydroxide:

$$Fe^{2+} + 2OH^- \rightleftharpoons Fe\,(OH)_2 \quad \text{.....(2.2)}$$

In an oxidizing environment ferrous hydroxide is oxidized to ferric hydroxide, which is more insoluble, and appears as oxides of iron found in iron rust.

The electrons released at the anodes are conducted to the cathode and are responsible for cathode reactions which are more complicated than anode reactions. In the absence of oxygen, in acid media, the following reaction takes place:

$$2e^- + 2H^+ = H_2\,(gas) \qquad .....(2.3)$$

In neutral or alkaline media and absence of oxygen, the reaction is:

$$2e^- + 2H_2O = H_2\,(gas) + 2OH^- \qquad .....(2.4)$$

As in the above reactions hydrogen gas evolves, this type of corrosion is known as **hydrogen evolution type corrosion**.

In the presence of dissolved oxygen, in acid media, the reaction taking place is

$$4e^- + 4H^+ + O_2 = 2H_2O \qquad .....(2.5)$$

In neutral or alkaline media the reaction is :

$$4e^- + 2H_2O + O_2 = 4OH^- \qquad .....(2.6)$$

It is noteworthy that electrode potentials of reactions (2.5) and (2.6) are different. In the former case it is +1.229 V and in the latter it is +0.401 V. The electrode potential of reaction (2.1) is -0.440V. This means that corrosion of iron in the presence of dissolved oxygen in acid medium will generate an emf of 1.669 V whereas in neutral or alkaline medium only 0.841 V.

Only in a few cases hydrogen peroxide is formed thus:

$$2e^- + 2H_2O + O_2 = H_2O_2 + 2OH^- \qquad .....(2.7)$$

In many cases several of these reactions take place simultaneously but dominant one in a particular medium is the one mentioned above. Corrosion process involving oxygen is known as **oxygen absorption type corrosion.**

Although theoretically it is possible to predict the rates of these reactions, in actual practice it has not been possible to predict the exact rate of corrosion reaction because there are several polarizing (i.e. opposing) reactions. For example, if an insoluble corrosion product is formed as a porous and non-adherent film on the metal surface, the polarizing effect is small. On the other hand if an insoluble product is deposited as an impervious, thin, continuous and adherent film, the corrosion reaction is completely stopped because polarization is complete. This condition is known as the *passivity* of metals. It is this process which is responsible for the corrosion resistance of some of the common metals and alloys like aluminium, tin, stainless steels, chromium, nickel, etc., on the surfaces of which oxide films are formed. If the environment is changed the passivity may disappear as happens when aluminium is used in sea water or stainless steel is in contact with hydrochloric acid solution.

The primary products of the electrochemical reactions are usually soluble substances. Sometimes insoluble products are later produced but generally they are porous and loose. Thus electrochemical reactions are more dangerous than direct chemical attack. The insoluble primary products of electrochemical reactions generally form impervious and adherent coatings and they prevent corrosion but such products are rarely formed.

To summarize, for electrochemical corrosion the three essentials are: 1. formation of anodic and cathodic areas; the anode corrodes by forming ions; at the cathode reduction takes place, 2. electrical contact between anode and cathode for the conduction of electrons, and 3. electrolyte, usually provided by the presence of moisture. The difference in electrode potentials is responsible for the formation of anode and cathode. The active part of the metal becomes anode and it gets corroded. Cathode part is always safe and does not corrode.

2.3 Factors Influencing Corrosion

It is clear from the foregoing that surface or part of the equipment which becomes anodic gets dissolved or corroded. It is, therefore, necessary to know where the anodic areas are formed so as to safeguard against eventual failure. In many cases it is not easy to know the exact spots but the following illustrations will go a long way in locating the corrosion spots.

When more than one metal is used in the structure, with the help of the galvanic series it is easy to find the metal which is most Anodic. And it is this metal that will corrode, others will be saved. In an equipment using zinc and iron parts, zinc will corrode and not iron because zinc is anodic to iron. Iron will corrode in contact with copper because it is anodic to copper. The same principle applies when impurities are present. For example, impure zinc dissolves in acid faster than pure zinc because of the cathodic impurities. From the electrode potentials an idea of relative corrosion tendencies can be formed. For example, iron in contact with copper will corrode faster than in contact with lead. In drawing conclusions the importance of the polarizing effects, already discussed, should not be lost sight of.

A little more complicated is the case of a single metal being used, corrosion may take place due to the formation of concentration cells. A metal may be in contact with a solution whose concentration is not uniform, as can happen in the storage tank of a solution. A more common cause is dissolved oxygen whose concentration is not the same throughout the entire mass of the solution. This is known as **differential aeration**.

Differential aeration is quite a common cause of corrosion at many unsuspected spots. In the section on 'Corrosion Reactions' eqn. 2.5, 2.6 and 2.7 illustrate what happens at the cathode in the presence of oxygen. Of these the most common reaction is eq. 2.6 which is:

$$4e^- + 2H_2O + O_2 \rightarrow 4OH^-$$

This shows the part played by oxygen. The cathode is the surface at which electrons are consumed. Oxygen helps consumption of electrons. Therefore, the surface which is more exposed to oxygen is cathodic to the surface less accessible to oxygen. Consequently, the surface on which oxygen concentration is lower gets corroded. Common examples of this type of corrosion are illustrated in Fig. 2.4. When there is a crack, in a metal there is lower concentration of oxygen in the crack hence it becomes anodic and corrosion proceeds in the crack increasing its size. The surface covered with dirt is also less accessible to oxygen and is anodic. In the same way there is less oxygen where the wires of a fence cross and also under a rivet head. When once corrosion starts on a metal surface at certain spots they become less open to oxygen with the result that corrosion proceeds faster at those very spots making deeper and deeper cavities. This is known as **pitting**. Pitting is easy to explain on the basis of differential aeration. The surface under the rust or scale is less accessible to oxygen, therefore, is anodic and corrodes. As corrosion proceeds, a cavity or pit is created and as it becomes deeper its bottom becomes still less accessible to oxygen which makes it still more anodic.

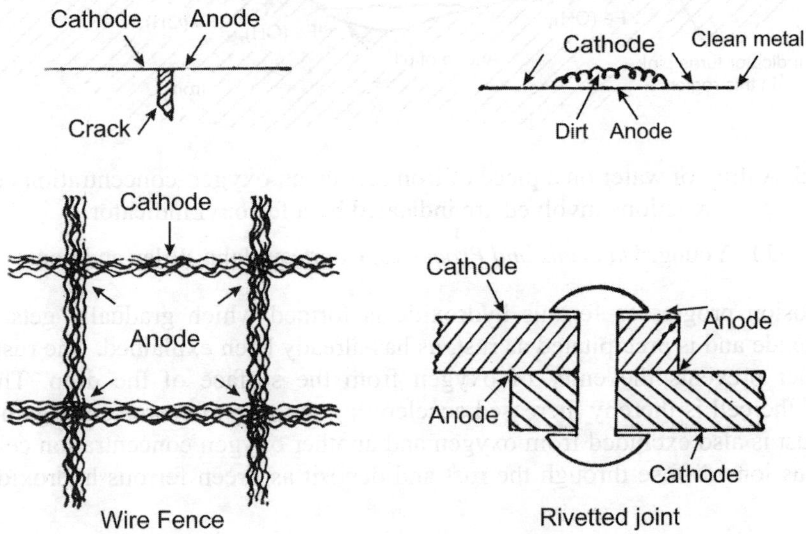

Fig. 2.4 Illustration of corrosion by differential aeration.

Corrosion by differential aeration has also been explained as being due to the initial formation of a thin metal-oxide film on the surface. A galvanic cell is now set up between the metal-oxide surface, which is anodic, and the metal surface. Therefore, the part which has higher oxygen concentration being cathodic is saved. The phenomenon has also been attributed to the formation of a protective hydroxide layer at the cathode. It seems thal corrosion due to differential aeration is quite a complicated process and several factors together are responsible for it.

A drop of water on an iron or steel plate (Fig. 2.5) is an interesting illustration for the study of the formation of an oxygen concentration cell. The concentration of oxygen at the center of the drop is less than at its surface because as oxygen is used up it is replaced more easily at the surface of the drop. Thus, there is a difference in oxygen concentration from the centre to the surface of the drop and consequently, the centre becomes anodic and the surface region becomes cathodic.

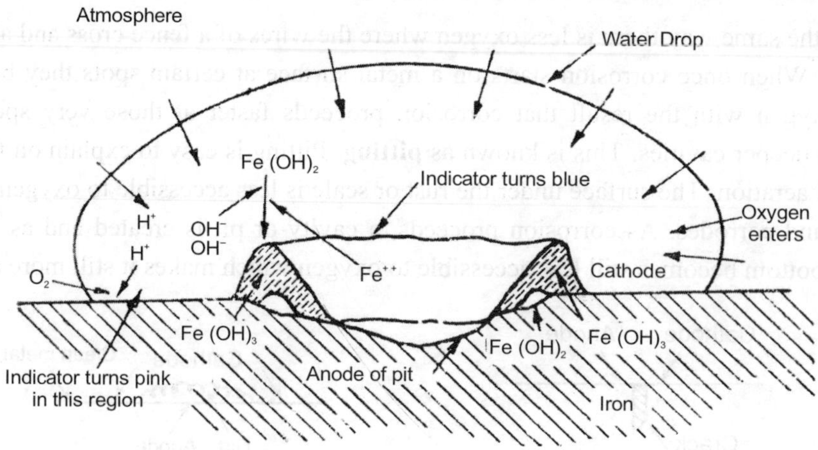

Fig. 2.5 A drop of water on a piece of iron sets up an oxygen-concentration cell. The reactions involved are indicated by a ferroxyl indicator

(J.F. Young, *Materials and Processes*, Courtesy, John Wiley and Sons)

As corrosion progresses ferrous hydroxide is formed which gradually gets oxidized to ferric hydroxide and is precipitated as rust, as has already been explained. The rust around the anode further prevents the entry of oxygen from the surface of the drop. The electrical potential of the cell is thereby increased accelerating the rate of corrosion. Again the surface under the rust is also excluded from oxygen and another oxygen concentration cell is formed. Some ferrous ions diffuse through the rust and deposit as green ferrous hydroxide under the red rust.

The reactions taking place in the drop can be observed by adding ferroxyl indicator solution prepared by mixing solutions of phenolphthalein (which indicates the presence of excess hydroxyl ions) and potassium ferricyanide (which indicates the presence of ferrous ions). For demonstration purposes it is helpful to add a trace of sodium chloride to the indicator so that corrosion may be accelerated and the experiment may be easily shown. On placing a drop of this solution on a polished iron surface a concentration cell is quickly formed because the oxygen is used up at the centre. The centre becomes anode and the solution in the centre turns blue because of ferrous ions. Cathode is formed at the edge of the drop and the solution there turns pink because of hydroxyl ions.

Water line corrosion is quite common in ships, iron tanks used for storing water, etc., in which a portion of the metal surface is under and the other is above water. The concentration of oxygen dissolved in water is greater at the water surface than deeper down. When a steel tank is regularly used for filling water up to a certain level the metal at the water surface is in contact with water having a higher concentration of oxygen than the metal under the surface. This causes the formation of a concentration cell and the anode is the lower portion and the cathode is at the water level. Because of the poor conductivity of water the ions just below the water level are more readily available for reaction, therefore, the metal corrodes just below the water level. Ships sunk for many years deep under water are known to be free from corrosion whereas water line corrosion of ocean-going ships is a practical problem.

Soil corrosion depends on salts content, moisture content degree of aeration and presence of microorganisms leading to the formation of electrochemical cells (concentration cells) and the corrosion occurs at anode and e^- are conducted by the electrolytes of salts in the water of soil. Generally less aerated areas become active places and form amodes to corrode the metal in soil. Such cases are seen in various types of soils because of formation of differential aeration concentration cells on metal pipes in the ground.

Certain *physical factors* like surface conditions and internal or external stresses also affect the corrosion process. A rough surface is more likely to collect dirt than a smooth and polished surface. As has already been explained, the surface under dirt is anodic and will corrode. Many structures are known to have failed due to *stress corrosion*. The internal stresses may have been produced during manufacture or fabrication or heat treatment. The external stress is due to service conditions. From thermodynamic calculations it has been shown that the atoms under stress are at higher energy level. Consequently the points under stress **(anode)**, as well as the grain boundaries (Inter granular corrosion), are anodic and there is a large stress-free surface which is cathodic. The difference in potential, however, is very very small but the path of stress is very narrow, hence corrosion is concentrated in a narrow path which is anodic with a large cathodic surface, and helps to open up cracks or crevices which widen due to the stress. The cracks have been noticed in areas under tensile stress and not under compressive stress. Stress corrosion appears quite suddenly and its reason is totally different from that of creep failure which is due to elongation under steady load. **Caustic embrittlement** of iron and steel, which is sometimes the cause of boiler failure, is similar to stress corrosion. Where caustic concentration cells are formed which one acts as anode (corrodes) and other part acts as cathode.

Another way the anodic areas are formed is by *scratches* on a protective coating which is cathodic to the metal. When a scratch is formed on the common 'tin sheet', which is in fact a mild steel sheet coated with tin, severe corrosion is noticed at the scratch. Tin is cathodic to iron and when it covers iron completely, iron does not come in contact with the corroding atmosphere. When a scratch is formed, the small iron surface exposed is anodic with respect to the large cathodic area around it. Therefore, the entire intensity of corrosion is concentrated on the area exposed by the scratch which often results in the formation of a hole in the sheet. In galvanised iron sheets, which are iron sheets coated with zinc, the case is the reverse. Zinc is anodic to iron and it is zinc that continues to be attacked even when a scratch is formed because the exposed iron surface is cathodic to it. Therefore care has to be exercised while handling metals protected by coatings.

When an organic coating like paint, is scratched, often the edges of the coating become loose and oxygen can enter to a small extent. If this happens, there will be corrosion inside the scratched edges due to the lower concentration of oxygen as compared to the concentration at the exposed surface.

In certain alloy systems corrosion takes place due to preferential attack of the environment, because one constituent gives rise to anodic and another to cathodic areas. This is known as preferential attack corrosion. In brass, zinc is anodic to copper and in corroding environments removal of zinc takes place. In cast iron there are quite large areas of graphite which are cathodic to iron. This helps in the removal of iron; another type of preferential corrosion is intergranular corrosion. It is due to impurities or inhomogeneties at the grain boundaries of the metals and alloys.

2.3.1 Factor Affecting Corrosion of Pure Metal

We encounter many factors affecting corrosion depending on nature of metal used, the design of the article, environmental conditions like temperature, contact of other materials, etc. Here we are mainly dealing with pure metal without any protective cooting and any special design of the article. The corrosion rate depends on the following conditions :

Nature of metal: The alkali metals are highly corrosive on account of their tendency to easily donate an electron and attain a noble gas configuration. The corrosion of metals decreases as we go from left to right across the periodic table and from top to bottom.

Galvanic series: The corrosion situation is more complex if more than one metal is present within the system. Which metal will undergo oxidation is determined by the 'standard electrode potential', which is measured against an arbitrary zero created by a 'standard hydrogen electrode' gives standard electrode potentials for many of the metals used in electronic packaging the more positive the electrode potential, the more 'noble' the metal is and hence the less prone to oxidise and hence corrode.

When dissimilar materials are connected together, the less noble metal will corrode relative to the more noble one. Thus, given a corrosive environment, one would expect the tin-lead of the joints to be corroded before the copper of the tracks, and this is indeed usually the case.

Standard oxidation electrode potentials for common metals used in electronic packaging.

Metal	Metal ion in equilibrium	Electrode potential at 25 °C (V)
Gold	Au^{3+}	1.498
Palladium	Pd^{2+}	0.987
Silver	Ag^{+}	0.799
Copper	Cu^{2+}	0.337
Lead	Pb^{2+}	–0.126
Tin	Sn^{2+}	–0.136
Nickel	Ni^{2+}	–0.250
Iron	Fe^{2+}	–0.440
	Fe^{3+}	0.331
Chromium	Cr^{3+}	–0.744
Aluminium	Al^{3+}	–1.662

Over voltage: When a metal which is at a high position in galvanic series (eg Zn) is placed in a dilute acid it undergoes corrosion with deposition of a film on its surface and evolution of hydrogen gas. So the initial rate of corrosion is high but decreases after a while due to salt film and H_2 film surrounding the metal. So an over voltage (a raised voltage due to new conditions developed) reduces the corrosion rate. Presence of ions like Cu^{2+} in the present example lead to deposition of Cu metal acting as minute cathodes and hence reduction of overvoltage and increase in corrosion.

Purity of metal: Impurity in metals cause heterogeneity and hence small localized galvanic cells and therefore an enhancement of corrosion. For example if Zinc metal is contaminated with impurities like Pb or Sn which are at a higher level than Zn, they use Zinc as sacrificial anodes and hence more corrosion is seen in comparison to pure metal.

Nature of oxide film: In aerated atmosphere, practically all metals get covered with a thin film of metal oxide. The ratio of the volumes of the metal oxide to the metal is known as specific volume ratio. Greater the specific volume ratio, lesser is the oxidative corrosion rate. Specific volume ratios of Ni, Cr and W are 1.6, 2.0 and 3.6 respectively suggesting tungsten has least corrosion. Further the corrosion depends on nature of oxide film. Metals like Al have a firm oxide film in comparison to Fe and hence Al in less corrosive (follows Pilling – Bedworth rule). The iron oxide is porous in nature and this leads to extension of corrosion to inner surface.

Nature of corrosion product: If the corrosion product is insoluble in the medium (eg: Pb SO_4 in H_2SO_4 medium) or interacts with the medium to form another insoluble product, then the corrosion product acts as a physical barrier and prevents further corrosion. Similarly the volatile corssion products expedite the corrosion rate. For example molybdenum oxide (M_oO_3) is volatile and results in excessive corrosion.

2.3.2 Influence of Environmental Condition on Corrosion Rate

It has already been explained that in spite of the rates of individual corrosion reactions being predictable, it has not yet been possible to predict the rate of corrosion in an actual situation because it is often a result of many complicated and not very well understood factors. However, there are several common environmental conditions whose effects in general are worth taking note of.

Cathode and Anodic areas: One of the factors that greatly affects the corrosion rate is the *relative proportion of anodic and cathodic areas.* If there is a large cathodic area with only a small anodic area, intense corrosion at the anodic area takes place, as happens when a protective coating, like tin, nickel or chromium on iron, breaks at a few points.

Effect of pH: In the corrosion reaction described H^+or OH^- are also involved. Therefore the effect of pH is obvious. It can be easily seen from the chemical equation for a reaction the direction in which it will shift by change in concentration of H^+ or OH^-. As a general rule, acids are more corrosive than neutral or alkaline solutions. Exceptions to the rule are amphoteric metals, like aluminium, zinc and lead, which form anions as well as cations. In acid solutions these react quickly like other metals but in alkaline solutions they form complex ions and go into solution. Consequently these metals most resistant to corrosion in neutral solutions. In the case of other mctals the corrosion rate is higher in acidic solutions than in alkaline solutions. The corrosion of iron or steel is quite slow in alkaline solutions and gradually increases as acidity increases and becomes rapid when the pH value is below, say, 4.5 to 3. This is the reason why the pH value of water for boiler feed or for cooling systems is kept in the alkaline range.

Effect of Oxidant: *Oxidizing agents* may increase or decrease corrosion rates. Systems handling water, e.g., boilers, heat exchangers, etc., are faced with serious corrosion problems due to dissolved oxygen. In high pressure boilers less than 0.005 mg/l of dissolved oxygen is permissible. The oxidizing agents oxidize materials and are themselves reduced at cathodes and, thereby support cathode reaction and promote corrosion. It is noteworthy that on some metals, like stainless steels, aluminium and magnesium, a thin film of oxide is deposited on the surface rendering the metal resistant to corrosion and makes it passive. Such metals and alloys are more resistant in oxidizing environments. Monel metals (Ni-Cu alloys) corrode rapidly in the presence of air whereas 18-8 stainless steel (Cr-Ni-Fe alloy) corrodes rapidly in the absence of air. It has been reported that in an experiment using 5% sulphuric acid at 30° the corrosion rate of Monel metal was 240 mdd (mg per sq. decimeter per day). When the acid solution was saturated with air, but on exposure to the same solution free from air corrosion was 40 mdd. Under similar conditions the corrosion rates of stainless steel were 2 and 300 mdd. This illustrates the sharp difference in the behaviours of Monel and stainless steel both of which are known for their resistance to corrosion. Hence, care needs to be taken while using materials

Effect of Temperature: *The effect of temperature* on the corrosion rate is complicated because of the fact that it affects the various factors in different ways. The rate of chemical reaction increases, with rise in temperature but the solubilities of gases, like oxygen which affect corrosion, decrease. The temperature may affect the protective coatings in different ways. In general, the rate of corrosion due to oxygen or oxidizing agents is decreased with rise in temperature but the rate of hydrogen type corrosion is increased.

Effect of water / humidity: A mild steel equipment handling water gets corroded at a faster rate as the temperature increases, because of a higher rate of chemical reaction, reaching the maximum rate at 80°. Above this temperature the effect of lower solubility of oxygen in water becomes dominant and the rate of corrosion falls sharply. Similar results have been reported for the corrosion of Monel metal in dilute sulphuric acid in the presence of air. The effect of temperature on the corrosion of zinc in distilled water is interesting. Up to 50° the protective coating formed by the corrosion protects zinc. Above 50^0 the protective coating gives way and rapid corrosion takes place up to about 65°. Above 65° the effect of decreasing solubility of oxygen becomes pronounced and the corrosion rate falls.

Other factors: A number of *other factors* also affect corrosion rates. A liquid flowing with high velocity past a metal surface reduces the tendency to form a protective film and may even remove films already formed. It may even cause erosion. Consequently, at higher velocities the corrosion rates are higher. The dissolve salts may increase or decrease corrosion rates. The corrosion of stainless steel and of aluminium is increased in the presence of chloride ions as these react with the protective oxide films. In general, dissolved salts increase corrosion rates as they make the solution a better conductor, thereby facilitating diffusion of ions.

Corrosion Prevention by Inhibition: Mention may be made here of *inhibitors*, i.e., substances that hinder the reactions when added in small amounts to the environment, a number of organic as well as inorganic substances have been suggested. Some of these are polarizers and act by retarding cathodic reactions. Their exact action is varied e.g., formation of an insoluble salt, oxide on the metal surface, absorption on the metal surface, etc.

2.4 Corrosion Control

Corrosion Protection means not allowing corrosion reactions to take place. In most practical cases the cost of a metal or alloy or of fabrication is such that the most resistant material cannot be used and some steps have to be taken to prevent corrosion, e.g., noble metals do not corrode but they cannot be used for common purposes. In some cases protection can be done directly, e.g., by painting, and in some case indirectly, e.g., by attaching blocks of magnesium to underground pipelines. The various measures taken for corrosion protection can be studied under: 1. design and material selection, 2. cathodic protection and 3. protective coatings.

2.4.1 Design and Material Selection

Design consideration opens up a vast field of study. From the corrosion point of view it means proper selection and proper use of the metals. Sometimes even a small amount of an impurity in the common metals and alloys may cause severe corrosion, as for example, 0.003% iron dissolves in magnesium forming a solid solution. If iron is present in even slightly greater amounts it appears as small particles which are highly cathodic to the surrounding magnesium surface and cause very severe corrosion around these particles. This type of corrosion is known as *pitting*. Similar is the case of lead as an impurity in zinc diecasting alloys (whose solubility is about 0.002%). In such cases it is necessary to use high purity metals.

Sometimes a difficulty like the above can be avoided by alloying. Iron may be an impurity in aluminium. The difference between their potentials is about 0.8 volts which causes severe corrosion of aluminium. If small amounts of chromium or manganese are added, alloys like AI-Cr-Fe or AI-Mn-Fe are formed whose potentials are very close to that of aluminium.

In case a metal is under external stress, *heat treatment* like annealing, which relieves the stress, will help corrosion resistance. It is rather difficult to generalize the effect of heat treatment on corrosion. A quenched steel has one phase, viz. martensite. If it is tempered at low temperatures, carbide and ferrite phases appear and rapid corrosion takes place. High temperature tempering causes the carbide phases to come together, thereby reducing the number of galvanic couples. Annealing, which reduces the corrosion rate in many cases, increases corrosion of cutlery grades of ferrite steel. The reason is that annealing causes the precipitation of chromium carbide and thus reduces the amount of carbon in the neighbouring areas which get corroded because they are anodic to chromium carbide.

A major factor in the corrosion failure of a component or structure is a faulty design. The design may satisfy all the basic mechanical requirements but it may incorporate features that tend to promote corrosion. This will lead to either early failure or unnecessary maintenance throughout its life. A simple modification in design in many cases may protect them from corrosion to a large extent.

There are many equipments and structures which are meant to be exposed to corrosive environments, e.g., equipment handling acids, a bridge, and hull of a ship. In such cases often slight changes in their designs can be made so that there is no accumulation of corrosive material at a particular place.

Fig. 2.6 shows the design of drain pipe in a tank for chemicals. Fig. 2.6(a) illustrates a poor design because some solution will be left after draining and the life of equipment will be reduced because of corrosion. Fig. 2.6(b) shows a very slight modification in the design but the effects of the modification are far-reaching. In this case the solution can be drained completely. Cause of leaking drain pipes, taps, fittings, etc., can be traced to such defects in the design.

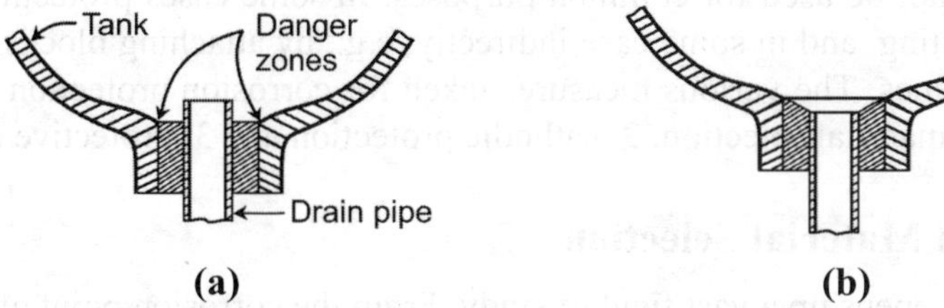

Fig. 2.6 Drain pipe in a tank for chemicals, (a) Poor design, (b) Correct design

(L. P Mondolfo and O. Zmeskal, Engineering *Metallurgy*, Courtesy, McGraw Hill Book Co., Inc.)

Fig. 2.7 shows an electrical box. The design shown in Fig 2.7(a) is such that rain water collects at the top and also seeps between the bolt and the housing and remains there because of capillary action. The rain water at the top is not drained off and as water evaporates, concentration of the salts increases with the result that corrosion takes place. If the housing and the bolt are of different materials, as is quite often the case, the formation of galvanic couples is still easier. Fig. 2.7(b) shows how the defects can be corrected by slight modifications in the design.

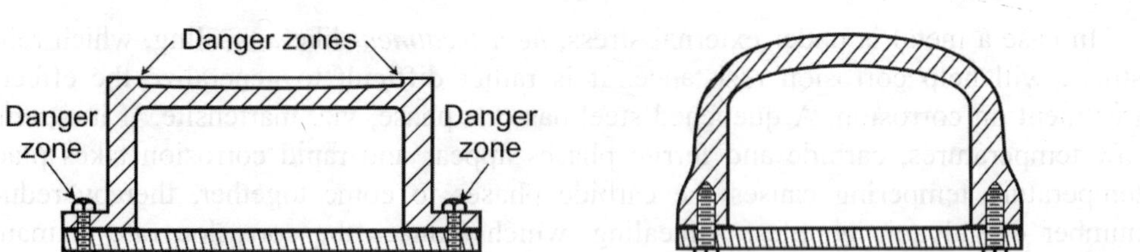

Fig. 2.7 Electrical box (a) Poor design, (b) Correct design.

(L. F. Mondolfo and O Zmeskal, *Engineering Metallurgy*), (Courtesy, McGraw Hill Book Co., Inc)

In the tropical countries roofing of corrugated iron sheets is quite common. Fig. 2.8 (b) shows a very bad way of joining the sheets as the rain water cannot be drained away. Fig. 2.8 (a) shows how the sheets should be assembled so that corrosion is not facilitated.

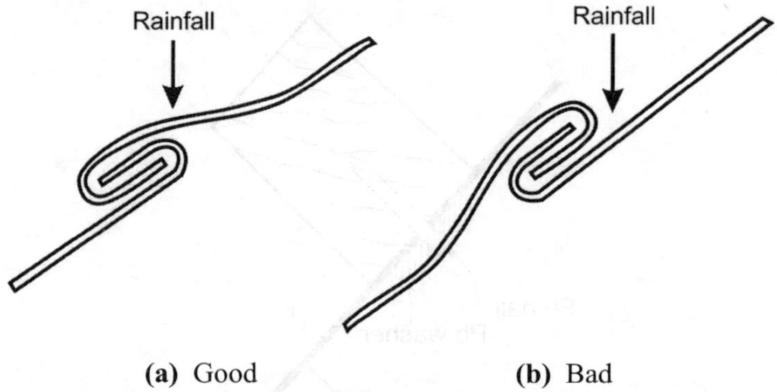

Fig. 2.8 Roofing sheet. (a) correct assembly (b) wrong assembly leading to rapid corrosion.

(L. F. Mondolfo and O Zmeskal, *Engineering Metallurgy*), (Courtesy, McGraw Hill Book Co., Inc)

Materials like wood, paper, cloth, asbestos, magnesia, etc., which absorb and retain moisture, should not be used in contact with metals. Care should be exercised in the use of soft metal (like lead) washers because of the formation of galvanic couples. It is better to use rubber or plastics washers because they do not absorb moisture and moreover they insulate the metals. Sharp corners and projections should be avoided as they cause accumulation of materials.

The use of more than one metal in equipment often becomes unavoidable. This opens up the danger of formation of galvanic couples. Since the rate of corrosion depends upon, besides difference in potentials, the relative sizes of the anodic and cathodic areas it is desirable that the smaller parts be made of the noble metals. For example, a steel bolt in a copper equipment is undesirable because the effect of corrosion will be concentrated on the small steel bolt. Engineering News (1915) describes the failure of the yacht "Sea Call" whose hull plates were made of Monel metal. These were fastened with steel rivets to the steel frame. Within three months the Steel rivets were destroyed and the frame became so weak that the vessel had to be scrapped. If more than one metal has to be used in the construction, much of the corrosion problem can be avoided if (i) the metals are so chosen that they have as little difference in their potentials as possible, (ii) the smaller part is of nobler metal, and (iii) the metals are electrically insulated from each other.

A note of warning may be sounded. Corrosion is a complicated problem with many aspects whose effects are difficult to assess. Undue haste should not be exercised in changing designs that have stood the test of time. For example, aluminium sheets fastened by iron nails using lead washers are known to stand for years, Fig. 2.9. Electrochemically, it is undesirable because the difference between the electrode potentials of aluminium and lead is about 2 volts. The resistance to corrosion in this case is attributed to an insulating film of aluminium oxide which prevents the formation of the galvanic couple.

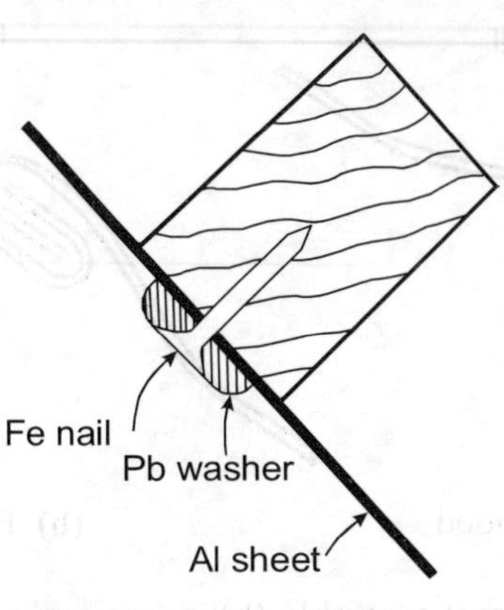

Fig. 2.9 Method of fastening roofing sheet.
(L. F. Mondolfo and O Zmeskal, *Engineering Metallurgy*),
(Courtesy, McGraw Hill Book Co., Inc)

Stray currents cause anodic corrosion, especially in areas where electric traction is used. Because of defective insulation stray currents pass through metallic conductors, like poles and pipes, supported on the ground. The point at which the current leaves the conductor becomes anodic and gets corroded. This can be checked by proper insulation.

2.4.2 Cathodic Protection

Interest in cathodic protection has greatly increased in recent years. It can be defined as a method of protecting metals and alloys from corrosion by making them completely cathodic and no site of the metallic object is allowed to act as an anode. Obviously, there cannot be a cathode without an anode. Therefore, to achieve this, an auxiliary anode is provided in the corroding medium which is connected to the structure to be protected through an external source of current. This is illustrated by the simple diagrams shown in Fig. 2.10.

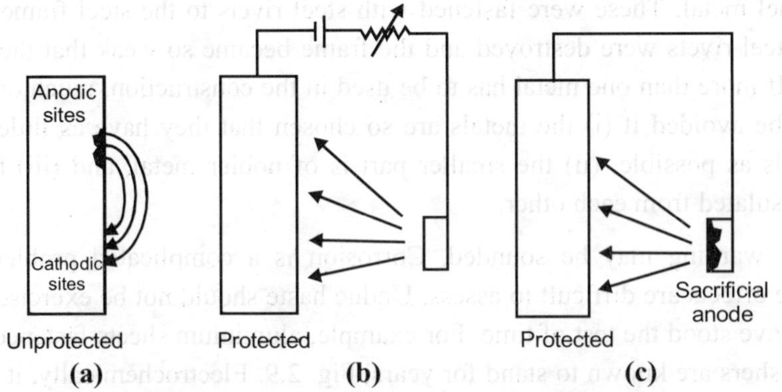

Fig. 2.10 Cathodic protection metal (a) piece being corroded (b) It can be protected by applying an impressed current or (c) by connecting a sacrificial anode.

The cathodic protection is applied either by using **sacrificial anode** or by applying **direct current** from a rectifier or a battery. In the former case a more active metal, like zinc, magnesium, aluminium or their alloys, than the metal to be protected is used as the anode which gets corroded in the environment and the metallic object which has now been made a cathode is protected. (See Fig. 2.11). The electrons for protection in this case, are supplied by the dissolution of sacrificial anode. This method is advantageous in the sense that it requires low installation cost and minimum maintenance cost after installation. In the impressed current method the auxiliary electrode is made of noble metal and is kept in contact with corroding environment. The anode, therefore, remains unaffected and electrons are supplied by rectifier or battery.

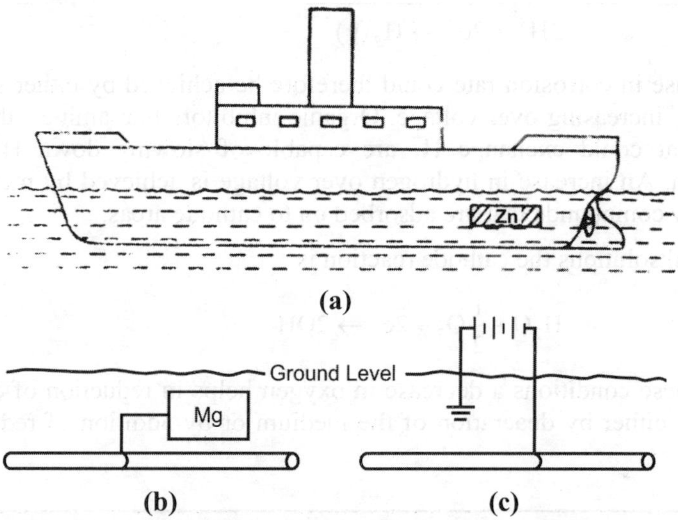

(a) (b) (c)

Fig. 2.11 Applicatin of cathodic protection. (a) Protection of ship's hull below waterline by fixing zinc plates, (b) Protection of underground pipe-line by attaching magnesium plates. (c) Protection of underground pipe-line by applying a small impressed voltage. In all these cases the part to be protected has been made a cathode.

Although, the method of cathodic protection appears to be an ideal method for the prevention of corrosion, there are some demerits in this method. Firstly, this method is expensive as it requires very high current, especially in the case of impressed current method. Secondly, it is difficult to maintain uniform current over the entire metal surface and as a result localized corrosion may occur. Finally, as the metal under protection is made cathodic, hydrogen evolution reaction may occur increasing the pH of the environment which may cause the corrosion of base metal. The continuous production of H_2 at cathode may also lead to hydrogen embrittlement.

Impressed current cathode Protection: An impressed current is applied in opposite direction to nullify the corrosion current, Fig. 2.11 (c). This converts the corroding metal from anode to cathode. The impressed current is usually derived from a D.C source like a battery or rectifier. Here, usually, the insoluble anode like iron scrap is buried in the soil in the corrosion medium and connected to the D.C anode. To increase the electrical contact with the soil a backfill like coke or gypsum is added.

Use of Inhibition: "A corrosion inhibitor is a substance which when added in small quantities to the moist environment effectively decreases corrosion rate".

(i) *Anode Inhibition:* High oxygen anions of transition elements like chromate, phosphate, tungstate, etc., inhibit the corrosion by forming a sparingly soluble salt with the metal ion of corroding element. This sparingly soluble salt forms a protective film on the metal surface there by reducing corrosion rate. This method is not fool proof since several uncovered areas of the metal are left unprotected.

(ii) *Cathode Inhibition:*

(a) In acidic solution the main cathodes reaction is evolution of hydrogen.

$$2H^+ + 2e^- \rightarrow H_2(g)$$

A decrease in corrosion rate could therefore be achieved by either slowing down H^+ ion or by increasing over voltage. Organic inhibitors like amines, thiourea, micelles, soaps that could exchange H^+ are capable of slowing down H^+ and hence the corrosion. An increase in hydrogen over voltage is achieved by metallic arsenic and antimony compounds that are adsorbed on to cathode areas.

(b) In neutral solutions the cathode reaction is

$$H_2O + \frac{1}{2}O_2 + 2e^- \rightarrow 2OH^-$$

Under these conditions a decrease in oxygen helps in reduction of corrosion. This is achieved either by deaeration of the medium or by addition of reducing agents like Na_2SO_3.

2.5 Protective Coatings

The need to apply protective coatings arises from the practical difficulties in using the most resistant metal and avoiding all the various factors that support corrosion. These difficulties are due to cost and availability of materials and fabrication problems. After taking all due care in the selection of the materials and using them in the best possible manner it often becomes necessary to cover the metal surface with a coating which is intended to protect it from corrosion. Such coatings are called protective coatings. These may be of metallic, inorganic or organic materials. Most of these coatings act by cutting off the contact between the metal and the corrosive surroundings. Before any coating can be applied, it is necessary to prepare properly the metal surface.

2.5.1 Surface Preparation and Methods of Application of Coatings on Metals

If the metal surface has not been properly prepared, the protective coating will not stick to the surface and not only will the purpose be defeated but still more severe corrosion may results at points exposed to the corrosive environment. The surface imperfection, besides physical defects, is generally due to the presence of two kinds of substances: oil or grease, and rust or scale. Oil or grease is removed by washing with soap or alkali solution in water. Several patented compositions containing alkalis, phosphates and silicates are marketed for removing oil. The use of solvents, mostly chlorinated hydrocarbons, has become quite wide spread in the U S A.

Rust or scale is generally removed by dipping the material in dilute (5 to 10%) sulphuric acid and by mechanical rubbing with a wire brush. Large equipment or structures like bridges are cleaned by the use of a wire brush or by sand blasting. Another method of cleaning is by applying flame from a kerosene burner. The old paint, oil or grease burns and the scale is removed because of sudden expansion of the metal.

A surface may also be cleaned by making it a cathode and passing high density current through it when the evolved hydrogen gas does the cleaning.

If the surface is not smooth, it is made smooth by mechanical polishing or by electropolishing. The clean surface catches dirt and gets corroded quickly. Hence it should not be left in that condition and should be quickly protected by applying a protective coats on the surface by selecting suitable method of application from the following methods.

2.5.2 Metallic Coatings

The metals used for coatings may be placed under two categories: (1) metals which are anodic to the metal to be protected, i. e. metals standing above it in the galvanic series, or (2) cathodic metals, which stand below it in the galvanic series. Metals like zinc, magnesium, cadmium and aluminium, when applied on iron and steel fall under the first category. These being anodic get themselves corroded, saving the iron which is to be protected. As explained under 'Formation of Anodic Areas', in such cases even if the surface is scratched, the metal to be protected does not get corroded. These metals may be applied on the surface by hot dipping or spraying.

Metals like tin, nickel and chromium on iron and steel surfaces fall in the second category. These are cathodic to iron or more resistant to corrosion and offer protection only so long as the surface is completely covered. If the coating breaks or is applied in a manner so that pinholes are left, the points at which the iron surface is exposed gets very severely corroded because they are anodic to the large cathodic surface. Hence great care has to be exercised while applying these. Generally, the cathodic metal coating is applied by electroplating because by this method a more uniform coating is produced which completely covers the surface.

2.5.3 Anodic Coatings & Hot dipping method:

Galvanizing Zinc coating on iron by hot dipping, is the most widely used metallic coating. The metal sheet is cleaned by alkali wash followed by pickling in dilute sulphuric acid and finally washed with water. It is then passed through a bath of molten zinc at about 450°. Smaller articles like bolts, nuts, etc., are kept in a perforated container and dipped in molten zinc and the excess of zinc is removed by centrifuging. A layer of an alloy of zinc on iron is formed to which the outer layer of zinc sticks The article should not be dipped in zinc at a very high temperature or for a long period as a thick layer of alloy is formed which is brittle.

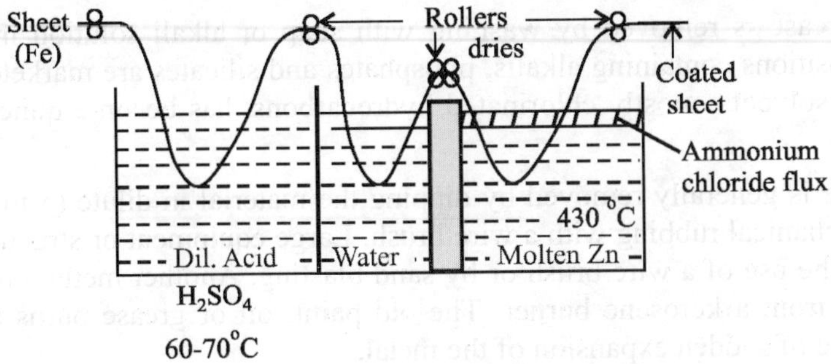

Hot dipping

Galvanizing is widely used for protecting iron exposed to the atmosphere, as is the case with roofs, wire fences, pipes and articles fabricated from galvanized sheets like buckets, tubes, etc. Galvanized ware is not used for keeping eatables because of the solubility of zinc. The popularity of galvanizing is due the low cost of zinc, easy application and the anodic protection offered by the zinc.

Cathodic Coatings: ***Tinning*** of mild steel plates is done mostly for the requirements of the foodstuff industry. Tin is a nobler metal than iron, therefore, it is more resistant to chemical attack. As has already been explained, if the coating formed does not cover the surface completely during use and leaves iron exposed, more rapid corrosion of iron will take place.

A cleaned iron sheet is passed through a bath of molten flux, like zinc chloride, then through molten tin and finally through a suitable vegetable oil (to protect the hot surface against oxidation). Lastly it is passed between rolls to adjust the thickness of the tin layer, which may be about 0.002 mm thick.

2.5.4 Electroplating

"Electroplating is the process of deposition of coated metal on the base metal by passing direct current through an electrolyte solution of the metal".

Nickel and chromium plating on steel are expensive but are commonly used for costlier equipments. Both the metals are cathodic to steel and care should be taken that the coating is continuous, uniform and firmly adherent to the surface, otherwise rust spots appear. Nickel coating deteriorates more quickly than chromium. Chromium gets covered with a strongly adherent oxide film which is extremely stable. Chromium plating is more difficult and is more expensive. During plating, hydrogen evolves and makes the metal brittle. The layer should be thin otherwise it cracks. Steel is first plated with copper which is nickel plated and is then highly polished. Finally chromium plating is given. If due care is taken such a plating should remain bright for a long time. Generally steel is made as cathode over which metal coating is done. Electrolytic plating is well known industrial activity. Graphite is the anode and steel becomes cathode in an electrolytic bath containing the salt of coat forming metal in solution of a buffer and a flux to help the process of electroplating on steel. The Ni, Cr, and Cu salts are used in solution to coat (Ni, Cr and Cu) metal as cathodic coat on steel. When current is passed through the electrolytic bath, the metal ions (Ni^{2+}, Cr^{3+} or Cu^{2+}) move from anode and go to cathode (steel) and get deposited on cathode as protective coat on steel. By selecting suitable concentrations of electrolyte, buffer and current density, uniform thick protective layer can be obtained.

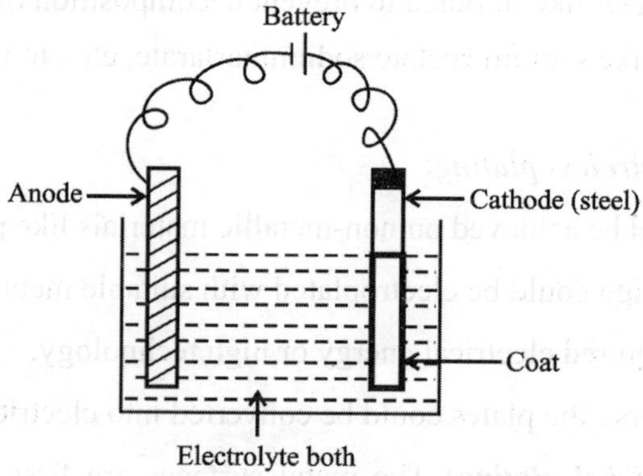

(Ni salt for Ni plating) (Cr salt for Cr plating) (Cu salt for Cu plating)

Objective of electroplating

(i) *On metals:* (a) To increase corrosion resistance of plated metal (b) To improve physical appearance & hardness (c) To increase commercial value by giving suitable color.

(ii) *On non-metals:* (a) For enhancing decoration of plated non-metals surfaces like plastic, photo frames, etc. (b) For creating conducting surface on the non-conducting material.

2.5.5 Electroless plating

Electroless plating is a technique of depositing a noble metal on a catalytically active surface of less noble metal by the usage of a suitable reducing agent without employing any electrical current.

$$M^{n+} + \text{reducing agent} \rightarrow \underset{\text{deposited on the activated surface}}{M} + \text{oxidized product}$$

The pretreated catalytically active surface is the driving force behind this reaction. The process involves two steps (i) preparation of active surface (ii) electroless plating.

(i) *Preparation of active surface of the object to be plated:* This is achieved by adopting any one of the following process;

  (a) by etching or cleaning surface using acid treatment

  (b) by electroplating a thin layer of any supportive metal coating.

  (c) by treating with stannous chloride followed by dipping in palladium chloride. This treatment produces a thin coat of palladium and is very useful for coating on the plastics.

(ii) *Electroless plating bath:* This bath contains;

  (a) the electrolytic solution of the metal to be coated.

(b) a reducing agent like formaldehyde, hypophosphite, etc.

(c) complexing agent like tartarate, citrate, etc. used to improve quality of deposit.

(d) a stabilizer like thiourea to prevent decomposition of plating bath solution.

(e) buffer like sodium acetate sodium tartarate, etc., to maintain the pH of bath during plating.

Advantages of electroless plating:

1. Plating could be achieved on non-metallic materials like plastic and wood.
2. Intricate design could be electroplated with suitable metal.
3. Does not required electrical energy or high technology.
4. The insulators / the plates could be converted into electrical conductors.

1. ***Electroless nickel plating:*** The metal surfaces are first degreased. Metals and alloys containing Al, Cu, Fe can be nickel plated without any activation of surface. Stainless steel is generally activated by dipping in dilute H_2SO_4. Metals like Mg are activated by coating Zn or Cu over it.

   Non metallic plastics and wood articles are first dipped in $SnCl_2$ solution containing HCl followed by dipping in $PdCl_4$ solution. On drying a thin layer of Pd is coated on the surface.

   Composition of coating bath: Coating nickel solution – Ni Cl_2 (20g/L), reacting agent – Na H_2 PO_2 (20g/L), buffer – Sodium acetate (10g/L), complexing agent – Sodium succinate (15g/L), optimum pH = 4.5, optimum temperature = 93°C

   Reactions

   Cathodic reaction $Ni^{2+} + 2e \rightarrow Ni$

   Anodic reaction $H_2PO_2^{2-} + H_2O \rightarrow H_2PO_3^- + 2H^+ + 2\overline{e}$

   Overall reaction $Ni^{2+} + H_2PO_2^{2-} + H_2O \rightarrow Ni + H_2PO_3 + 2H^+$

   Below 70 °C the plating is very slow and above 93 °C the bath solution starts the composing. Therefore the optimum temperature of 93° C is maintained.

Applications:

1. *Electroless Ni plating* is used on hydraulic compressor parts, pumps and pressure vessels. Many plastic articles like knobs of electronic gadgets, imitation jewellery, bottle caps, car trims, etc. are also nickel plated using this technique.
2. *Electroless copper plating:* Degreasing is done with ammonia treatment. Pretreatment & etching is done using dilute H_2SO_4 solution. Metals like Fe, Co, Ni can be directly copper plated.

Composition of coating both:

Coating solution – Cu SO_4 (12g/L)

reducing agent – formaldehyde (8g/L)

Buffer – NaOH (15g/L) + rochelle salt (14g/L)

Complexing agent – EDTA (20g/L)

Optimum pH = 11.0

Optimum temperature = 25°C

Reactions:

Cathodic reaction – $Cu^{2+} + 2e^- \rightarrow Cu$

Anodic reaction – $2HCHO + 4OH^- \rightarrow 2HCOO^- + 2H_2O + H_2 + 2e^-$

Overall Reaction – $Cu + 2HCHO + 4OH^- \rightarrow Cu + 2HCOO^- + 2H_2O + H_2$

Applications: Typical application of electroless plating of Cu lies in making Printed Circuit Boards (PCB) used for making many electronic gadgets. In particular electroless Cu plating derives an advantage over electroplating where the plating has to be carried out through holes in multilayed boards.

2.5.6 Metal Claddings

Coating of a thin uniform homogeneous layer of a coating metal on a base metal an fused so that it strongly binds permanently either on one side (as in copper clades in cooking vessels) or on both sides (as in alclad "duroallumin" sandwich between two Al layers of pure 99.8% Al) and passed through rollers under heat and pressure. All corrosion-resistant metal (NI, Cu, Ag, Au, pt) and alloys *steels Ni alloys) can be used as cladding materials. Base metals on which cladding is done are mild steel, aluminium, copper, nickel and other alloys Duralumin is very light metal alloys used in airfracft industry.

Terne plate is the name given to the plate which has a coating of an alloy of tin and lead.

Metal spraying methods are there to protect the "base metal" the molten/powdered coating metal is sprayed on the base metal with the help of a gun/aspirator to give a continuous thin protective film/coat.

In some cases metal oxide powders are taken in a revolving heating drum in which the "base metal" is thoroughly immersed acid airtighted. This causes diffusion/cementation of the coat with the base metal to protect from corrosion. Where ZnO is used, it is known as Sherardizing. When Al_2O_3 and Cr_2O_3 are used, it is known as chromizing and when only Al_2O_3 is used, it is known as colorizing. All these give thin coat to protect the base metal from corrosion.

2.6 Questions

1. Define corrosion.
2. Explain passivity.
3. Difference between galvanic and electro chemical series.
4. Explain pitting corrosion.
5. Explain galvanic corrosion.
6. What is the chemical formula of the black rust?
7. How Fe corrodes in neutral or alkaline medium.
8. 'Rusting of iron is more in saline water', explain.
9. Differentiate galvanic and concentration cell corrosion.
10. Corrosion of gases will depend on which factors.
11. Write important applications of protective coatings.
12. Define anodizing.
13. Explain cathodic protection.
14. What is corrosion? Discuss different methods of corrosion control.
15. What is cathodic protection? How is it done by using sacrificial anode method?
16. Explain why an iron pipe does not rust when connected to zinc metal but rusts rapidly when connected to copper metal.
17. Write a brief note on dry corrosion
18. What is metallic corrosion? Write a note on water line corrosion
19. Explain the factors affecting chemical corrosion
20. Explain the steps involved in pre-treatment of a metal surface prior to being given a protective coating
21. Write short note on
    (a) Cathodic protection against corrosion
    (b) Anodizing
    (c) Electroplating
22. How would you explain the corrosion of zinc rod partially immersed in a solution of sodium chloride.
23. List any three methods employed for corrosion control and explain their features.
24. Describe the following metallic coatings
    (a) tinning (b) hot dipping
25. Explain the following methods of corrosion control
    (i) proper designing (ii) protective coatings
26. (a) What is the difference between pitting and stress corrosion?
    (b) Describe the important factors that influence the process of corrosion.

27. (a) Explain the process of differential aeration corrosion.
    (b) What is sacrificial anode and how does it work?
    (c) Write a note on metallic coatings that prevent corrosion.
28. (a) What is electrochemical corrosion?
    (b) Explain the process of galvanizing and tinning.
29. Explain the difference between wet and dry corrosion.
30. (a) Explain the electrochemical theory of corrosion.
    (b) Discuss the important factors influencing corrosion.
31. Explain the process of
    (i) tinning (ii) electro plating.
32. (a) What is galvanic corrosion and how it can be prevented?
    (b) Explain the phenomenon of stress corrosion with suitable examples.
33. (a) Discuss the phenomenon of
    (i) metal cladding (ii) electroplating used to prevent corrosion.
    (b) Write a note on underground/soil corrosion.
34. What is the role of surface preparation in corrosion prevention and how it is done?
35. (a) Explain the process of
    (i) tinning (ii) galvanizing.
    (b) Describe the method of anodic protection.
36. (a) Explain the process of pitting corrosion and differential aerationcorrosion.
    (b) What are the important factors that influence the corrosion phenomenon?

2.7 Objective Questions

1. Which of the following acts as corrosion inhibitor by removing oxygen from aqueous solutions.
   (a) $ZnSO_4$ (b) Na_2CrO_4 (c) N_2H_4 (d) $NaNO_2$
2. Galvanizing is the process for coating iron with
   (a) Ni (b) Zn (c) Cu (d) Sn
3. During galvanic corrosion the more active metal acts as
   (a) anode (b) cathode
   (c) anode as well as cathode (d) none of the above
4. The principle in electroplating is
   (a) electrolysis (b) oxidation
   (c) reduction (d) anodization
5. The metal which is protected by a layer of its own oxide
   (a) Cu (b) Fe (c) Au (d) Al

6. During galvanic corrosion, the most noble metals act as
   (a) anode (b) cathode
   (c) anode as well as cathode (d) corroding metal
7. Corrosion of zinc metal containing an impurity of copper is called ______ corrosion.
   (a) water-line (b) galvanic
   (c) drip (d) crevice
8. The type of corrosion occurring in a wire fence is ______ corrosion
   (a) galvanic (b) differential aeration
   (c) inter granular (d) water line
9. A metal will be easily corroded if it is
   (a) less acidic (b) more acidic
   (c) less basic (d) more basic

2.8 Fill in the Blanks

1. Evolution of hydrogen type corrosion occurs in _____ environment.
2. The metal ______ (higher/lower) in electrochemical series undergoes corrosion.
3. _____ is an example for stress type of corrosion.
4. Presence of CO_2 in air _____ the rate of corrosion.
5. The smaller the grain size of metal ______ will be the corrosion.
6. Corrosion is a reverse of _____ process.
7. If the cathodic area is small and anodic area is large the rate of the corrosion will be _____.
8. _____can be used as cathodic inhibitor.
9. The purpose of adding chromium in steel is to prevent _____.
10. An acid which is used in the phosphate coatings is _____.
11. Chromate coatings are mainly applied to _____ Metal.
12. In chromium electroplating electrolyte used is _____.
13. An acid which is used in electroplating of chromium is _____.
14. Coating of tin over iron is example for _____ coating.
15. Coating of zinc over iron is example for _____ coating.
16. In Anodic coatings the metal which is used for coating is _____ to the base metal.
17. _____ is an example for cathodic coating.
18. _____ can be used as sacrificial anodic protection.

3 Polymers

(Plastics, Rubbers, Conducting Polymers and Liquid Crystals)

3 Polymers

(Plastics, Rubbers, Conducting Polymers and Liquid Crystals)

3.1 Polymers

3.1.1 Introduction

Although the large number of macromolecules, containing few hundred to few thousand atoms) all linked to one another firmly, have evolved along with the universe, their science remained obscure and emerged suddenly into lime light, to name this technology era as *polymer or plastic age*. Their actual importance may be realized from the statistics that half the American chemists and chemical engineers are employed in industries concerned with polymeric materials. A common man can readily realize their importance by attempting to recognize the polymeric materials around him and within him. The naturally occurring macromolecules, viz., cellulose, wool and rubber are used to drape ourselves, to live, to walk and to roll on them. Majority of the food items contains polysacharides, proteins and fibers and the very existence of life and its replication is the result of biopolymers called DNA and RNA. Similarly most of the synthetic materials existing around us in the form of plastics, fibers, gums, laminates, paints, gadgets, etc., are all different forms of macromolecular materials. Some of the synthetic polymers have gone to the extent of replacing ailing hearts and other very sensitive organs.

Polymers, often called as *macromolecules*, are very large molecules formed by linking together small, simple chemical units called *monomers*. The resulting structure comprises of repetition of relatively small molecular fragments usually referred to as *repeating units*. For example, polyethylene is a polymer made up of inter linking of a number of ethylene monomer molecules. In this process the double bonds of ethylene molecules open up and bind with the adjacent C_2H_4 unit. Here (C_2 H_4) is the repeating unit.

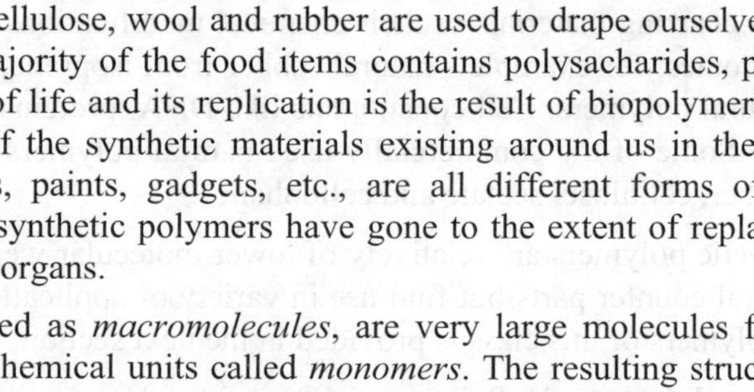

Ethylene A monomer — Polyethylene A polymer — Repeating unit

The physical and chemical properties of the polymer are much different from the monomer. In fact polyethylene behaves much like saturated hydrocarbons with their inactivity towards chemical reactions characteristic of double bonds like facile bromination.

Although polymeric materials were discovered by early chemists, polymer science may be thought to have entered this new era largely because of untiring efforts by Staudinger through his macromolecular hypothesis presented in 1920, which earned him the 1953 Noble Prize. He proposed and substantiated the molecular formula of polystyrene, rubber and other polymers. This soon paved way for commercial exploitation of polymers like PVC, Nylon, Neoprene, Styrene -Butadiene synthetic rubber, etc. Ziegler and Natta (1955) developed new polymerization catalysts with stereo-regulating powers and obtained the second Nobel Prize of polymer chemistry in 1963. Flory elucidated the mechanisms of polymerization processes and added one more Noble Prize to the field in 1974. As a result of these discoveries the polymer industry currently produces around 30 billion pounds of polymer materials annually.

3.1.2 Classification

The polymeric materials can be broadly classified on the basis of a physical verification of origin, elements involved in the backbone chains and physical properties.

(a) Natural and Synthetic Polymers : Depending on their origin, polymer can be grouped into Natural and Synthetic polymers. The natural polymers are either naturally occurring materials or their chemical modifications. In spite of discovery of synthetic routes, for example - natural rubber from isoprene, these materials still fall under natural polymers. Cell constituents like DNA, proteins etc., also fall under this category. Some of the commercially used natural polymers are cotton, silk, leather, wool, rubber, cellulose, acetate and cellophane.

The synthetic polymers are relatively of lower molecular weight when compared to their natural counter parts but find use in variety of applications. Classifications and uses of polymers of this class is provided in the next section.

(b) Organic and Inorganic Polymers : Organic polymers, as expected, have carbon back bone structures while inorganic polymers usually have back bone structures made out of non-metals other than carbon. Elements like Si, Ge, N, P, As, O, S, Se etc., with limited catenations (inter linking) usually make the back bone of inorganic polymers. In either case the side valencies of the back bone elements are satisfied by elements like hydrogen, halogens, oxygen, nitrogen, etc. Almost all the polymers of biological origin and majority of the synthetic origin belong to the group of organic polymers. Therefore, when we mention simply polymers, we generally mean organic polymers. As we already stated, organic polymers have a variety of applications and does not fit in any simple frame work. However, inorganic polymers have some specific and unique applications, where organic polymers are not of much use. The typical and most useful polymers of this class are glass and silicones (-Si-O-Si-back bone). Depending on the size and cross linking silicones may be fluids, greases, resins or rubber. Silicone rubbers retain elasticity over a wide range of temperatures and silicone lubricants are mixed in mineral oils.

(c) Thermosetting and Thermoplastic Polymers : This classification depends on the differences in physical properties on heat treatment. Polymers that soften on heating and regain original physical properties on cooling are known as Thermoplastics. On the contrary, Thermosetting plastics are deformed irreversibly on heat treatment. Linear chain polymers fall under the first category of thermoplastics. The process of heating, reshaping and retaining the same on cooling can be repeated several times on this class of polymers. Polyethylene, PVC, Nylon, etc., are typical examples of thermoplastic polymers. In case of cross linked polymers, heat treatment brings changes in the cross linkages and converts them into infusible mass, are thermosets which generally formed by condensation polymerisation.

(d) Fibers, Elastomers, Plastics and Liquid Resins : This nomenclature is based on the ultimate form which in turn depends on dominant physical property. Fibers, elastomers, plastics and liquid-resins differ in the extent of the crystallinity and physical state and are referred as crystalline, semi-crystalline, amorphous and liquid respectively.

(a) ***Fibers*** **:** Being highly crystalline, possess high tensile strength. Because of this property these materials are drawn into fibers used in weaving and making ropes. eg: Nylon, Terelene.

(b) ***Elastomers*** **:** These polymers possess only partial crystallinity and therefore can be stretched to one and half times its original length, but regain original shape on releasing. Because of their rubbery characters they are also used as synthetic rubbers. They exhibit good strength and elongation. eg., Silicone rubber, Neoprene, Natural rubbers.

(c) ***Plastics*** **:** Polymers that can be shaped into hard and tough utility articles by the applications of heat and pressure are called plastics. Eg : PVC, Teflon, Polymethylmethacrylate, Polystyrene, etc.

(d) ***Liquid Resins*** **:** Polymers existing in liquid state are used as adhesives, lubricants, sealants and resins. Eg. Some Silicones, Polysulphide Sealants, Epoxy resins.

3.1.3 Types of Polymerization

W.H. Carothers in 1929 categorized the polymerization reactins into *Addition polymerization and Condensation polymerization*.

3.1.4 Condensation Polymerization – Step Growth Polymerization

Condensation polymers are those in which the repeat unit the polymer lacks certain atoms present in corresponding monomer(s). Two like or unlike monomers join each other by the elimination of small molecules, such as water, to give the polymer chain.

When the monomers of the same kind join together, the polymer obtained is called monopolymer. Nylon-6 is formed by self condensation of ε-aminocaproic acid

$$2n\,H_2N-(CH_2)_5-COOH \rightarrow -[HN-(CH_2)_5-CO-HN-(CH_2)_5-CO]_n-OH + n\,H_2O$$

ε-aminocaproic acid Nylon-6

Polymerization of two or more different monomers gives a copolymer. Nylon-6,6 is a best example of condensation polymers. It is a copolymer, formed by the combination of a dicarboxylic acid and a diamine.

Polyamide (Nylon 6,6)

$$n\,H_2N\,(CH_2)_6-NH_2 \;+\; n\,HOOC-(CH_2)_4\text{-}COOH$$

Hexamethylene diamine Adipic acid

$$\downarrow$$

$$[\text{-}NH\text{-}\,(CH_2)_6-NH-CO\,(CH_2)_4-CO-]_n + 2\,n\,H_2O$$

Nylon

Mechanism

A general mechanistic scheme for condensation polymerization involves a simultaneous combination of any two monomers to give dimers, which in turn combine to give tetramers and so on and finally grow into a polymer in a step-wise manner.

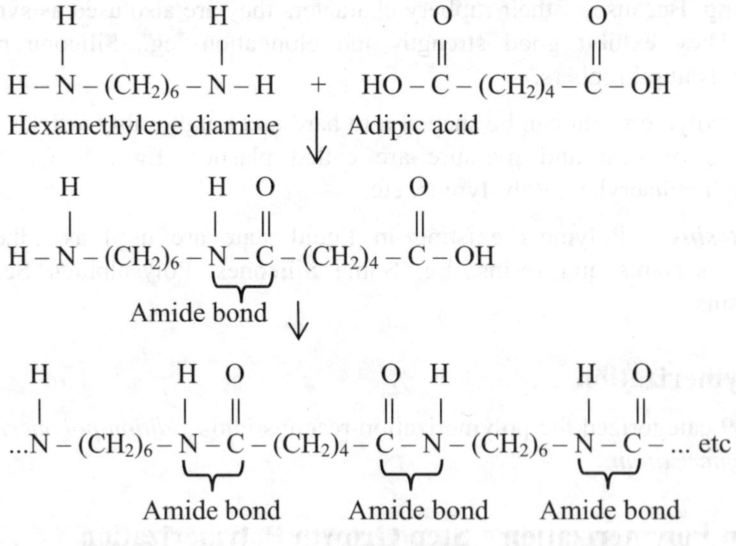

Poly (Hexamethylene adipamide) or NYLON – 6,6, is a thermoplastic polymer.

Molecular weight of condensation in polymers is determined by the number of reactive functional groups present on the monomer. Monofunctional monomers give only low molecular weight condensation products. Bi and poly functional monomers give high molecular weight (longer chain) condensation polymers. Others most useful condensation polymers are Polyesters (trade name Dacron) and Polyurethane (trade name PUF).

Polyester (Dacron or terylene) is another typical example of a condenser polymer is a thermoset polymer.

$$n\,HO-CH_2-CH_2-OH + n\,H_3CO-OC-C_6H_4-CO-OCH_3$$

Ethyleneglycol Dimethylterephthalate

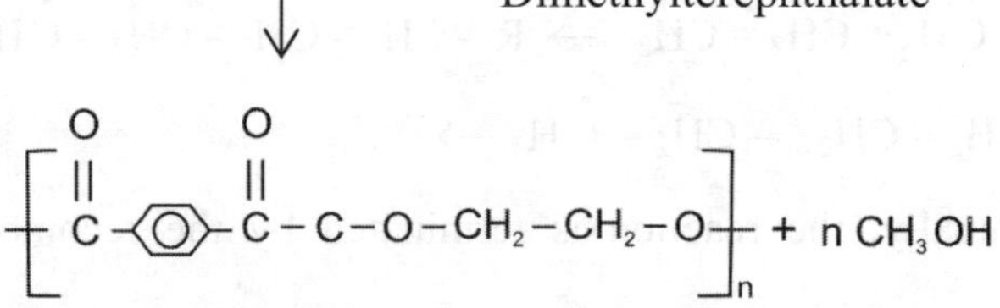

Polyester (Polyethyleneterephthalate)

Polymers having more than one atom in their back bone structure are called *heterochain polymers or heteropolymers*. Because Condensation polymers have functional groups as a part of polymer back bone, they invariably have more than a single element in the backbone structure and therefore are always heterochain polymers.

3.1.5 Addition Polymerization – Chain Growth Polymerization

Addition polymers are those in which the repeat unit and monomer(s) have same composition. Additional polymers are formed by simply adding monomer units together *without undergoing any loss of atoms or groups*.

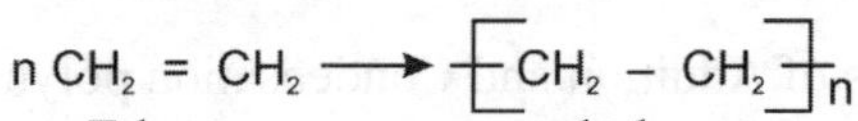

Ethene polythene

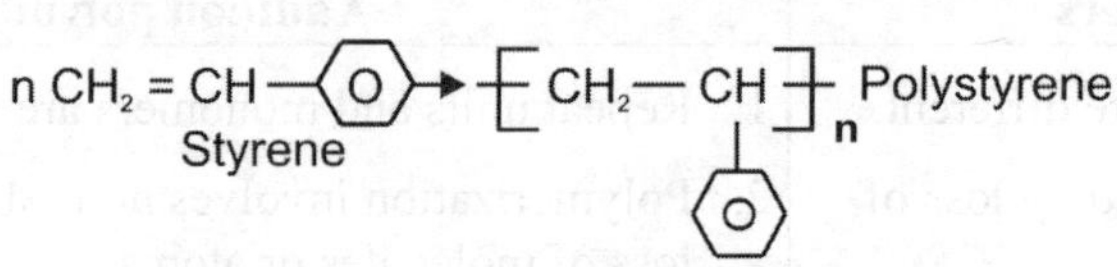

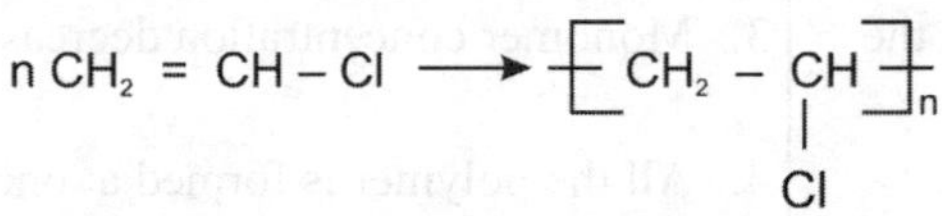

Vinylchloride polyvinylchloride

Addition polymerization is usually initiated by a radical or a cation or an anion. Most of the vinyl monomers undergo addition polymerization. For example ethylene polymerizes into polyethylene at high temperatures (100-200 °C), high pressures (100-300 atm) and in the presence of benzoyl peroxide.

Mechanism of Chain Growth Polymerization

Initiation : Homolytic cleavage of initiator leads to formation of free radicals. Thus Benzoyl peroxide decomposes to give benzoyloxy free radicals C_6H_5-CO-O$^\cdot$ (denoted by R$^\bullet$).

$$2\,R\text{-}O\text{-}R \xrightarrow{heat/h\nu} 4\,R^\bullet + O_2$$

Propagation : Benzoyloxy radical readily combines with ethylene, causing opening of double bond and yielding another highly reactive radical. This combines further with another ethylene molecule and this process goes on till a very long chain of CH_2 groups in built.

$$\dot{R} + CH_2 = \dot{C}H_2 \rightarrow R - CH_2 - \dot{C}H_2$$

$$R - CH_2 - \dot{C}H_2 + CH_2 = CH_2 \rightarrow R - CH_2 - CH_2 - CH_2 - \dot{C}H_2$$

$$\rightarrow R(-CH_2 - CH_2)_n - CH_2 - \dot{C}H_2 \rightarrow .....$$

Termination : Eventually, the reaction is terminated by the recombination of final free radicals.

$$R\left[CH_2 - CH_2\right]_{n-1} CH_2 - \dot{C}H_2 + R\left[CH_2 - CH_2\right]_{m-1} CH_2 - \dot{C}H_2 \longrightarrow R\left[CH_2 - CH_2\right]_{n+m} R$$

Here *n* gives the total monomer molecules combined to give the polymer.

The above mechanism suggests a gradual but fast disappearance of monomer to give a long polymer chain. Since the process is akin to interlinking of rings one by one to give a chain, addition polymerization is also called *Chain polymerization.*

The distinguishing features of addition and condensation polymers is given in Table 3.1.

Table 3.1 Distinguishing features of Addition and Condensation polymers

Condensation polymers	Addition polymers
1. Repeat units and monomers are different.	1. Repeat units and monomers are same.
2. Polymerization is accompanied by loss of atoms.	2. Polymerization involves no molecules or loss of molecules or atoms.
3. Almost all the monomer disappears in the first 10% of the reaction.	3. Monomer concentration decreases steadily.
4. Molecular weight of the polymer rises throughout reaction.	4. All the polymer is formed at once and molecular weight of the polymer changes little throughout
5. Low molecular weight polymers are formed. Longer reaction times are essential for obtaining high molecular weight polymers.	5. High molecular weight polymers are formed all at once.
6. Mainly heterochain polymers often have functional group atoms as a part of chain.	6. Mainly homochain polymers often have carbon chains.
7. Monomer needs to be bi- or poly- functional.	7. Monomer could be mono- functional.

3.1.6 Mechanism of polymerization

There are three main types of mechanisms encountered in polymerization reactions:

(a) ***Free radical Mechanism.*** This type of mechanism, as discussed under chain polymerization, operates in the presence of peroxide catalyst, light or heat energy.

(b) ***Anionic Mechanism***. This type of mechanism is brought into play by the use of coordination catalysts i.e. These are like transition metal ($TiCl_3$, Z_2Br_3 etc.) with organometal compounds ($Al(CH_3)$, $Al(R)_3$) Zeigler-Natta catalysts. Here the catalyst transfers a negative charge to the monomer giving a *carbanion*, which undergo initiation, propagation and termination to yield the polymer.

(c) ***Cationic Mechanism.*** In this type of mechanism, the polymerization is initiated by protonic acids, metal halides and carbonium salts e.g. tertbutyl perchlorate, which dissociates to give carbonium ion.

3.1.7 Polymer Branching

There are three main types of polymer branching :

(a) ***Linear Chain*** exhibited in the case of polymerizations of bifunctional groups (Fig. 3.1 a).

(b) ***Branched Chain*** present in the case where branches grow from the parent chain (Fig. 3.1 b).

(c) ***Cross linked Chain*** shown in the case of polymerizations from bi and tri-functional groups (Fig. 3.1 c).

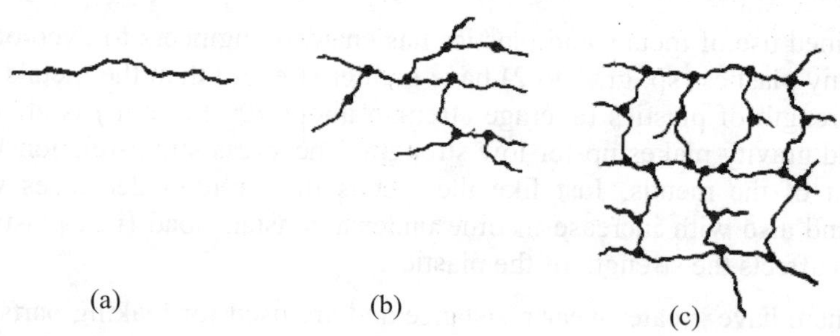

Fig. 3.1 Formation of plastics (a) Linear polymer, (b) branched polymer, (c) cross linked polymer.

3.2 Plastics

A substance that can be easily formed or moulded into a desired shape is plastic. A plastic material can be permanently formed into a desired shape at some stage in its manufacture or processing by means of mechanical force, or under combined effect of mechanical force and heat. In most cases the temperatures required are not very high and steam heating is adequate. The common plastics are organic high polymers 'which possess the property of plasticity at some stage during their fabrication, and which are substantially rigid after fabrication'.

The origin of modern plastics may be traced back to the discovery of nitrocellulose in 1846. Articles made of nitrocellulose were displayed in 1862. The history of industrial plastics begins with the preparation in 1869 of celluloid from cotton linters, wood pulp, camphor and nitric acid. The second landmark in industrial plastics was the preparation of casein plastics in 1890 from formaldehyde and sour milk. Real incentive to the industry came in 1909 from the synthesis of Bakelite: named after its discoverer Dr. L. H. Bakeland. It is prepared by the reaction between phenol and formaldehyde. During the last half century, a large number of plastics have been developed and are manufactured on a very large scale.

In the manufacture of plastics raw materials like coal, petroleum, cellulose (prepared from wood and cotton), salt, sulphur, limestone, air, water, etc, are used. These are subjected to several processes of chemical decomposition and recombination giving intermediates which are then made into plastics.

The Indian plastics industry is of recent origin. The production of phenol formaldehyde started in 1946, polystyrene and polyethylene in 1957, polyvinyl chloride in 1961 and cellulose acetate in 1963. The development of the plastics industry depends to a considerable extent on the petro-chemical industry. As the latter is firmly established in the country the production of plastics will increase.

3.2.1 Plastics as Engineering Materials

The chief engineering advantages of plastics over other engineering materials result from their light weight low fabrication cost, low thermal and electrical conductivities, high resistance to corrosion and solvents and transparency. In addition they have a decorative effect. A further advantage is that the plastics can be used in conjunction with metals.

The combined use of metals and plastics has enabled engineers to overcome many design problems. Many plastics (sp. gr. 1 to 2) have greater strength than the metals per unit weight. The tensile strength of plastics (average strength about 700 kg/ cm^2) is about one fourth of aluminium and gravity makes up for low strength. The stress-strain-relationship of plastics is similar to that of the metals. Just like the metals the strength decreases with increase in temperature and also with increase in time under a constant load (i.e., plastics show creep). Humidity also affects the strength of the plastics.

Some of them have greater wear resistance and are used for making parts that rub against other parts, e.g., bearings, gears, etc. Remarkably low coefficient of friction of nylon against metals enables it to be used for making bearings and gears. Another advantage is that plastics reduce noise and vibration in machines. Plastics parts can be easily and cheaply made with accurate dimensions. Complicated shapes can be produced by moulding. Plastics, particularly thermosetting plastics, have a tendency to shrink.

Most plastics cannot be used at elevated temperatures, e.g., thermoplastics generally cannot be used above 100° and thermosetting above 150°. Plastics being bad conductors of heat are useful for making handles used for hot objects, and for the same reason in cold climates the plastics feel warm to the touch. Most plastics are inflammable, but some are fire-proof.

Most plastics are electrical insulators and find large scale use in the electrical industry.

In general, plastics are resistant to chemicals. Thermosetting plastics are more insoluble in organic solvents than the thermoplastics.

Most plastics are clear and transparent. They can be given beautiful and brilliant colours. Optical clarity combined with strength makes them suitable for making wind screens for automobiles, aircrafts, etc.

Many failures in the usage of plastics materials in engineering have occurred because of inadequate knowledge of their properties under working conditions. Many of the plastics have poor weather resistance and their properties deteriorate and dimensional changes take place. Generally they are soft, have low strength and low modulus of elasticity. It is difficult to repair parts made of plastics.

3.2.2 Types of Plastics

Plastics have been classified in several ways. From the engineering point of view it is most satisfactory to classify them as : 1. thermoplastics and 2. thermosetting plastics. Some of the more common plastics are classed as under:

Types of Plastics

Thermoplastic Resins	*Thermosetting Resins*
Acrylics	Alkyds
Cellulose derivatives (e.g., celluloid cellulose acetate)	Melamines
Polyamides (e.g., Nylon)	Phenolics (e.g., Bakelite)
Polyethylenes	Polyesters (e.g., terylene)
Polystyrenes	Silicones
Polyvinyls	Ureas

The behaviour of a few plastics is in between that of the two extreme types but most plastics fall under one or other of these types. The load-carrying capacity of thermoplastics is appreciably affected by the working conditions, because the material becomes soft and plastic with an increase in temperature. They have higher compressive strengths than thermosetting plastics materials, although the actual strength of the latter is usually greater than that of the former.

3.3 Thermoplastic Resins

As already explained, the thermoplastics soften or melt when heated. On cooling they become rigid. On reheating they again soften or melt and can be given the desired shape by moulding. A Resin is a pure polymer and plastic is a compounded resin. Thus these plastics are repeatedly heated and cooled without decomposing, to get back same chemical properties but physical structure may change. Their behaviour can be compared to that of the common sealing wax which softens on heating and takes the desired shape. On reheating it softens again and a new shape can be given. The thermoplastic resins consist mostly of linear polymers. Heating or moulding operations do not change the chemical structure or the molecular weight of the material. The thermoplastics are smaller molecular weight compounds than the thermosetting plastics. In them the inter-molecular forces are weaker than in thermosetting plastics. This explains why these swell or dissolve in suitable solvents. This property enables them to be mixed with plasticizers which keep the material soft and pliable.

On the basis of the raw materials used, the thermoplastics may be subdivided into the *cellulosic* and the *non-cellulosic* types.

3.3.1 Cellulose Resins

The important cellulose derivatives are its nitrate, acetate, acetate-butyrate, propionate and methyl cellulose. Strong alkalis decompose most of the cellulosic plastics. Weak acids and weak alkalis affect them slightly. Cellulosic plastics are very good electrical insulators. They have considerable strength and toughness. Most of them have poor fire resistance. Rayon is regenerated cellulose in the form of filament. It has low gravity and great tensile strength.

Cellulose molecule, $(C_6H_{10}O_5)_n$ has a large chain of $C_6H_{10}O_5$, groups joined together, each of which has three hydroxyl groups which may be esterified or etherified Cellulose derivatives in general and some other thermoplastics are not workable. Substances like camphor, glycolesters, phthalates, etc. known as *plasticizers*, are added to these plastics. The plasticizers improve the flow properties of the plastics. They enable the plastics to be worked at lower temperature, make them resistant to water and temperature change, and improve toughness, flexibility and non-inflammability. Plasticizers usually decrease the strength.

Cellulose acetate is made by first preparing cellulose triacetate by acetylating cellulose, e.g., purified cotton linter is treated with acetic anhydride, glacial acetic acid and sulphuric acid. When the acetate content is about 62.5% it is hydrolysed, precipitated, washed and dried. The product is kneaded in a mixer with a plasticizer. Cellulose acetate plastics are stable, transparent and slow burning. They are not very resistant to moisture, but are tough, have high impact strength, and are available in attractive colours. On account of its mechanical strength, toughness, wear resistance, ease of fabrication and transparency, cellulose acetate has found wide application in the automotive industry. Cellulose acetate plastics can be more easily moulded into the desired shape than cellulose nitrate by extrusion and especially injection moulding. Cellulose acetatebutyrate has similar properties but has better resistance to moisture and better dimensional stability than the acetates.

Cellulose nitrate is quite hard and brittle. It can be used as a plastic material only after incorporating a plasticizer like camphor. Rods, sheets and tubes of cellulose nitrate are available in various colours. Cellulose nitrate plastics are highly inflammable, unstable at higher temperatures, sensitive to sunlight and are affected by acids, alkalis and organic solvents like alcohols, ethers and ketones. In spite of these drawbacks, they are extensively used because of their toughness, strength, dimensional stability and water resistance. They are used mostly as sheets or blocks and are shaped by sawing, turning, drilling or pressing the pre-cut solid blocks. They cannot be processed by injection moulding as they do not withstand the temperature required for the process. Unlike most other plastics materials, granules of cellulose nitrate are not used for moulding because they do not weld together.

Ethyl cellulose is the most important of the cellulose ethers. It is chemically stable and is resistant to alkalis and to dilute acids. It has a low softening temperature (120°C). Ethyl cellulose plastics are very tough, strong, resistant to moisture and are good insulators. They have good dimensional stability and do not distort on heating. Thin sheets of this material are widely used for packaging. Because of its low softening point, it is easily fabricated by compressing, extrusion and injection moulding. Ethyl cellulose is largely used in coatings on adhesives.

3.3.2 Non-Cellulose Resins

The second type of thermoplastics is derived from *non-cellulosic materials*. Plastics of this type are in general very tough, chemically inert, non-inflammable or slow-burning and have good electrical and water resistant properties.

Polyacrylates : A large quantity of non-cellulosic plastics are derived from ethylene. Acrylic esters are prepared from ethylene by converting it first to ethylene chlorohydrin, then ethylene cyanohydrin which on esterification gives the acrylates :

$$\underset{\text{Ethylene}}{CH_2 = CH_2} \xrightarrow{HOCl} \underset{\text{E. Chlorohydrin}}{CH_2OH.CH_2Cl} \xrightarrow{NaCN} \underset{\text{E. cyanohydrin}}{CH_2OH.CH_2CN}$$

$$CH_2OH.CH_2CN + \underset{\text{Methanol}}{CH_3OH} \xrightarrow{H_2SO_4} \underset{\text{Methylacrylate}}{CH_2 = CHCOOCH_3} + NH_3$$

The esters obtained in such a manner are polymerized to give the plastics called polyacrylates. The **acrylates** stand as a class because they are water-clear transparent. Age, weather and sunlight have little effect on their transparency. Because of these properties, numerous applications follow, e.g., aircraft and cockpit enclosures, windshields, 'Plexiglas', tubes, sheets, rods, etc.

The acrylates also have toughness, strength, flexibility, low specific gravity and permanence of dimensions. Acrylates are also avaibable as staple fibres (e.g., orlon).

3.3.3 Teflon: Tetrafluoroethylene (PTFE)

It is prepared by addition or chain polymerization of water-emulsion mixture of bifunctional tetrafluro ethylene monomer under pressure in a vessel in the presence of benzoyl peroxide catalyst.

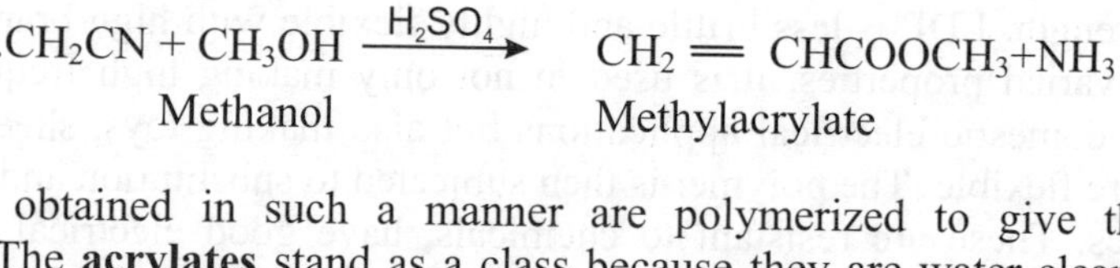

Teflon belongs to the *fluorocarbons*. These have very good chemical, electrical and thermal resistance and can be used from –70° to 250° temperature range. Because of these properties, the fluorocarbons are used for handling hot and corrosive liquids and as insulators in equipment using high voltage and high frequency electricity.

It has high softening temperature (350 °C) and high density (2.1 to 2.3 g cm^{-3}) but can not be dissolved to give a molten state. It sinters at 350 °C to form a very viscous, opaque mass that can be moulded by applying high pressure.

3.3.4 Polyethene (PE)

These are prepared by polymerizing, bifunctional ethylene monomer under high pressures and moderately high temperature. It crystallizes very easily from solution and has highly symmetrical chain-structure. The crystallinity varies from 40 to 95% depending on the degree of branching which in turn governed by the operating pressures in the process of polymerization as explained in 3.1.5.

$$nCH_2 = CH_2 \longrightarrow [\!-CH_2-CH_2-\!]_n$$

High pressure causes high and low pressure causes low branching structures in the polymer products resulting in three grades of polymers with varied densities. 1. Type-I is a low density polymer with 0.91 to 0.925 g cm^{-3}. 2. Type-II is a medium-density with 0.925 to 0.940 g/cm^3 and Type-III is a high-density with 0.94 to 0.965 g/cm^3 densities. High density polymer (HDP) is completely linear molente with a high softening point of 130 °C, brittle and has less impact strength. LDP is less brittle and highly flexible with high branching strength. Because of these varied properties, it is used in not only making high frequency insulator parts, kitchen and domestic electrical applications but also making toys, sheets, tubes, pipes and bags as they are flexible. The polymer is then subjected to substitution and other reactions to give the plastics. These are resistant to chemicals, have good electrical resistance, low specific gravity, low water absorption and flexibility. They are much cheaper than the other plastics.

Their resistance to chemicals falls drastically at higher temperatures. They are adversely affected by sunlight. Above 60° oils, greases, aromatic hydrocarbons, etc., affect polyethylenes.

3.3.5 Polystyrene (PS)

These are prepared from bifunctional styrene monomer (i.e., phenylethylene) as explained in 3.1.5.

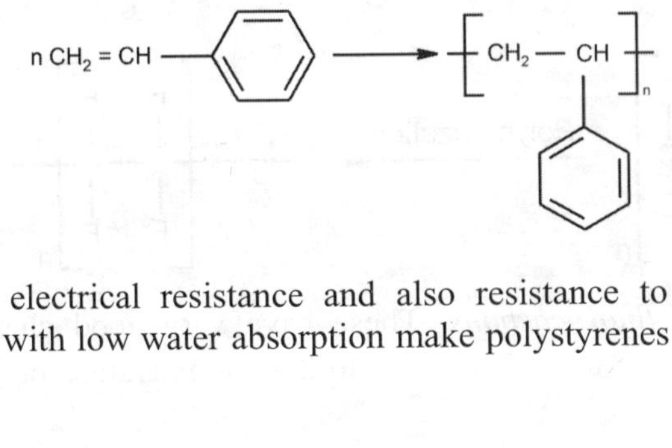

They have high electrical resistance and also resistance to acids and alkalis. These properties combined with low water absorption make polystyrenes very suitable materials for making electrical and electronic appliances. They also have permanence of dimensions and are hard. This can be used in the manufacture of Buna styrene rubber.

3.3.6 PVC

These are prepared from bifunctional vinylchloride monomer as explained in 3.1.5. *Vinyl* plastics are numerous. They are basically derived from acetylene. The vinyl group, $CH_2{=}CH-$ is very reactive and readily gives acetal, acetate, alcohol and chloride. These materials, on polymerization give the corresponding plastics. In general, these plastics have high electrical resistance and resistance to chemicals and to corrosion, combined with flexibility. These properties make them very suitable for coating wires and cables. They are strong, machinable, do not absorb moisture and the shapes of the finished products do not change. Vinyl plastics have a wide range of properties, e.g., from soft and flexible to hard rigid. High softening temperature of 140 °C.

$$n\,CH_2=CH-Cl \longrightarrow \left[-CH_2-\underset{\displaystyle Cl}{\underset{|}{CH}}- \right]_n$$

The largest use of vinyl resins is in the manufacture of safety glass and helmets. They are also used in large quantities in the manufacture of chemical resistant fabrics and for electrical insulation. Platicized PVC is used in manufacturing of plastic cloth, belts and thermal insulating foam.

3.3.7 Nylons

The condensation of diamines and dibasic acids gives linear products. Properties can be varied greatly (from hard and tough products to soft and rubbery products) by varying the nature of the acid and the amine. The use of nylon, which is a *polyamide*, in the textiles is very well-known. It is tough, and heat and abrasion resistant. It is also moulded to produce gears and bearings.

These are prepared by poly condensation of diamines and dibasic acids as explained in 3.1.4.

$$2n\,H_2N-(CH_2)_5-COOH \longrightarrow \left[-HN\,(CH_2)_5-\overset{\displaystyle O}{\overset{\|}{C}}-NH-(CH_2)_5-\overset{\displaystyle O}{\overset{\|}{C}}- \right]_n$$

Nylon-6 + H_2O

Polyamide (Nylon 6,6)

$n\ H_2N-(CH_2)_6-NH_2 \quad + \quad n\ HOOC-(CH_2)_4\text{-}COOH$

Hexamethylene diamine   Adipic acid

↓

$[-NH-(CH_2)_6-NH-CO\,(CH_2)_4-CO-]_n + 2\,n\,H_2O$

Nylon-6, 6:

Among the Polyamides the most widely known are the **nylons** of which nylon 6,6 is the most common. It is prepared by the condensation reaction of hexamethylene diamine with adipic acid. It is a copolymer formed by polycondensation of two or more monomers as explained in 3.1.4. It has remarkably low coefficient of friction against many metals which makes it a good material for making bearings and gears. It is self extinguishing when it comes in contact with a flame. It has low water absorption and good strength which does not deteriorate even when used for a long period of time.

Nylon can be shaped by casting, extrusion, blowing and moulding giving products of excellent durability and toughness.

3.4 Thermosetting Resins

As already explained, the thermosetting resins are produced when the molecular weight and chemical structure of the resin changes on heating. When the material is moulded, additional cross links are formed between the long chains and the molecular weight increases. In the formaldehyde resins the chemical reaction is condensation, in the glyptal resins it is esterification. In all such cases cross linking is very important. In the partially set state these are usually soluble in solvents but when cross linkings are formed, the plastics acquire some of their characteristic properties, e. g., they dissolve or swell less in solvents and soften less on heating, and hard, tough, and highly insoluble products are obtained. The thermosetting resins have direct covalent links between the long chain molecules, giving a threedimensional molecule. In the fibres like nylon, which are thermoplastics, there are strong intermolecular forces of attraction which bind the molecules, and the material consists of many smaller (as compared to thermosetting resins) molecules held together by these forces. This accounts for the differences in the behaviour of the two groups of plastics. These are summarized.

Thermoplasts	Thermosetts
1. Heat & cooling: Softened on heating and hardened on cooling by regaining original properties. These can be reshaped and used.	Cannot be softened as they are potentially set. If we attempt to soften them, they decompose and can not get back its structure. Hence can not be reshaped and used.
2. Formation: Generally by addition polymerization except nylon.	Condensation polymerization is seen largely in these formation.
3. Size & Shape: Small molecular weight compound with linear structures	Large molecular weight compounds with three dimensional network
4. Linkages: Little or less cross-linked structures.	More highly cross-linked structures.
5. Bondages: Weak vander walls forces or H-bonds	Highly cross-linked structures strong covalent bonds are responsible for strength
6. Generally soft weak and less brittle	Usually hard, strong and more brittle
7. These can be reclaimed and reshaped from the waste	Cannot be reclaimed and reshaped from waste.
8. Generally soluble in organic solvents	Not soluble in almost all organic solvents.
9. Curing: By cooling	By applying heat and pressure.
10. Examples: PE, PS, PVC and Teflon	Backlite and polyester

3.4.1 Phenolic Resins - Bakelites

Phenol and formaldehyde react to give active monodtrimethylol monomers which in turn involve in polycondensation to give various linear or cross-linked polymer structures as shown below.

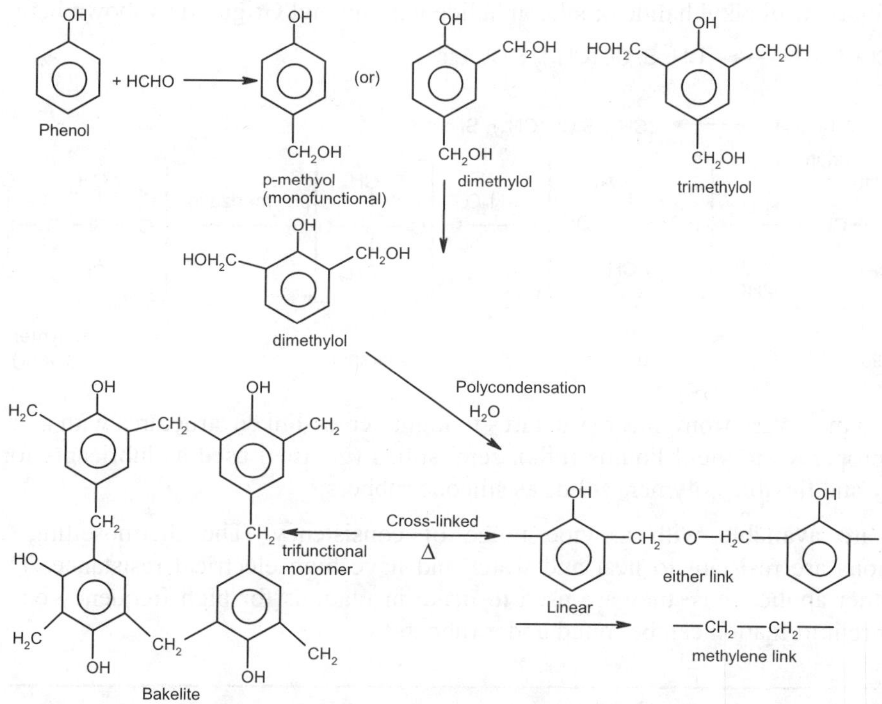

Resins of this group are more widely used than those of any other group. The most important ones are formed by the condensation of phenol and formaldehyde as shown below. The properties of the product depend upon several factors, like the concentrations of reactants, the catalyst, the time and temperature of the reaction. Phenol-formaldeyde resin, known as **Bakelites**, have outstanding electrical and mechanical properties and are resistant to heat, water and many chemicals. Strong caustic alkalis and concentrated Sulphuric and nitric acids decompose them. They are hard and infusible and retain their shapes. Usually various fillers and fabrics are combined with the resins for different purposes. To increase mechanical and dielectric strength and lightness, wool floor is commonly used as a filler. The resulting product is used where light material is desired, as for aircraft, automobile and telephone industries. Resistance to chemicals and heat is increased by using asbestos as a filler. Moulding sheets, which have many applications, are made by impregnating paper with the resin. Thus by a suitable filler the range of usefulness of phenol-formaldehyde plastics is greatly increased.

Phenol also forms a condensation product with furfural. The product has properties similar to phenol formaldehyde plastics and can be used under difficult and complicated moulding conditions which require a long time of flow. Thus articles of thick sections can be made out of it.

3.4.2 Silicones

These have silicon-oxygen chain to which are linked various organic groups.

These are prepared by polycondensation of the active monomers of monofunctional/bifunctional/trifunctional alkyl silicon chlorides which are prepared by reacting silicon with alkyl halide or silicon halide with methyl Grignard as shown here

1. $CH_3Cl + Si \longrightarrow CH_3\,Si\,Cl, (CH_3)_2\,SiCl_2$ etc

2. $Si\,Cl_4 + \underset{\text{grignard}}{CH_3MgCl} \longrightarrow (CH_3)_3SiCl, (CH_3)_4\,Si$ etc

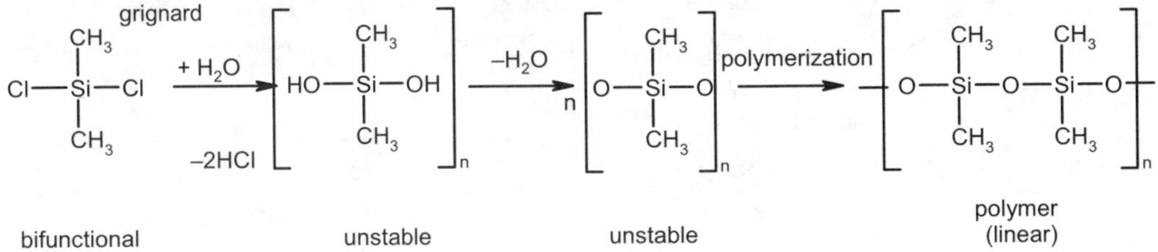

The polymers vary from linear structures to highly cross-linked structures with a variety of physical properties to yield liquids (oils), semi solids (greases) used as lubricants for heavy machinery and flexible polymers solids as silicone rubbers.

They are available with a wide range of consistency. The thermosetting silicone compositions are resistant to heat and water and have high electrical resistance. Therefore besides other applications, they are used to make insulations for high frequency equipment. (Some more information can be found under rubbers).

3.5 Compounding of Plastics

The materials described above are called resins or binders and are the chief ingredients of plastics objects. They have plasticity or binding property but need other ingredients to be mixed with them for fabrication of useful shapes.

Almost invariably plasticizers have to be added to the thermoplastics because the thermoplastics are inflammable and lack flexibility and toughness. Plasticizers reduce these defects and also improve the flow-property of the plastics. The plasticizer should dissolve the resin at the working temperature and should not evaporate. Plasticizers are not used with thermosetting resins because the two are incompatible and a plasticizer gradually sweats out.

Thermosetting resins generally lack elasticity and they crack under stress, To remove this defect *fillers* or *extenders* like wood flour, cotton or cloth fibres, asbestos, mica and clay are added, These are usually cheaper than the resins and the amount added may be more than that of the resin. Fillers impart some very useful properties e.g., wood flour reduces the specific gravity and absorbs shocks, cotton and other fibers impart strength and toughness, asbestos and mica are especially useful for plastics for electrical appliances.

To give decorative effect, *dyes* and *Pigments* are added. To prevent the plastics from sticking to the moulds lubricants like oils, waxes and soaps are added.

Fabrication of Plastics

To give suitable shapes to plastics, several methods of fabrication are in use. The method chosen depends upon the type of the plastics, i.e., thermoplastics or thermosetting plastics, and the shape of the finished product.

Compression Moulding

This is the most widely used method, both for thermosetting and thermoplastic resins. The quantity of the resin, filler and other ingredients placed in the mould should be such that the volume of the compact mass is slightly more than the volume of the cavity of the mould. The mould is usually preheated. It is closed after placing the ingredients and heat and pressure are applied, then the ingredients soften and fill the cavity of the mould. For thermoplastic resins, the mould is cooled before releasing the pressure, otherwise the object will distort. Thus because of alternate heating and cooling, compression moulding is not the most suitable method for thermoplastics. The thermosetting resin has to be kept in the mould for a period of time sufficient for the completion of the chemical reactions and the formation of the permanent shape. Cooling of the mould is not necessary. The temperature ranges from 100 to 200° and the pressure from 100 to 500 kg/cm^2.

Cold Moulding

It is mostly used for phenol-formaldehyde resins for electrical fixtures. The products are of inferior strength and surface finish. The moulding composition is shaped by pressure alone. In the case of a thermosetting resin, the article is cured, i.e. heated in an oven. The chief advantage of the method is that the output is large because the time taken in the mould is short, say, 10 seconds.

Injection Moulding

It is the most widely used method for moulding thermoplastic resins. In compression moulding of thermoplastics the difficulty is that the mould has to be cooled before removing the object from it. In injection moulding use of hot plastics in a cold mould removes this difficulty. The material is passed through a heating chamber where it softens. See Fig. 3.2. It is then forced by a ram into a cold mould and left there to become hard. The process is usually made automatic and the time required is 10 seconds to 1 minute. *Jet moulding* is a modification of injection moulding so as to make it suitable for thermosetting plastics. As the material is forced through the nozzle, intense heat is applied by induction heating equipment. The material hardens in the mould. The advantage of this process is a high rate of production.

Injection and jet moulding methods are similar to compression moulding methods but the latter method is slower. However, more complex shapes can be formed and a variety of raw materials can be used in compression moulding.

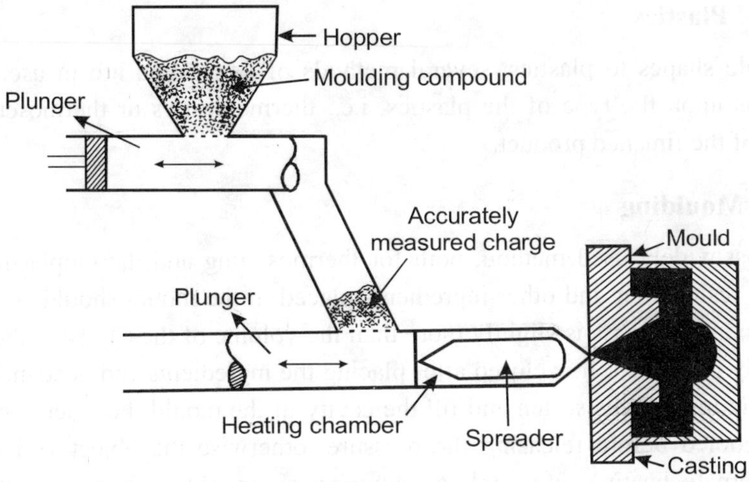

Fig. 3.2 Injection moulding.

Transfer Moulding

This method is almost exclusively used for thermosetting plastics and has been developed to make compression moulding applicable to more complicated shapes and permits the use of delicate inserts. The material is heated in a chamber to melt it and then is forced into a hot mould. The mould is opened after the plastics has set.

Extrusion Moulding

It is similar to injection moulding and is applicable to thermoplastics only. The material is forced by a screw drive into a heated chamber where it softens, and is forced out through a die. The finished product coming out from the die is cooled by blowing air or spraying water. Extrusion moulding is used for making continuous shapes like rods, tubes, sheets, etc.

Casting

Casting is a very simple method but is only used to a limited extent. The mould is usually made of lead. The molten resins are poured in the mould and cured at about 70° for several days. The mould is removed, remelted and a new mould is again made. This method is used for making special shapes. The products are water-white in colour, free from internal stresses and can take high polish. The cost of production is high.

Blowing

This method is similar to that used in the glass industry. The thermoplastic material is softened and then blown by air or steam into a closed mould.

3.6 Fibers

Fibers are a class of materials that are continuous filaments or discrete elongated pieces. They are highly crystalline form of polymers. They are biologically very important in both plants and animals. The ligaments etc. that hold the tissue material are basically fibers. Fibers are used for making textiles, utilities, ropes, strings, etc.

Fibers are classified into two categories depending on their origin, viz., natural and synthetic.

Natural fibers: Nature fibers include those produced by plants, animals and geological material. They are environment friendly and biodegradable. They are:

(i) *Vegetable fibers:* They are basically cellulosic material and include cotton, jute, etc. Plant fibers found application by even early man and are used for making textiles, ropes, mats, paper, bags, etc.

Dietary fiber is an essential component of our food and its deficiency could result in cancer.

(ii) *Wood fibers:* The strength of a plant is due to presence of wood fiber. Wood pulp is used in making paper and wood fibers like jute are used for making bags.

(iii) *Animal fibers:* They are largely made of protein. Pure silk, wool, hair are animal fibers. We move our limbs using fibers present in them. Spider silk is used for making special bullet proof jackets.

(iv) *Mineral fibers:* Asbestos is a typical example of mineral fiber. Mica and other minerals are also used as fibers.

Synthetic fibers: Synthetic fibers can be produced in large qualities and are cheaper than some of the natural fibers like pure silk. Polyamide nylons, polyesters, PVC, phenol formaldehyde resins, polyethylene are often used for making textiles. An account of nylon, PVC etc., are already given and we now give an account of polyesters.

Polyester: This category of polymers have ester linkages in the main chain. It takes 18% of market share of synthetic polymers and only third after polyethene (33.5%) and polypropylene (19.5%) the production in the year 2008 is around 50 million tones out of which 80% (40 million tones) is used exclusively for making textiles.

Synthesis of polyesters is achieved by polycondensation of purified terephlthalic acid (1, 4 – benzene dicarboxyne acid) (PTA) and monoethylene glycol (MEG). The general equation could be written as a reaction between di acid and a diol.

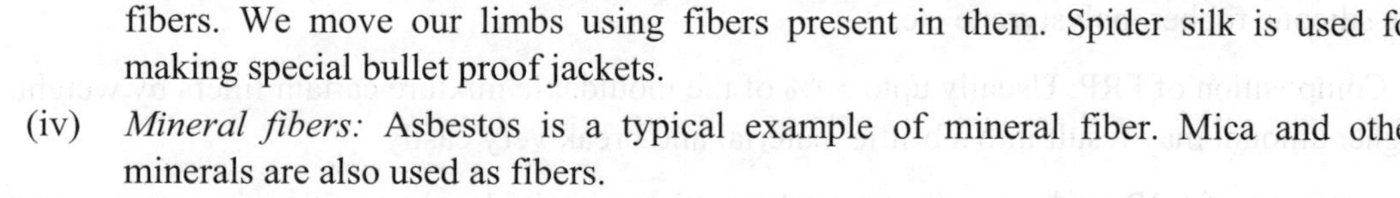

The enterfication is catalized by the presence of mineral acid like HCl.

3.7 Reinforced Plastics

Generally pure polymers are soft and less used, where as the compounded polymers with plasticizers are hard, elastic and largely used.

The combination of plastic material and solid fillers will give hard plastic with good mechanical strength and impact resistant, known as reinforced plastic. The fibre polymers impregnated with solids / fillers to impart mechanical strength and hardness without loosing plasticity are known as fibre reinforced plastics (FRP). Various fillers are known to impart characteristic property to the plastic material. For example, fillers like carborandum, quartz and mica are added to impart hardness and strength. Barium salt fillers are impervious to X-rays. Asbestos fillers added to plastics provide heat and corrosion resistant for FRP. A number of such fillers are available to impart one or other advantage property for FRP in the market. These are 1. Woodflour 2. Saw-dust 3. Ground cork 4. Paper pulp 5. Corn husk 6. Cotton fibres 7. Kelvan fibres 8. Nylon fibres 9. Polyester fibres 10. China clay 11. Marble powder 12. Punice powder 13. Metal oxides 14. SiC 15. Carbon 16. Graphite 17. Metal powders of Al, Cu, Pb etc. 18. Asbestos 19. Mica 20. Glass fibres.

Nature of polymers used for FRP: Both thermoplastics and thermosett plastics are used such as PE, PP, Nylon-6, PET; polyester, melamine. The rubber materials of natural or synthetic like silicone rubber, polystyrene etc.

Composition of FRP: Usually upto 50% of the mouldable mixture cartain fillers by weight. Higher amount may result into a brittle material and break very easily.

A variety of FRP are known in the market which contain the above fillers in varied extent imparting gradual increase in a specific property as cited herein.

1. Addition of carbon black to natural rubber gradually increase the tensile strength of rubber upto 40% and such rubbers are used in the manufacture of tyres of various vehicles.
2. Addition of ehina clay improves the insulation property of polymers like PVC, Teflon.
3. Calcium carbonate addition to PVC, generally used for the insulation purposes of tubings, seat covers, wires and cables.
4. Asbestos filled FRP are largely employed for electrical appliances for insulation.
5. FRP possess very good shock and thermal resistances, processability, mouldability, dimensional stability and repairability.

Applications and uses: The reinforced plastics with above properties are suitably applied for making variety of novel products. Some of those and their uses are mentioned below:

1. For making door handles, battery cases, exhauster fans, engine cooling fans etc., in automobiles using the base polymers like polyethene, polypropylene and nylon-6.

2. For making nose cones, pistol grips and riffle bullets in defence industries using the base polymers nylon-6 and polystyrene.
3. For making computer tape, insulators, wires and cable insulation, switch, gear parts, spools; exhaust fans etc., as electrical appliances in electrical industries using the base polymers like PET, nylon, PP and styrene acrylonitrile copolymer.
4. For making shuttle and textiles using nylon-6 as base polymer in textile industries.
5. For making the consumer goods like doors, windows, tables, chairs, alimarahs, cameras, housings etc., using polypylene (PP) and Acrylobuitadiene styrene blended plastics.
6. For making other miscellaneous items like water meters and chemical pump housings, tubings, seat coverings, etc., using the base polymers such as PVC, polypropylene and nylon-6.

3.8 Elastomers–Rubbers

Polymers having molecules arranged in the form of zig-zag or spring-shaped chains are commonly called elastomers. Most of the commercial rubbers are thermoplastic materials of this type. In recent times the distinction between rubbers and plastics has become very obscure, because a large number of copolymers now being used have structures between rubber and simple thermoplastic resins.

3.8.1 Natural Rubber

Natural rubber is a high molecular weight hydrocarbon polymer represented by the formula $(C_5\ H_8)_x$. It is obtained from a milky emulsion, called latex, by tapping the bark of the tree *Hevea brasiliensis*. This tree was originally found along the Amazon river in South America and its seedlings were planted in Ceylon.

Rubber plantations flourish best in the equatorial climate. In India about 95% of the plantations are in Kerala. The remaining are in Tamil Nadu, Karnataka. Goa, Assam and Maharashtra States. The total production nearly 2 lac tonnes.

When the trees are 6 years old spiral grooves are cut in the bark, the sap flows slowly and is collected in cups. It has about 35% rubber. The latex is diluted with water and filtered to remove coagulated rubber, barks etc., and acetic or formic acid is stirred in it when the rubber coagulates. Rubber is separated from the liquid by squeezing between rolls and is then washed with water.

The sheets so obtained are dried either in the air when 'crepe' rubber is obtained, or in hot smoke when 'smoked' sheet rubber is obtained.

The product so obtained is a high polymer of isoprene,

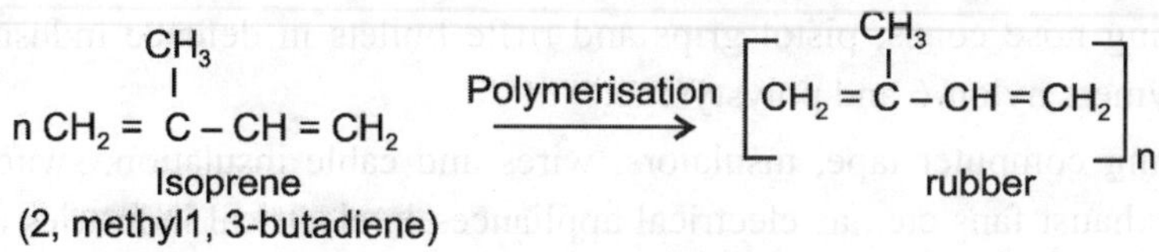

In a normal position the molecule is like a coiled helix. It uncoils in tension. When tension is removed, the molecule recoils and returns to the original coiled state. This accounts for the elasticity of rubber. If rubber is stretched too much, one molecule chain slips over another and a permanent deformation takes place. In this process of uncoiling and recoiling some energy is lost. This gives rubber the property of absorbing shock and reducing vibrations in machines. Raw rubber has some serious defects, e.g., it becomes soft and sticky with heat and stiff and brittle in cold. It is not very tough and deteriorates through slow atmospheric oxidation giving a bad small. These defects are removed by vulcanization.

3.8.2 Vulcanization

When raw rubber is passed between hot rolls both physical and chemical changes take place and the mass becomes plastic. Before vulcanization proper substances are added in order to control properties like strength, toughness and hardness and to make it resistant to abrasion, chemicals, solvents, oxygen, light and oils. Some fillers like barytes ore, cotton, etc. are also added. Zinc oxide and zinc soaps are added to decrease the vulcanization time and are called accelerators. The vulcanized rubber deteriorates with age, hence some age-resisters, mostly organic nitrogen compounds, are also added. Antioxidants are used to give resistance to oxygen. Carbon black increases abrasion resistance and is the important reinforcing agent. Raw rubber, when in a plastic state during hot rolling, is mixed with these ingredients and with sulphur, which is the most important vulcanizing agent. Soft rubber is obtained when 0.5 to 5% sulphur is used. Addition of over 30% sulphur produces hard rubber or ebonite. For uniformity of composition the material is passed many times between the rolls. This operation, called milling, requires considerable power and the temperature rises. Rubber is then cut into pieces of suitable size, placed in moulds, and formed into the desired shape and vulcanized in a single operation by the simultaneous application of heat and pressure. The rubber flows and fills the cavities and sets by vulcanization giving the desired shape. The moulds are opened and the finished products are obtained.

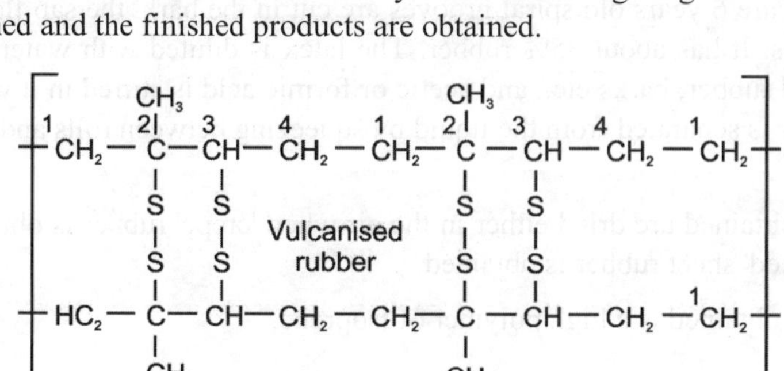

Vulcanization involves complex and not too well understood, chemical reactions. The helix-like molecules of unvulcanized rubber can be stretched but they slide over each other and raw rubber deforms readily, i.e., does not regain the original shape. In vulcanization, cross links are formed between the long unsaturated hydrocarbon chains of the original molecules. The cross links provide anchorage and prevent inter-molecular movements with the result that when the vulcanized rubber is subjected to stress, elastic strain is produced and it regains the original shape on removing the stress. The cross links may be directly from carbon to carbon or through a sulphur atom or through an oxygen atom or through all the three. Thus the vulcanized rubber is of higher molecular weight than the raw rubber and is less unsaturated.

On vulcanization, the properties of rubber change considerably. For instance, the unvulcanized rubber deforms, very readily becomes plastic and soft on working between hot rolls, and becomes viscous in hydrocarbons like benzene; but vulcanized rubber is elastic, crumbles between hot rolls and retains its form in many solvents. Besides bringing about these changes in properties, vulcanization enables the manufacturer to manipulate the material to produce the desired shapes.

3.8.3 Artificial Rubber

Strictly speaking, synthetic rubber is not produced in the sense that the synthetic product differs in chemical composition and properties from those of the natural rubber. 'Synthetic' rubbers are only rubber-like products. Therefore the term 'artificial' rubber is more appropriate than 'synthetic' rubber. The development of the automobile industry brought a revolution not only in the petroleum industry but also in the rubber industry. Rubber from the plantations could not meet the demand and attempts were made to synthesise natural rubber. As this did not succeed, rubber-like materials were synthesised. Further, due to several specific and specialized applications, need was felt to retain e!astomeric properties over a wider range of condition than natural rubber has. As natural rubber is produced in tropical regions, the industrialized countries felt the urgency for having their own resources of rubber. With the development of the petrochemical industry, it is planned to increase the artificial rubber production. In India in 1980 its production was 28,000 tonnes.

At present the following five principal types of rubber are in large scale use :

Buna S. (SBR, GRS) Rubber : It is a copolymer of butadiene., $CH_2 = CH - CH = CH_2$ (75% by weight) and styrene, $C_6H_5 - CH = CH_2$ (24% by weight). In the early days of its synthesis sodium was used as the catalyst, hence the name, bu (for butadiene), na (symbol, Na, for sodium) and S (for styrene). It is also called GR-S (Government Rubber Styrene) or SBR. Buna S rubber can be combined with natural rubber, thus combining the useful properties of both, and is used for the manufacture of tyres belting, shoe roles, coated fabrics, floor tiles, tank lining, carpet backing, etc. This rubber has lower strength and flexibility than the natural rubber.

An emulsion of butadiene and styrene is made in water with the help of soap-like emulsifying agents. A catalyst is added and the mass is heated under pressure when a latex-like liquid is obtained. It is then steam distilled to remove the uncombined hydrocarbons. Rubber is coagulated and processed in a manner similar to the natural rubber using sulphur as the vulcanizing agent. Buna S rubber is synthesised more than the other types of rubber.

At Bareilly, U. P., Synthetics and Chemicals Ltd. have an artificial rubber plant of 20,000 tonnes a year capacity. The plant annually uses about 70 million litres of ethyl alcohol, which is supplied by the various distilleries, and 9 million litres of benzene obtained from the coke oven plants of the steel industry. Alcohol is dehydrogenated in the presence of a catalyst, to give acetaldehyde. Acetaldehyde mixed with more alcohol is passed over another catalyst, butadiene is obtained :

$$CH_3\,CH_2\,OH \longrightarrow CH_3\,CHO + H_2$$

$$CH_3\,CHO + CH_3\,CH_2\,OH \longrightarrow CH_2 = CH - CH = CH_2 + 2H_2O$$

Dehydration of ethyl alcohol gives ethylene which, with benzene, in the presence of catalysts, gives styrene:

$$C_6\,H_6 + CH_2 = CH_2 \longrightarrow C_6\,H_5\,CH_2\,CH_3$$

$$C_6\,H_6\,CH_2\,CH_3 \longrightarrow C_6\,H_5\,CH = CH_2 + 2H_2O$$

Copolymerization of butadiene and styrene gives Buna S rubber which has a structural unit of

$$-[CH_2 - CH = CH - CH_2 - CH_2 - \underset{\displaystyle C_6H_5}{\underset{|}{CH}}]_n -$$

Buna N (GR-A and GR-N). It is a copolymer of butadiene 75% and acrylonitrile, (CH_2 = CHCN), 25% and its structure is similar to that to Buna S rubber, styrene being replaced by more polar acrylonitrile. The stable nitrogen component makes Buna N rubber highly resistant to solvents, oils and chemicals. It also has good resistance to abrasion, aging and to low temperatures and these properties increase with increase in acrylonitrile percentage. It is than natural rubber. The rubber is vulcanized with sulphur and is used for hose, conveyor belts, tank linings and for high altitude aircraft. It is also used in matching adhesives, oil resistant foams and automobile parts. Further, in the form of latex it is used for impregnating paper, textiles and leather. In India a plant with installed capacity of 2,000 tonnes of nitrile rubber was started in 1976 which produces about 1000 tonnes.

$$H_2C = CH - CH = CH_2 + CH_2 = CH - CN \rightarrow \left[H_2C - CH = CH - CH_2 - CH_2 - \underset{\displaystyle CN}{\underset{|}{CH}} \right]_n$$

Butyl Rubber (GR-I). 98% of this rubber consists of isobutylene which polymerizes to give a saturated polymer:

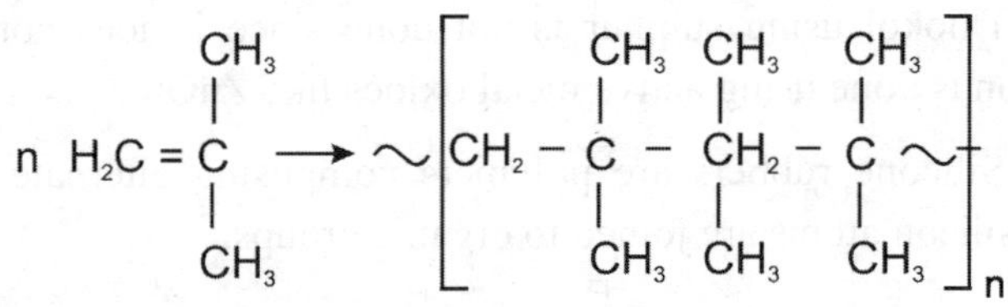

Polybutene (Butyl rubber)

Because it is saturated no cross links can be formed, hence it cannot be vulcanized. To overcome this difficulty, about 2% of butadiene is copolymerized with isobutylene. Butadiene provides centers for cross links and thus the rubber can be vulcanized. A plant for 20,000 tonnes per annum capacity of polybutadiene is working in India since 1978.

Butyl rubber is cheaper than other artificial rubbers. Another advantage of this rubber is that it is impermeable to gases, hence is satisfactory for making cycle and automobile tubes. It is also resistant to aging, ozone, acids or alkalis. Its other uses are for hose, tank linings and electrical insulation. A limiting factor of butyl rubber is that because of very low unsaturation it cannot be hardened much.

Neoprene (GR-M). It is a polymer of chloroprene :

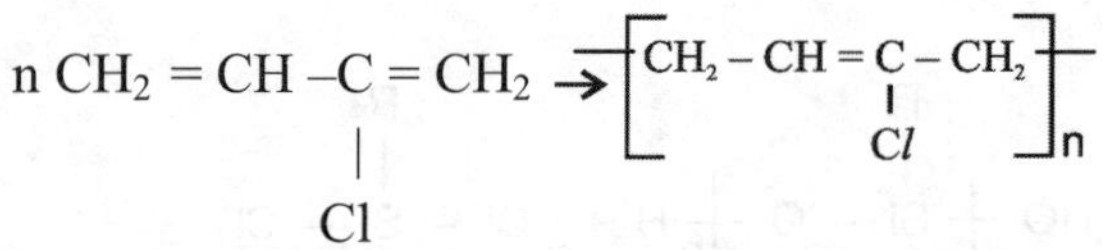

Starting materials for it are acetylene and hydrogen chloride. It is valcanized by zinc oxide or magnesium oxide. It has very useful properties and resembles natural rubber but is more expensive than rubbers. It has very good resistance to aging and to oils and is used for hose, wire insulation and belts.

Thiokol (GR--P). Several types of rubbers of this group are made by copolymerization of organic dichlorides, like ethylene dichloride, with inorganic polysulphides, like, Na_2S_4.

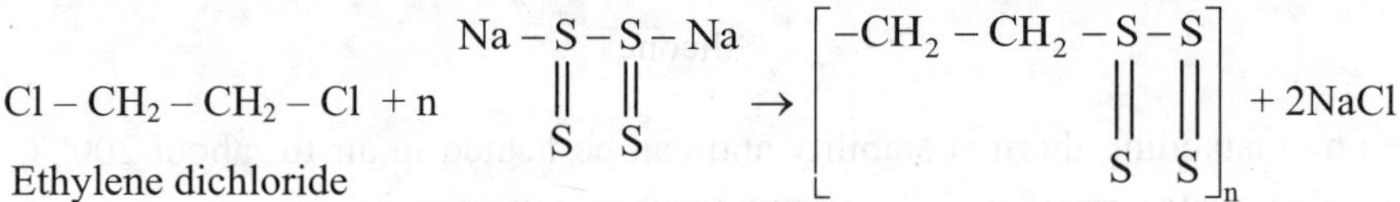

Ethylene dichloride

Sodium tetrasulphide

Thiokol

These rubbers neither resemble natural rubber in structure nor in properties they have very good resistance to solvents and can withstand cold but not heat. They are cheap and can easily be shaped, but have an unpleasant smell. Because of their being practically unaffected by petrol, fuel oils and lubricating oils they are mostly used for making hose pipe and tank linings for handling these materials.

Vulcanizing of Thiokol using sulphur is not done since it does not yield the hardened rubber. Vulcanization is done using active metal oxides like ZnO.

Silicone rubber *:* Silicone rubbers are polymers composing alternate silicon and oxygen atoms in which the silicon atoms are joined to organic groups.

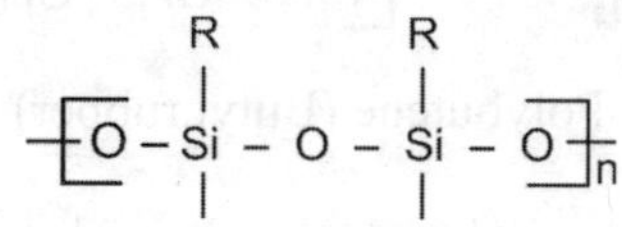

Silicone polymers were developed in 1930s during searches for heat resistant electrical insulating materials. Dicholrosilanes readily react with water to give silanols, which are unstable and condence to form silicones.

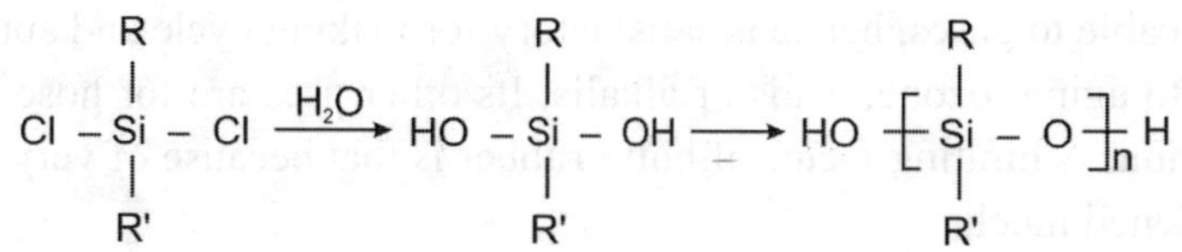

(Dialkyl silicon dichloride) Silanol

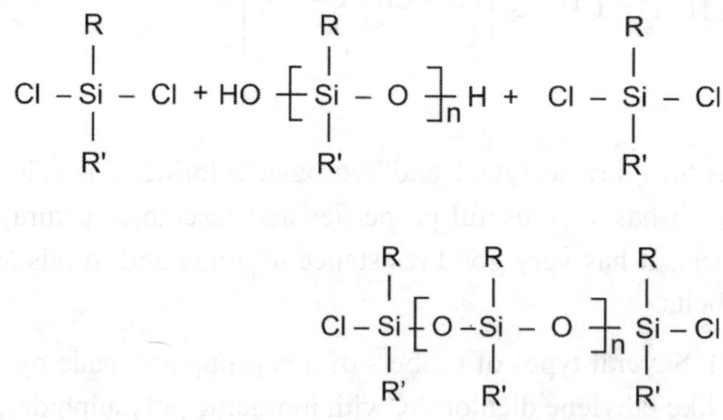

Silicone

Silicones have an outstanding thermal stability and can be heated in air to about 200 °C. Silicone rubber is also water repellent. So silicone rubbers are mainly used in making lubricants, electric blankets, protective coatings for fabrics, types for fighter aircrafts that could with stand high frictional heat and for making thermal impulsion for washing machines.

Polyurethane Rubber : These are obtained by condensation polymerization of short chain polyalcohols. The short chain hydroxyl terminated polymers like polyesters and polyethers are treated with diisocyanate to give the polyurethane rubbers

$$HO \sim\sim R \sim\sim OH + OCN - R' - NCO \quad \left[O - R - O - \underset{\underset{O}{\|}}{C} - NH - R' \right] NCO$$

Hydroxyl terminated Short polymer — Diisocyanate — Polyurethane

The properties of polyurethane rubber are decided by the short chain polymers and the final molecular length. Required strength for the rubber is achieved by vulcanisation using dissocyanate.

Urethane rubbers are highly resistant to oxidizing environments and to many organic solvents. However, they are attacked by acids and alkalies particularly in hot conditions. They form light forms and are often used as mattresses and insulators. They are also used in making surface coatings.

Rubber as an Engineering Material. Rubber has several properties which make it useful as an engineering material. Rubber has large deformability. Most structural materials, if deformed more than 10%, exceed their elastic limit but some rubbers can be stretched 10 times without rupture and regain almost the original shape after the removal of the stress. This constitutes both an advantage as well as a disadvantage depending upon the service requirement. Rubber has *low modulus,* i.e., low stresses produce deformation or change in shape. This property enables rubber to absorb shocks and to be used for cushioning and sealing. Because rubber regains its original shape after large deformation it can store energy more than most other materials. When rubber is subjected to cyclic (i.e., alternating) loading, as in automobile tyres which are cyclically stressed when the automobile is moving, not all the energy is recovered. This results in some *hysteresis loss* of energy which appears as heat energy and the rubber gets heated. This shortens the life of rubber. This loss of energy enables rubber to reduce vibrations. Rubber continues to deform under steady load with passage of time i.e., it has high creep. This behaviour limits the use of rubber as structural material.

Rubber is a bad conductor of electricity. Soft rubber is used as an insulator and hard rubber or ebonite is used as a high grade dielectric. Rubber is impermeable to water and to a large degree to gases.

Natural rubber is superior to synthetic rubber in mechanical, properties but is inferior in resistance to solvents and oils.

Table 3.2 Addition Polymers with commercial uses

Polymer name	Monomer(s)	Polymer	Use
Polyethylene	$CH_2 = CH_2$ (ethene)	$-CH_2-CH_2-$	Most common polymer. Used in bags, wire insulation, and squeeze bottles
Polypropylene	$CH_2 = CH$ | CH_3 (1-propene)	$-CH_2-CH-$ | CH_3	Fibers, indoor-outdoor carpts, bottles
Polystyrene	$CH_2=CH$ 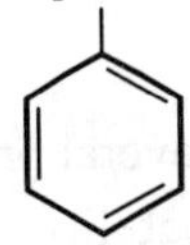 (styrene)	$-CH_2-CH-$ 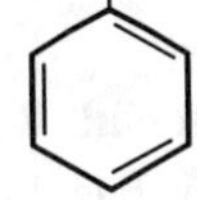	Styrofoam, molded objects such as tableware (forks, knives and spoons), trays, videocassette cases.
Poly (vinyl chloride (PVC)	$CH_2=CH$ | Cl (Vinyl chloride)	$-CH_2-CH-$ | Cl	Clear food wrap, bottles, floor covering, synthetic leather, water and drain pipe.
Polytetrafluoro-ethylene (Teflon)	$CF_2 = CF_2$ (tetraflouroethene)	$-CF_2-CF_2-$	Nonstick surfaces, plumbing tape, chemical resistant containers and films
Poly (methyl methacrylate) (Lucite, Plexiglas)	CO_2CH_3 | $CH_2 = C$ | CH_3 (methyl methacrylate)	CO_2CH_3 | $-CH_2-C-$ | CH_3	Glass replacement, paints, and household products

Table 3.3 *Contd...*

Polymer name	Monomer(s)	Polymer	Use
Polyacrylonitrile (Acrilan, Orlon, Creslan)	$CH_2 = CH$ | CN (acrylonitrile)	$–CH_2–CH–$ | CN	Fibers used in knit shirts, sweaters, blankets, and carpets
Poly (vinyl acetate) (PVA)	$CH_2=CH$ | $OOCCH_3$ (vinyl acetate)	$–CH_2–CH–$ | $OOCCH_3$	Adhesives (Elmer's glue), paints, textile coatings, and chewing gum
Natural rubber	CH_3 | $CH_2=C–CH=CH_2$ (2-methyl-1,3-butadiene)	CH_3 | $–CH_2–C=CH–CH_2–$	Rubber bands, gloves, tires, conveyor belts, and house hold materials
Polychlorprene (neoprene rubber)	Cl | $CH_2=C–CH=CH_2$ (2-methyl-1,3-butadiene)	Cl | $–CH_2–C=CH–CH_2–$	Oil and gasoline resistant rubber
Styrene butadine rubber (SBR)	$CH_2=CH$ 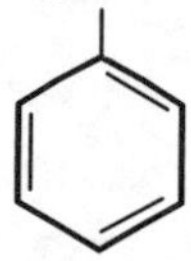 $CH_2=CH–CH=CH_2$	$–CH_2–CH–CH_2–CH– CH–CH_2–$ 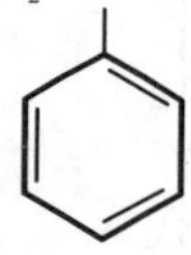	Non-bounce rubber used in tires

Table 3.3 Condensation Polymers with commercial uses.

Polymer name	Monomer(s)	Polymer	Use
Polyamides (nylon)	$HOC(CH_2)_nCOH$ (with C=O at both ends) $H_2N(CH_2)_nNH_2$	$–C(CH_2)_nC–NH(CH_2)_nNH–$ (with C=O at both ends)	Fibres, molded objects
Polyesters (Dacron, Mylar, Fortrel)	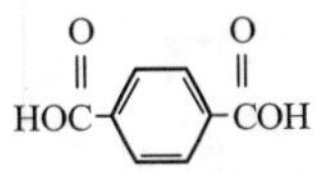 $HO(CH_2)_nOH$	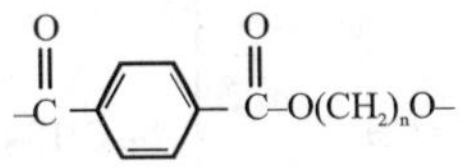	Linear polyesters, fibers, recording tape
Polyesters (Glyptal resin)	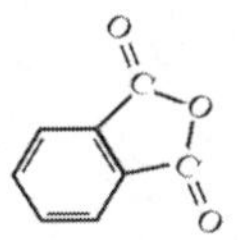 $HOCH_2CHCH_2OH$ (OH on central C)	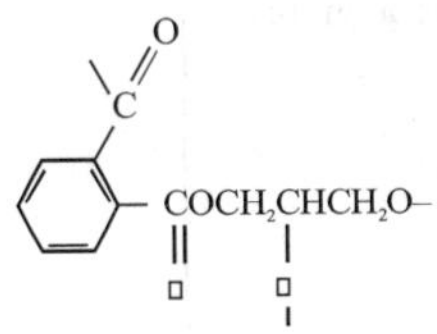	Cross-linked polyester, paints
Polyesters (Casting resin)	$HOCCH=CHCOH$ (with C=O at both ends) $HO(CH_2)_nOH$	$–CCH=CHC–O(CH_2)_nO–$ (with C=O at both ends)	Cross-linked with styrene and benzoyl peroxide, fiberglass boat resin, casting resin
Phenol-formaldehyde (Bakelite)	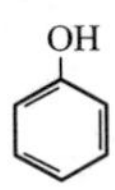 $CH_2=O$	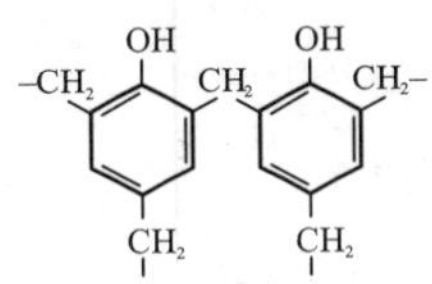	Mixed with filler, molded electrical cases, adhesives, laminates, varnishes
Cellulose acetate (cellulose is a polymer of glucose)	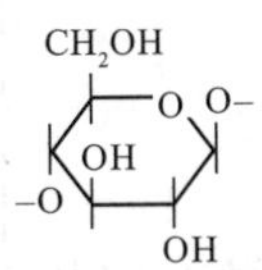	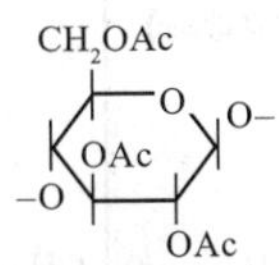	Photographic film
Silicones	$Cl–Si(CH_3)_2–Cl$ H_2O	$–O–Si(CH_3)_2–O–$	Water-repellent coatings, temperature-resistant fluids and rubber
Polyurethanes	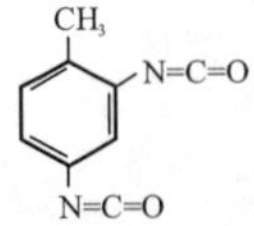 $HO(CH_2)_nOH$	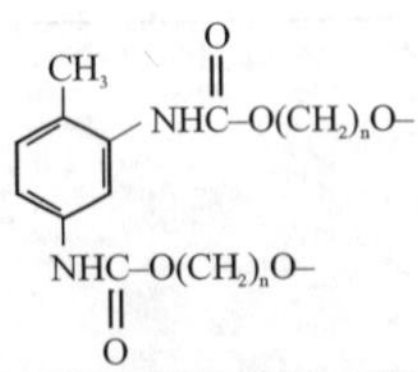	Foams, rigid and flexible, fibers

3.9 Conducting Polymers

Inherently, organic polymers with all carbon backbones are *insulators* and can be used as encapsulating materials for providing high resistivity to articles such as coatings for cables and electrical wirings and in electrical gadgets. It has been found that this resistivity decreases if composite of the polymer with carbon black or finely devided metal is fabricated, but the conduction in these cases takes place via the filler and not through the polymer which merely acts as a supporting matrix. Incorporation of a filler can also reduce the mechanical strength of the polymer. This scenario changed in 1970 due to pioneering work of Alan Heeger of the University of California, Santa Barbara, Alan MacDiarmid of the University of Pennsylvania, and Hideki Shirakawa of the University of Tsukuba in Japan – who shared the 2000 Nobel Prize in chemistry for this discovery. Polyacetylene was first synthesized, in 1958 by polymerization of acetylene, by Natta et al. as a *black powder*. This was found to be a semi-conductor with a conductivity between 7×10^{-11} to 7×10^{-3} Sm^{-1}, depending upon preparation process. This polymer was found to be not of much use due to its air instability. In 1967, when a postgraduate student of Hideki Shirakawa at the Tokyo Institute of Technology was attempting to synthesize polyacetylene, by mistake he took 1000 times more Natta catalyst and a *silvery shining thin film* was produced as a result of this mistake. Exposure of this form of polyacetylene to halogens increased its conductivity a billion fold. When it was weakly oxidized by compounds such as iodine it turned a golden colour and its conductivity increased to about 10^4 Sm^{-1}.

In the 1980's polyheterocycles (e.g., polythiophene) were first developed which were found to be much more air stable than polyacetylene, although their conductivities were not so high, typically about 10^3 Sm^{-1}. Since then it has been found that about a dozen different polymers and polymer derivatives (Fig. 3.3) exhibit conducting properties when doped with a weak oxidation agent or reducing agent.

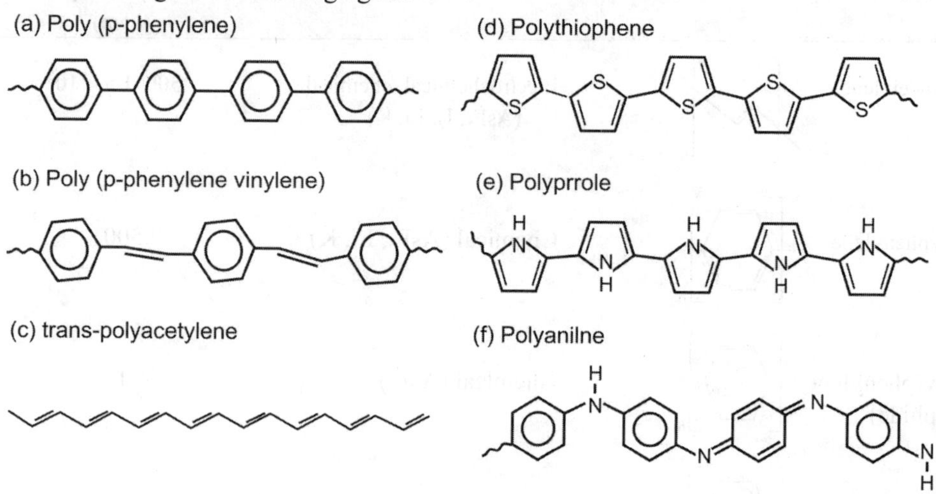

Fig. 3.3 Some Conducting Polymers-(*a-e*) are "stiff" linearly unsubstituted conducting polymers and (*f*) is a semi-flexible conducting polymer.

Conductive polymers include poly(acetylene)s, poly(pyrrole)s, poly(thiophene)s, poly(aniline)s, poly(fluorene)s, polynaphthalenes, poly(p-phenylene sulfide), and poly(para-phenylene vinylene)s. Classically, these compounds are also known as "blacks" or "melanins". The melanin pigment in animals is generally a mixed copolymer of polyacetylene, polypyrrole, and polyaniline.

Doping

Doping in silicon semiconductors is achieved by replacing some of the silicon atoms by electron rich (e.g., phosphorus) or electron-poor (e.g. boron) atoms to create *n*-type and *p*-type semiconductors, respectively. In contrast, in conducting polymers doping is achieved, through an oxidation or reduction process. The doping methods can be classified into (i) Chemical doping and electrochemical doping.

- ***Chemical doping****, involves exposing a polymer, to an oxidant (typically iodine or bromine) or reductant (far less common, but typically involves alkali metals).*
- **Electrochemical doping** *involves a polymer-coated working electrode suspended in an electrolyte solution in which the polymer is insoluble along with separate counter and reference electrodes. An electric potential difference created causes a charge from the electrolyte to enter the polymer in the form of electron addition (n-doping) or removal (p-doping).*

The different dopings of polymers is given in Table 3.4.

Table 3.4 Structures and Conductivity of doped conjugated polymers.
(Adapted from D.O. Cowan and F.M. Wiygul (1986))

Polymer	Structure	Typical methods of doping	Typical conductivity $(Scm)^{-1}$
Polyacetylene	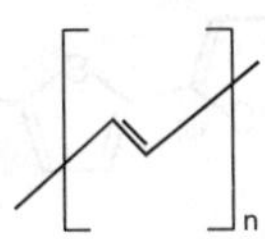	Electrochemical, chemical (AsF_5, I_2, Li, K)	$500–1.5 \times 10^5$
Polyphenylene	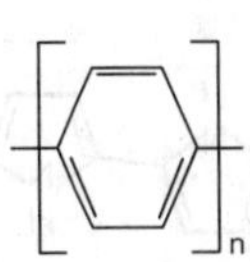	Chemical (AsF_5, Li, K)	500
Poly(phenylene Sulphide)	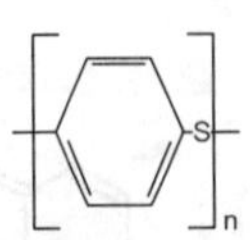	Chemical (AsF_5)	1
Polypyrrole		Electrochemical	600

Table 3.4 *Contd...*

Polythiophene	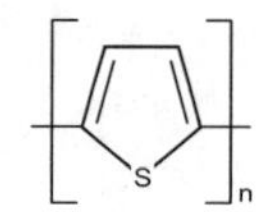	Electrochemical	100
Poly(phenyl-quinoline)	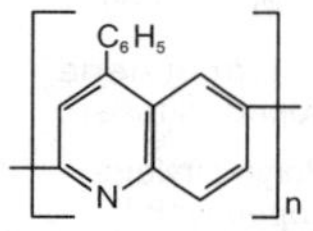	Electrochemical, chemical (sodium naphthalide)	50

Conjugation

Conduction in conducting polymers appears to be due to movement of electrons through the polymer and so referred to as *electroactive polymers*. Fully saturated chain such as polyetylene is 5 eV and in conjugated system like polyacetylene this gap decreases to 1.5 eV. Respective conductivities of these polymers are $\sim 10^{-17}$ S cm^{-1} and $\sim 10^{-8}$ S cm^{-1}. Ususally polyconjugated structures are either insulators or at the maximum semiconductors in their pure state but when treated with an oxidizing or a reducing agent can be converted into polymer salts with electrical conductivities comparable to metals. The generalized oxidation reaction that can be writen is

$$P_n \underset{\text{Red}}{\overset{\text{Ox/A}^-}{\rightleftharpoons}} [\,P_n^{+} A^-\,] \underset{\text{Red}}{\overset{\text{Ox/A}^-}{\rightleftharpoons}} [\,P_n^{2+}\ 2A^-\,]$$

where P_n represent a section of polymer chain.

Alternatively, after the first redox reaction, charge transfer complexes may form between charged and neutral segments of polymer

$$P_n \rightleftharpoons [\,P_n^{+} A^-\,] \overset{P_m}{\rightleftharpoons} [\,(P_n\,P_m)^{+} A^-\,]$$

The extended conjugation of a conductive polymer tends to give rise to fluorescence which has lead to the rapid development of polymer-based light emitting devices (OLEDs) and organic photovoltaic devices.

Conductivity

Conductive polymers are present in most mammal tissues where electrical conduction or transduction from light or sound are necessary, including the skin, eye, inner ear, and brain. Its electronic conductivity seems to be the underlying mechanism for absorption of light, and electron-photon interactions are exploited in hearing. A Logarithmic conductivity ladder of some of these polymers in relation to conductivity of Cu and Hg metals are shown below Fig. 3.3.

The conductivity is attributable to an extended conjugated backbone of continuous overlapping orbitals, for example, alternating single and double carbon-carbon bonds, which leaves a continuous path of overlapping *p*-orbitals (Fig. 3.4).

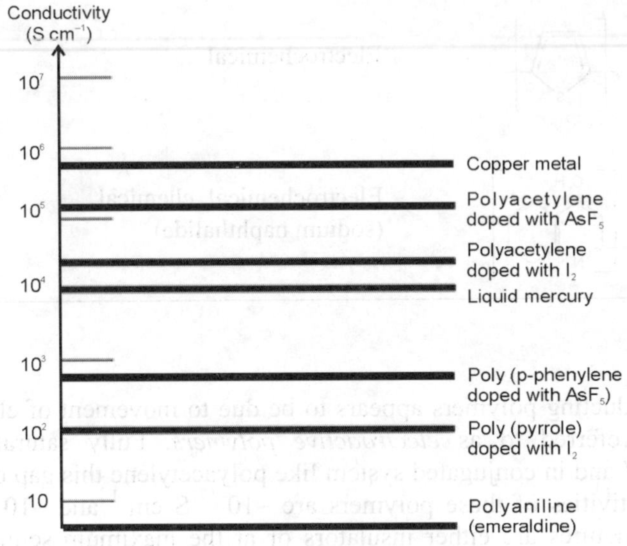

Logarithmic conductivity ladder locating some metals and conducting polymers

Fig. 3.3

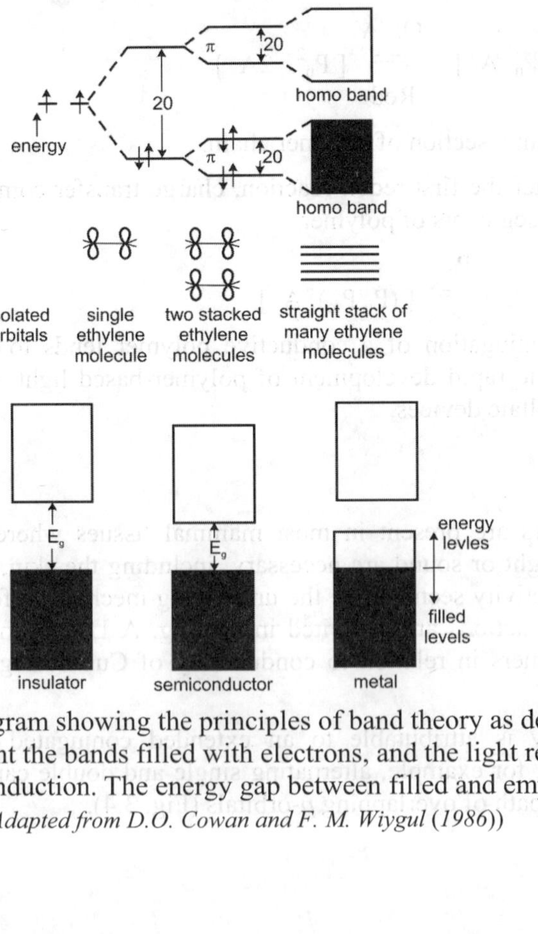

Fig. 3.4 Schematic diagram showing the principles of band theory as described in the text. The dark regions represent the bands filled with electrons, and the light regions the bands that are available for conduction. The energy gap between filled and empty states is E_g.

(Adapted from D.O. Cowan and F. M. Wiygul (1986))

This extended π-systems of conjugated polymer is highly susceptible to chemical or electrochemical oxidation or reduction. These alter the electrical and optical properties of the polymer, and by controlling this oxidation and reduction, it is possible to precisely control these properties. Since these reactions are often reversible, it is possible to systematically control the electrical and optical properties with a great deal of precision.

Preparation of Conductive Polymers

The normal methods of polymer synthesis like free radical initiated polymerisation, etc., do not give a polymer with suitable doping and required conjugation. So special methods are adopted to get required conductive polymer, some of which are presented here.

1. ***Polyacetylene :***

   Polymerisation of acetylene over Ziegler catalysts give polyacetylene which is predominately in *cis* form. On raising the temperature of this film gives more stable *trans* form. This polymer is infusible, insoluble and tends if become brittle on exposure to air. These problems are solved by Feast, adopting a two state process.

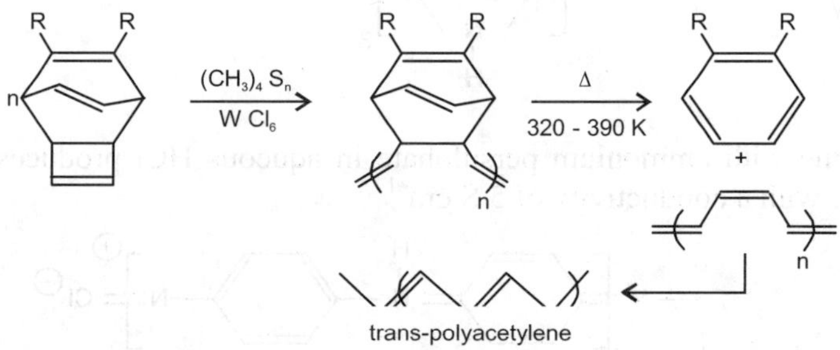

   Conductivity is magnified by doping. Exposure of the film to dry ammonia gives a polymer with conductivity of 10^3 S cm^{-1}. Controlled addition of p-doping agent like AgF_5, Br_2, I_2 or $HClO_4$ could boost to still higher conductivities.

2. ***Poly (p-phenylene) :***

   Polycondensation of p-dibromobenzene in the presence of Mg yields 'insoluble' oligomeric polymer material

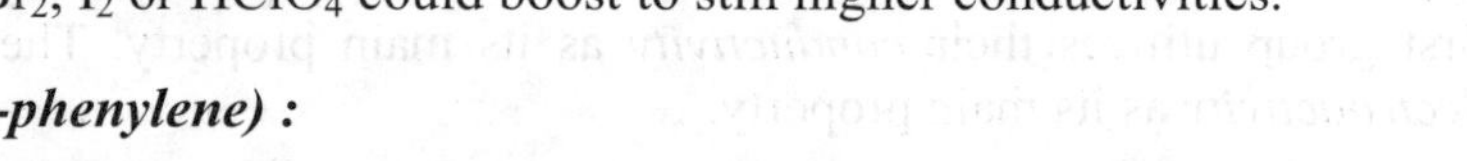

   This problem is solved by preparing first a precursor polymer which could be processed to suitable shape prior to use

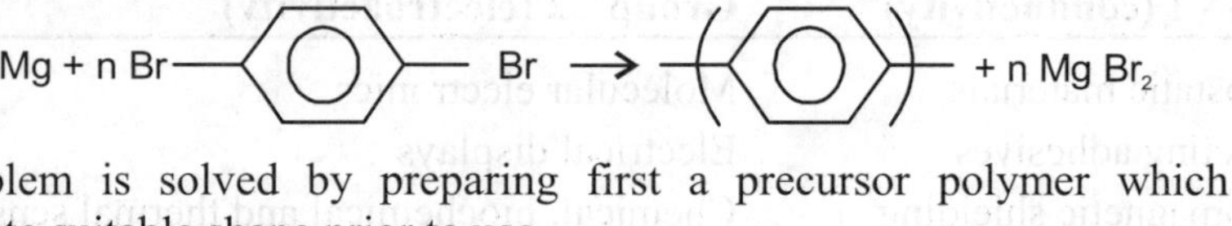

5, 6 - dihydroxycyclohex - 1, 3 - diene Precursor polymer Poly (p-phenylene)

   Doping is done as given for polyacetylene.

3. Poly pyrrole :

Electrolysis of a solution of pyrrole (0.06 M) and $(Et)_4 N^+ B F_4^-$ (0.1 M) in acetonitrile having 1% moisture leads to deposition of blue-black film of polymer.

The polymer has a conductivity of 10^2 S cm^{-1} and contains BF_4^- as do pant. The composition is given below

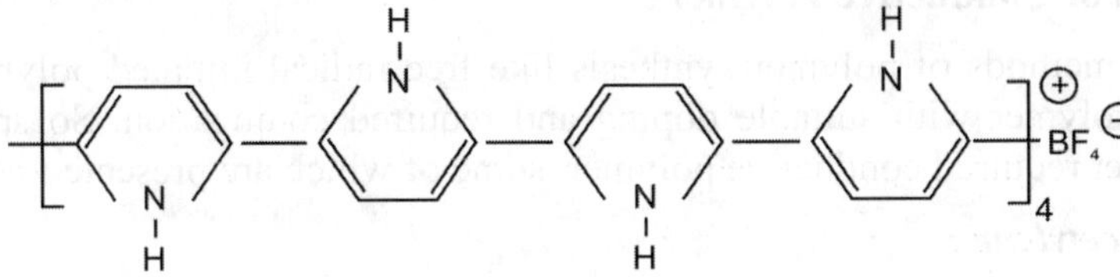

In the presence of Ag ClO_4 it gives a copper-bronze colored poly pynole per chlorate with following

Stoichiometry

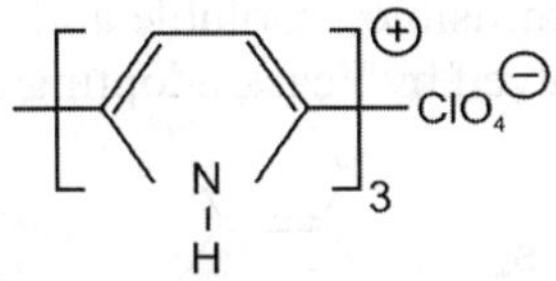

Polyaniline :

Reaction of aniline with ammonium persulphate in aqueous HCl produces polyaniline as a dark blue powder with a conductivity of 5 S cm^{-1}

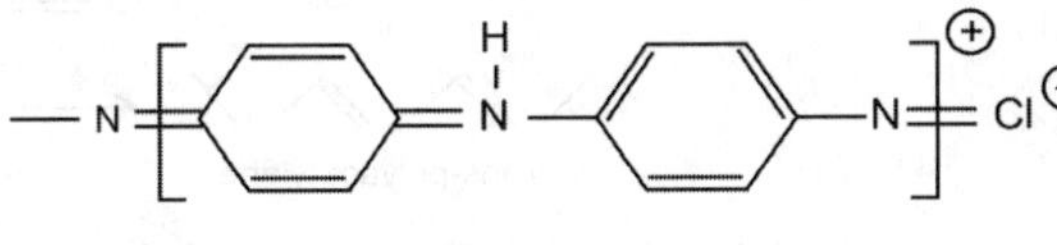

Classification of Conductive Polymers

The associated properties of conductivity and redox nature, group these polymers into two classes. The first group utilizes their *conductivity* as its main property. The second group utilizes their *electroactivity* as its main property.

The applications of these two groups are shown below :

Group - 1 (conductivity)	Group - 2 (electroactivity)
Electrostatic materials	Molecular electronics
Conducting adhesives	Electrical displays
Electromagnetic shielding	Chemical, biochemical and thermal sensors
Printed circuit boards	Rechargeable batteries and solid electrolytes
Artificial nerves	Drug release systems
Antistatic clothing	Optical computers
Piezoceramics	Ion exchange membranes
Active electronics	Electromechanical actuators
(diodes, transistors)	'Smart' structures
Aircraft structures	Switches

Most of these applications are in the development stage. However, conducting polymers have made it into a few small-scale commercial applications, such as antistatic coatings on photographic film and light emitting diodes in a display of maintenance information on an electric razor.

Application of Conductive Polymers

Group – I (*Conductivity*) Applications :

1. In the computer industry the sudden discharge of static electricity can damage microcircuits. Similarly in industries handling inflammable gases and explosives accumulation of static charge and a sudden discharge could lead to explosion. By coating an insulating surface with a very thin layer of conducting polymer it is possible to prevent the buildup of static electricity on its surface and therefore the damage.
2. By placing monomer between two conducting joints/surfaces and allowing it to polymerise it is possible to stick them together. This conducting adhesive eliminates the development of bad joints due to oxide films formed at the exposed surface.
3. Computer devices and cell phones often emit electromagnetic radiation in the microwave region. This leads to malfunctioning of neighbouring delicate electrical devises and LCD instruments. Normal plastic cases are transparent to these radiations while the conducting polymer coatings readily absorb these unwanted radiations.
4. Many electrical appliances use printed circuit boards. These are copper coated epoxy-resins. It is an expensive multistage process and the adhesion is generally poor. Conductive polymer coating on a polymer sheet has no complications and also has a very good adhesion.
5. It is possible to synthesize biocompatible conducting polymers. Due to the biocompatability these polymers they may be used to transport small electrical signals through the body, i.e. act as artificial nerves. Perhaps this could even pave way for brain implants.
6. Polymers have one tenth of density of metals and their maneuverability is also comparatively more. We are not far away from designing and using conducting polymers in our household wiring and electrical gadgets. This will be an added advantage in aircrafts where the load matters.

Group - 2 (*Electroactive*) Applications :

In the quest for improved performance, better processibility and novel applications a wealth of newer compounds with specific chemical architectures and electroactive properties have been synthesized. Better maneuverability is achieved through three parameters viz.,

- ✓ *greater main chain flexibility* (e.g., polyaniline),
- ✓ *side chain substitution* [e.g., poly(3-alkylthiophenes)],
- ✓ *fabrication from a soluble precursor polymer* [e.g., poly (*p*-xylene-dimethylsufonium chloride) to yield poly (*p*-phenylene vinylene)].

For example, side chain substitutions are now routinely used to enhance solvent solubility and fusibility so that various conventional polymer processing methods like casting and moulding may be done easily.

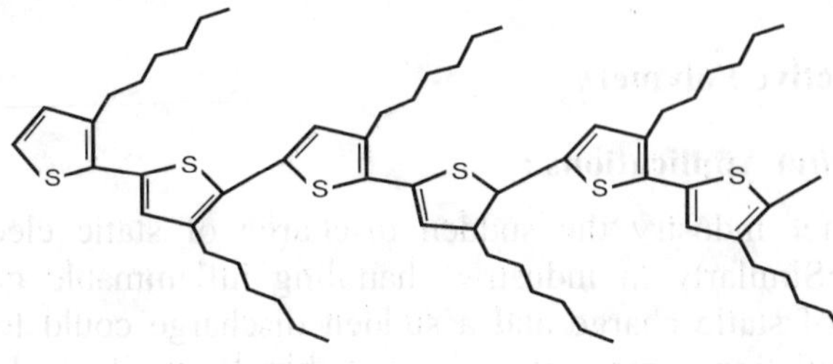

Further this may create new materials that exhibit enhanced electronic properties as compared to the linearly unsubstituted parent polymer. These side chain substituted conducting polymers also exhibit properties which do not exist in the unsubstituted hosts like pronounced *thermo- and solvato-chromism*, i.e. change of colour with temperature and solubility.

- **Smart/Intelligent Materials :** Smart materials are materials that 'remember' configurations and can conform to them when given a specific stimulus. These materials can respond to changes in electricity, heat, or magnetic waves. The electrical sensitivities make the conducting polymers used as smart or intelligent polymers. Changes in current values bring changes in the configuration of the conducting polymer. Taking polyacetylene as a typical example, normal preparation methods yield a random configuration while the one obtained by Shirakawa technique is a highly orderly transform exhibiting conductivity.
- **Gas Sensors :** The chemical properties of conducting polymers change their properties during reaction with redox reagents acting as dopants. Thus polypyrrole behaves as a quasi '*p*' type material. Its resistance increases in the presence of a reducing gas such as ammonia, and decreases in the presence of an oxidizing gas such as nitrogen dioxide. This is due to changes in electron holes caused by their reaction with adsorbed oxygen ions.
- **Biosensors :** The ability of triiodide (active ingredient is iodine) to oxidize polyacetylene is used as a means to measure glucose concentration. Glucose is oxidized with oxygen with the help of glucose oxidase in the solution phase. This produces hydrogen peroxide which oxidizes iodide ions present in the medium to form triiodide ions. These triiodide ions inturn bring changes in polyacetylene and its conductivity. Hence, conductivity exhibited by polymer is proportional to the peroxide concentration which in return is proportional to the glucose concentration of the solution phase.

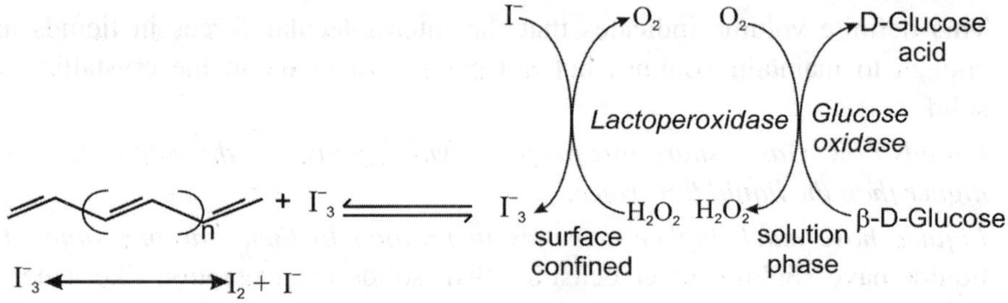

- **Polymer Rechargeable Batteries :** The polymer battery, such as a polypyrrole-lithium cell operates by the oxidation and reduction of the polymer backbone. During charging the polymer oxidizes anions in the electrolyte enter the porous polymer to balance the charge created. Simultaneously, lithium ions in electrolyte are electrodeposited at the lithium surface. During discharging electrons are removed from the lithium, causing lithium ions to reenter the electrolyte and to pass through the load and into the oxidized polymer. The positive sites on the polymer are reduced, releasing the charge-balancing anions back to the electrolyte. This process can be repeated about as often as a typical secondary battery cell. These prototype cells were found to be comparable to, or better than nickel-cadmium cells available now on the market.
- **Ion-exchange Membranes :** The charge residing on the conducting polymers make them useful as ion-exchange membranes.

3.10 Liquid Crystals

When the temperature of a gas is lowered, the kinetic energy of the gas molecules reduces considerably and the intermolecular attractive forces begin to play a vital role resulting in the liquefaction of gases. Any slight increase in temperature of the system pushes the molecules into gaseous state. On further cooling, the liquids acquire solid state, known for its orderliness. Therefore, a liquid can be thought to occupy an intermediate position between gas and solid states.

$$\text{Solid} \xrightarrow{+\text{ heat}} \text{Liquid} \xrightarrow{+\text{ heat}} \text{Gas}$$

The liquid and solid states, known as condensed state have comparable densities and are incompressible and differ only in the orderly arrangement, with liquids being more disorderly. The liquid and gaseous forms, in turn, differ largely in their densities and differ mainly in the magnitude of total intermolecular forces. All these characteristics of liquids are listed below.

(i) *Liquids have extremely small compressibilities and expandabilities.* This property is equally shared even by solids. The relative incompressibility of liquids is due to compactness of gas molecules.

(ii) *Liquids have definite volume but have no definite shape and always take the shape of its container. Solids are characterized by their definite volume and shape.*

This definite volume indicates that the intermolecular forces in liquids are strong enough to maintain volumes but not enough to maintain the crystalline shapes of solid.

(iii) *Liquids flow. Gases share this property with liquids, but the rate of flow of gases is higher than the liquid flow rates.*

(iv) *Liquids have much higher densities in relation to their gaseous states.* Generally liquids have slightly lower densities than solids but exception like water fulfilling reverse condition are also known.

(v) *Liquids show equal pressures in all directions. This property is common with gases.* Equal pressures indicate equal number of collisions of the molecules on the walls of the container.

(vi) *Liquids mix with other liquids forming both homogenous and heterogeneous systems. Gas mixtures are always homogeneous.*

Since school, we destinguish only three states of matter: solids, liquids, and gases. Solids, namely crystals, are highly ordered, the constituting atoms or molecules possessing positional as well as orientational order. On contrary, simple liquids like water or gases exhibit no order at all. The arising question was: How can biological systems, polymers, etc., be classified. Obviously they are highly ordered, but no crystalline solids. This was the beginning for O. Lehmann, F. Reinitzer, L. Gattermann, D. Vorländer, and others to search for new states of matter which are soft and ordered at the same time. The study of liquid crystals began in 1888 when an Austrian botanist named Friedrich Reinitzer observed that a material known as cholesteryl benzoate had two distinct melting points. In his experiments, Reinitzer increased the temperature of a solid sample and watched the crystal change into a hazy liquid. As he increased the temperature further, the material changed again into a clear, transparent liquid. Further Reinitzer discovered that cholesteryl acetate exhibits optical properties known only from crystals so far. Because of this early work, Reinitzer is often credited with discovering a new phase of matter – the liquid crystal phase.

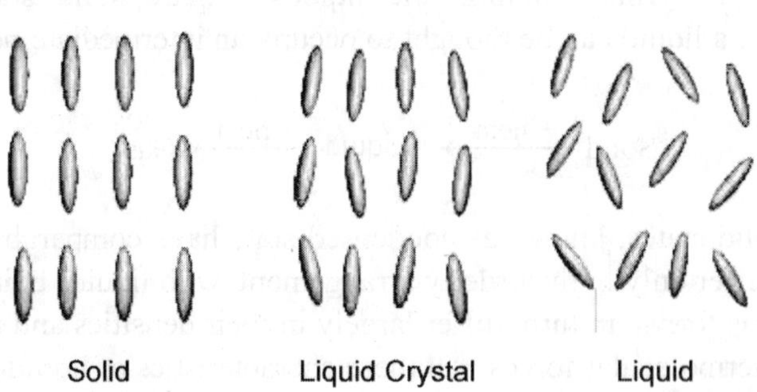

Fig. 3.4 Solid, Liquid and Liquid Crystal molecular arrangements.

Condensed matter which exhibit intermediate thermodynamic phases between the crystalline solid and simple liquid state are now called liquid crystals or mesophases ('middle' phases) and their molecular components are called mesogens (generators of mesophases) (Fig. 3.4).

Types of Mesophases

Liquid crystal materials are unique in their properties and uses. Liquid crystal materials generally have several common characteristics. Among these are a rod-like/disk like molecular structure, rigidness of the long axis, and strong dipoles and/or easily polarizable substituents (Fig. 3.5).

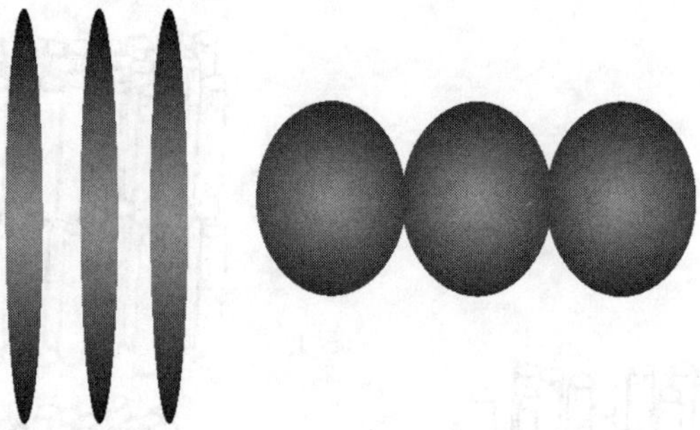

Fig. 3.5 Molecular shapes of Liquid Crystals.

They align themselves with a long range order due to anisotropic intermolecular forces. This fourth state of matter generally possess orientational or weak positional order and thus reveals several physical properties of crystals but flow like liquids and exhibit liquid properties like capillary rise, surface tension, etc.

If transitions between the phases are given ***by temperature***, they are called **thermotropic**. The temperature at which solid turns to mesostate is called **transition temperature** and the temperature at which mesomorphic state (mesophase) disappears to give liquid state is called the **melting point**. Thus for p-azoxy anisole the transition point is 118.3 ^{o}C and the melting point is 135.9 ^{0}C

In blends of different components phase transitions may also depend ***on concentration***, and these liquid crystals are called **lyotropic**. Lyotropic mesophases occur in concentrated solutions of rod-like molecules in an isotropic solvent, usually water, eg., sodium laurate in water, phosphotidyl choline in water. While thermotropics are presently mostly used for special technical applications, lyotropics are important for biological systems, e.g. membranes. In industry also we find their use as soaps, gels and colloids.

There are two types of thermotropic liquid crystals: discotics and rod-shaped molecules, Fig. 3.6.

Discotics are flat disc-like molecules consisting of a core of adjacent aromatic rings. Their molecules arrangement is akin to a pile of coins. Discotic liquid crystals can form columnar phases in which the coins are stacked into loosely defined columns. Disc-shaped molecules also form an orientationally ordered nematic phase which we are going to discuss shortly. The best demonstration could be made through stacked and orientational arrangement of Rs. 5 coins.

Rod-shaped molcecules have an elongated, anisotropic geometry which allows for preferential alignment along one spatial direction. They form both nematic and smectic phases. The best demonstration could be made through match sticks arranged randomly and in columns.

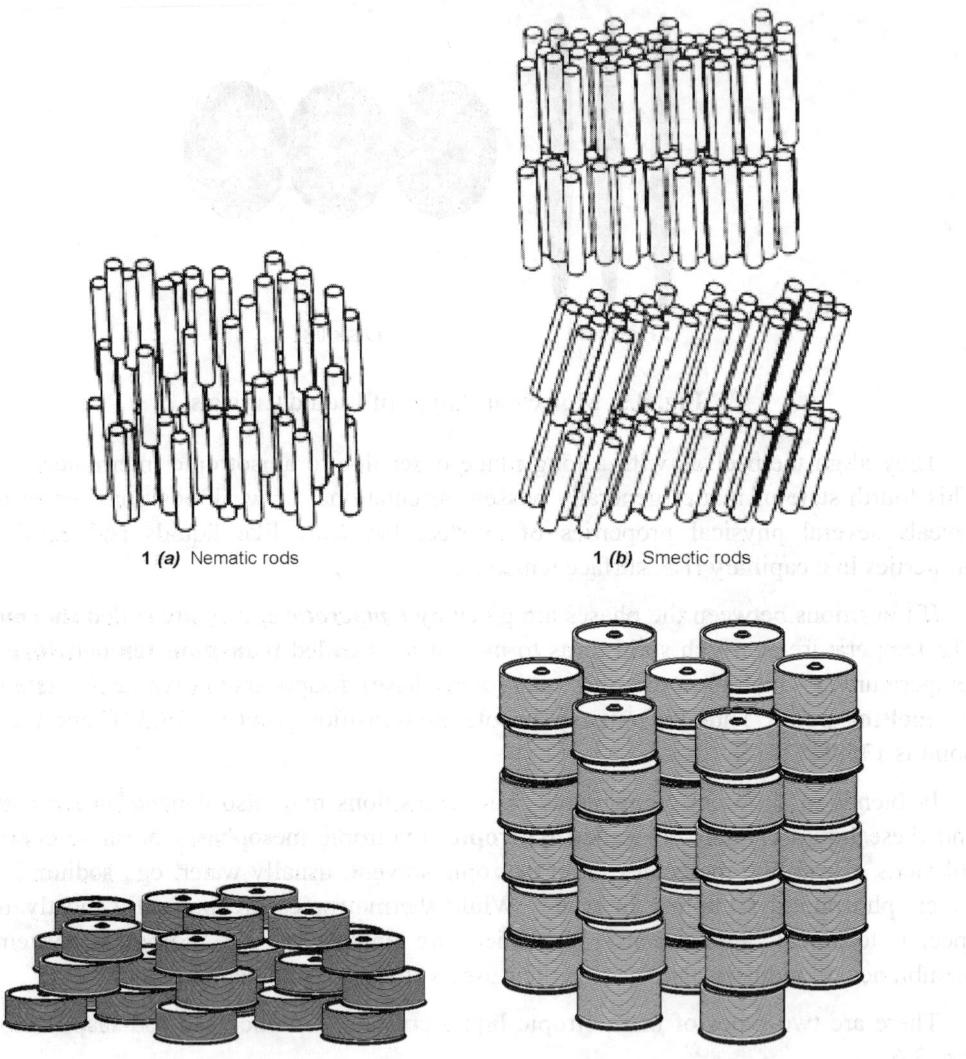

1 *(a)* Nematic rods — **1 *(b)*** Smectic rods

2 *(a)* Nematics disks — **2 *(b)*** Stack like - Columnar phase

Fig. 3.6 Liquid-crystal mesophases.

1. ***Rod-like molecules :*** **1(a)** Form nematic phase – have orientational order but no postional order **1(b)** Form smectic phase – have orientational and positional order in one dimension (that is, ordery in lines). **2.** Disc-shaped molecules **2(a)** Nematicdisks **2(b)** Stack-like columnar phase.

The distinguishing characteristic of the liquid crystalline state is the tendency of the molecules (*mesogens*) to point along a common axis, called the director.

This is in contrast to molecules in the *liquid phase, which are isotropic and have no intrinsic order.*

In the solid state, molecules have high long range order and have little *translational freedom.*

The characteristic *orientational* order of the **liquid-crystal state** is between the traditional solid and liquid phases (*middle*) state and this is the origin of the term **mesogenic.**

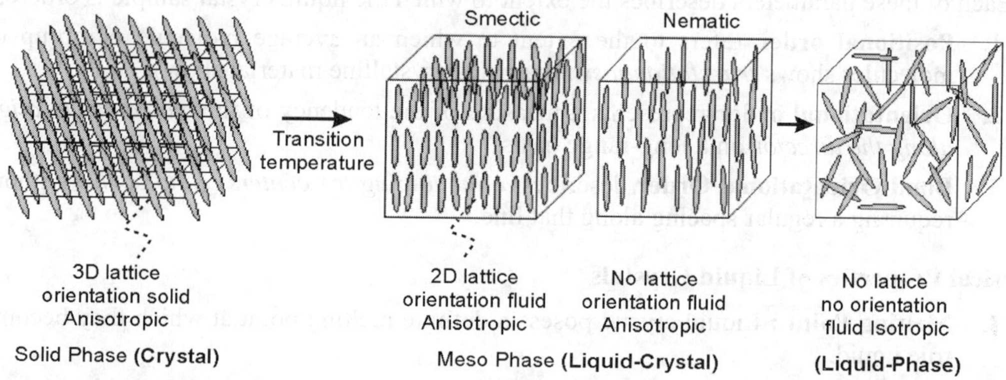

Fig. 3.7 Liquid crystalline mesophases between the solid and isotropic liquid phase.

To quantify just how much order is present in a material, an order parameter (S) is defined (Fig. 3.8).

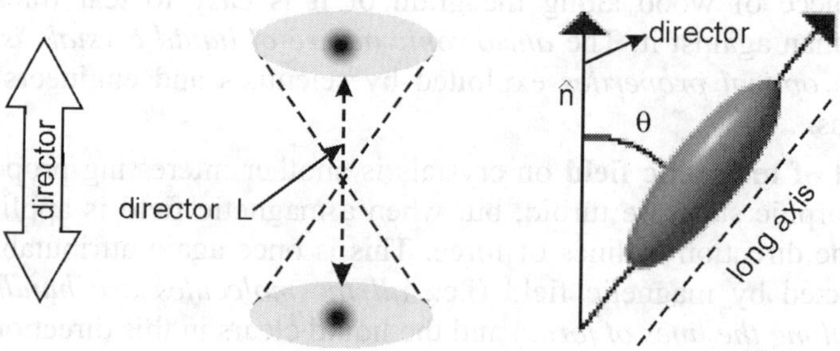

Fig. 3.8 Angle between director and long axis of a molecule.

Traditionally, the order parameter is given as follows :

Order parameter $S = \Sigma \cos \theta$

where theta (θ) is the angle between **the director** and the *long axis* of each molecule.

- In an isotropic liquid, the average of the cosine terms is zero, and therefore the order parameter is equal to zero.

- For a perfect crystal, the order parameter evaluates to one.
- Typical values for the order parameter of a liquid crystal range between 0.3 and 0.9, with the exact value depending on temperature as a result of kinetic molecular motion.

Characterization of Liquid Crystal Orders

The following order parameters characterize the liquid crystals :

1. Positional Order, 2. Orientational Order, 3. Bond Orientational Order

Each of these parameters describes the extent to which the liquid crystal sample is ordered.

1. **Positional order** refers to the extent to which an average molecule or group of molecules shows *translational symmetry* (as crystalline material show).
2. **Orientational order** represents a measure of the tendency of the *molecules to align along the director* on a long-range basis.
3. **Bond Orientational Order** describes *a line joining the centers of molecules* without requiring a regular spacing along that line.

Physical Properties of Liquid Crystals

1. **Melting Point :** Liquid crystal posess a definite melting point at which they become true liquid.
2. **Polymorphism :** Most liquid crystal compounds exhibit polymorphism, or a condition where *more than one phase is observed* in the liquid crystalline state.
3. **Anisotropy :** The tendency of the liquid crystal molecules to point along the director leads to a condition known as **anisotropy**. This term means that the properties of a material depend on the direction in which they are measured. For example, it is easier to cut a piece of wood along the grain or it is easy to tear mica sheet along a direction, than against it. The *anisotropic nature of liquid crystals is responsible for the unique optical properties* exploited by scientists and engineers in a variety of applications.
4. **The effect of magnetic field** on crystals is another interesting property. Substances in mesomorphic state are turbid, but when a magnetic field is applied they become clear, in the direction of lines of force. This is once again attributable to orientation order affected by magnetic field (i.e., *all the molecules are bundled in the same direction along the lines of force*) and the liquid clears in this direction.
5. **The viscosity** of liquid crystals is *small.* Where as that of liquids with similar molecular size is large.

Classification of Liquid Crystals

Meso phases are mainly of two types; 1. Nematic Phase, 2. Smectic phase. Each class could be further split into sub-classes.

Nematic Phases

In the simplest case *the molecules possess only orientational but no positional long range order*. Liquid crystals of that type are called nematic (from ***greek*** *nemos* = *thread*), the name of which has been given with respect to *thread-like textures* observed under polarizing microscope. The nematic liquid crystal phase is characterized by molecules that have no positional order but tend to point in the same direction (along the director). In the following diagram, notice that the molecules point vertically but are arranged

1. Orientational order
2. No positional order
3. No lattice
4. Uniaxial

with no particular order. Nematic phases are composed of rod-shaped molecular aggregates that are arranged with parallel but not lateral order. For example, imagine a large number of toothpicks put into a rectangular box and shaken. When you open the box, the toothpicks will be facing in about the same direction, but will have no definite spatial organization. They are free to move, but like to line up in about the same direction. This is a primitive model for nematic liquid crystals.

In nematics the distribution function is rotationally symmetric around the director, i.e. they are uniaxial.

The Smectic Phases

The word "smectic" is derived from the Greek word for soap. This seemingly ambiguous origin is explained by the fact that the thick, slippery substance often found at the bottom of a soap dish is actually a type of smectic liquid crystal. Molecules in this phase show a degree of translational order not present in the nematic.

In the smectic state, the molecules maintain the general orientational order of nematics, but also tend to align themselves in layers or planes. Motion is restricted to within these planes, and separate planes are observed to flow past each other. The increased order means that the smectic state is more "solid-like" than the nematic.

Many compounds are observed to form more than one type of smectic phase. As many as 12 of these variations have been identified, however only the most distinct phases are discussed here.

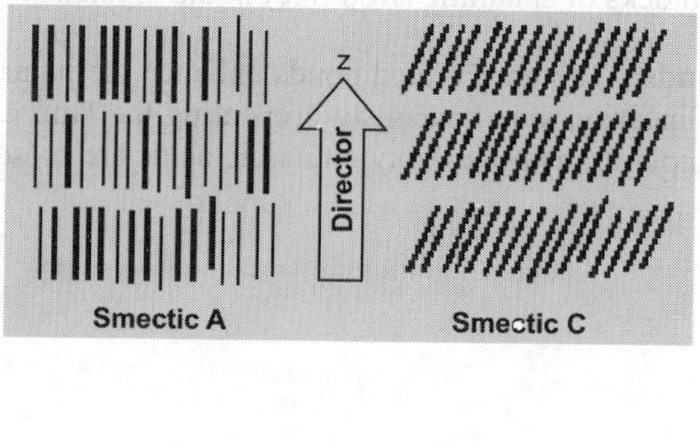

1. Orientational order
2. Positional (translational) order
3. 2D-symmetry
4. Uniaxial
5. More than one smectic phases possible

In the smectic-A mesophase the director is perpendicular to the smectic plane, and there is no particular positional order in the layer. Similarly, in the smectic-B, mesophase orients with the director perpendicular to the smectic plane, but the molecules are arranged into a network of hexagons within the layer. In the smectic-C mesophase, molecules are arranged as in the smectic-A mesophase, but the director is at a constant tilt angle measured normally to the smectic plane.

Chemical Nature of Liquid Crystals

The factors that control the physical structures and their aggregations are :

1. **Linear elongated molecules with molecular interactions lead to rod shaped structures :** Molecular interactions brought about by dipole-dipole and dipole-induced dipole interactions, hydrogen bonding and dispersion forces, are insufficient to bring about total 3D lattice symmetry of crystals but enough to bind them together leading to mesophases. In Fig. 3.9 the chemical structure for two of the most prominent representatives of rod-like molecules which exhibit liquid crystalline order are shown. 4-Methoxy bezyliden 4'-butylaniline (MBBA) was the first room temperature liquid crystal synthesized in 1969, whereas 4-pentyl 4'cyanobiphenyl (5CB) was the first member of the optically and chemically stable cyanobiphenyls, one of the most applicable family of liquid crystals, discovered in 1973. This is due to the interconnection of two rigid cyclic units. *The interconnecting group should cause the resulting compound to have a linear planar conformation.* Linking units containing multiple bonds such as –(CH=N)–, –N=N–, $–(CH=CH)_n–$,–CH=N–N=CH–, etc., are used since they restrict the freedom of rotation.

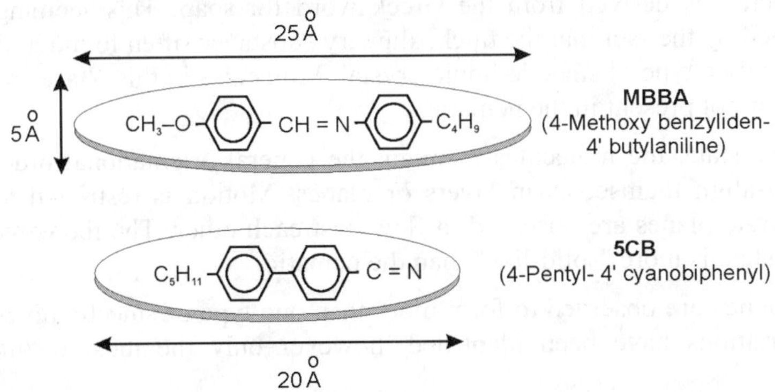

Fig. 3.9 Typical building blocks of calamitic (Rod like) liquid crystals.

2. **Presence of unsaturated bonds** : The unsaturated bonds enhance the polarizability, dipole character and overall linearity, simultaneously preventing the laminar bezene rings from twisting. For example, trans-p-n-alkoxy cinnamic acids are mesomorphic but its cis analogue is not.

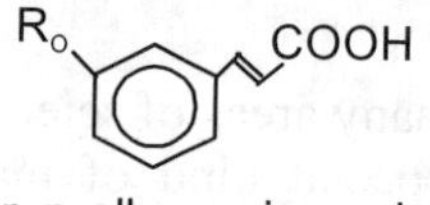

trans-p-n-alkoxy cinnamic acid

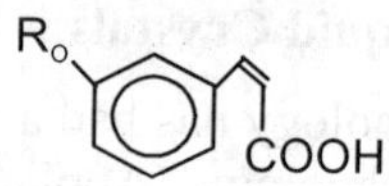

cis-p-n-alkoxy cinnamic acid

Similarly if the unsaturation is removed, the mesocharacter is destroyed due to twisting of chains leading to isotropy.

3. **Absence of bulky groups :** Bulky atoms like Br and bulky groups present in the chain other than as end groups do not allow aggregation

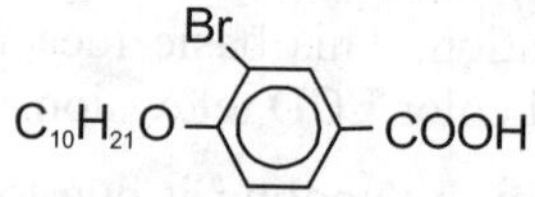

of mesogens. For example, inclusion of bromine atom on the benzene ring of 4-decyclo-3-benzoic acid prevent mesomorphic structure.

4. **H-Bonding :** Hydrogen bonding often destroys the mesomorphic character (i) by lengthening of molecular chains (ii) very strong bonds bringing crystallinity and (iii) by allowing crosslinking of chains and preventing formation of linear chains.

In case of carboxylic acids the mesomorphic character is destroyed due to H-bonding and lengthening of chains.

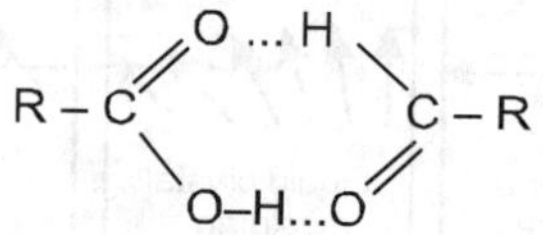

Fatty acids are not mesomorphic but their salts called soaps are strongly mesomorphic in nature.

Similarly 4-amino-4″-nitro-p-terphenyl (a), is not mesomorphic and crystalline with a melting point of 300 ^{0}C but becomes mesomorphic on substitution of a nitro group in ortho position to amino group (b).

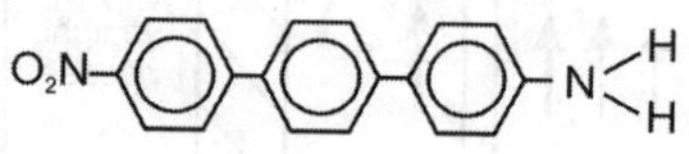

(a)

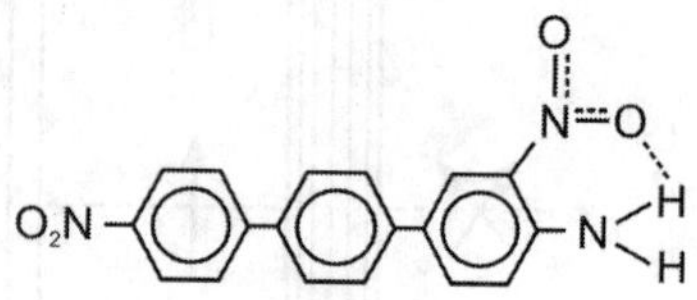

(b)

In some cases mesomorphism is brought about due to H-bonding. Thus cholesterol is not mesomorphic but it's ester shows liquid crystal property. The columnar structure is facilitated by the ester group.

Applications of Liquid Crystals

Liquid crystal technology has had a major effect on many areas of science and engineering, as well as device technology. Applications for this special kind of material are still being discovered and continue to provide effective solutions to many different problems.

1. **Liquid Crystal Display Devices :** The most common application of liquid crystal technology is in liquid crystal displays (LCDs). From the ubiquitous wrist watch and pocket calculator to an advanced VGA computer screen, this type of display has evolved into an important and versatile interface.

   A liquid crystal display consists of an array of tiny segments (called pixels) that can be manipulated to present information. This basic idea is common to all displays, ranging from simple calculators to a full color LCD television.

   A quartz watch stops it's display by cooling it due to formation of crystallites and turns blank on slightly increasing the temperature due to turning into liquid. This provides an easy way to demonstrate existence of liquid crystals in a quartz watch.

   A schematic diagram of functioning of liquid-crystal display is given in Fig. 3.10.

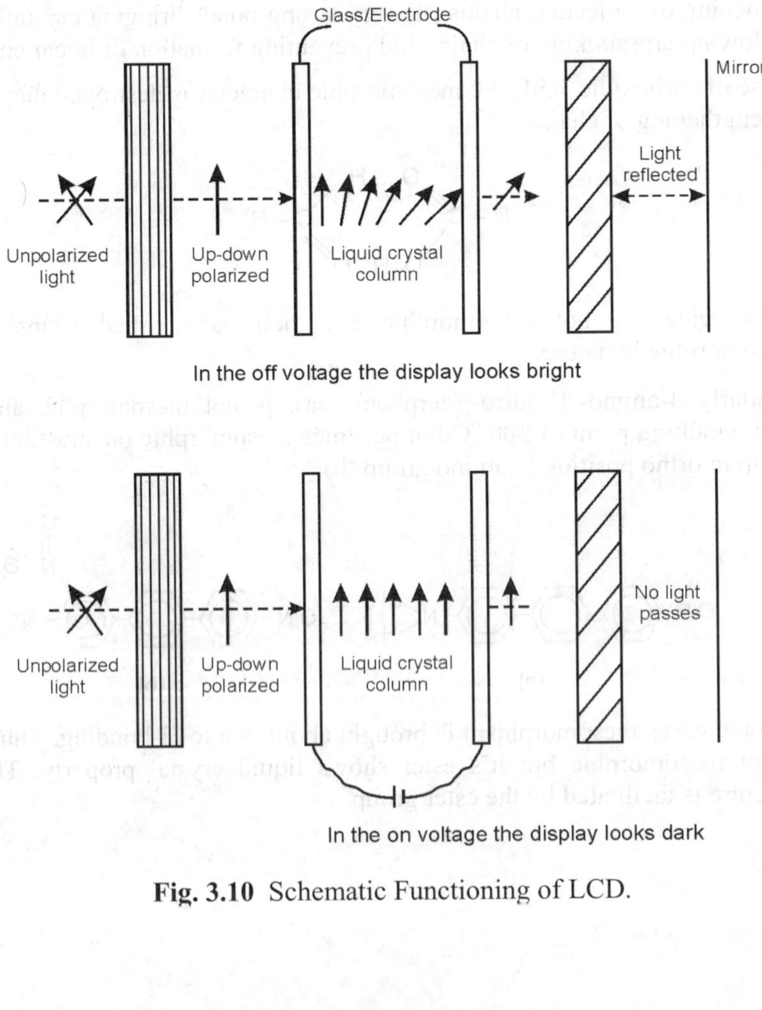

Fig. 3.10 Schematic Functioning of LCD.

Light travels through two polarisers twisted 90° and separated by a column of liquid crystal encased between two conducting glass sheets and falls on a mirror. In the journey of light, it is polarized by first polariser, say updown, progressively changes its direction in the liquid crystal solution and will be able to enter the rear polarizer, twisted by 90^0 or at an angle, and fall on the mirror to retrace it's path. When a small voltage is applied across the conducting glass sheets, the liquid crystal molecules line up and donot rotate the light. As a result it does not make through, back polarizer and the mirror gives *a dark* display. Thus we can generate any shape or number by selectively switching power across the display pixels. Colours can be introduced by including dyes in the liquid crystal. Normal nematic liquid crystals align their fields in few milli seconds but goback to their twisted state rather slow. Ferro electric liquid crystals meet this criteria and are widely used.

Why are liquid crystal displays important? The first factor is size. An LCD consists primarily of two glass plates with some liquid crystal material between them. There is no bulky picture tube. This makes LCDs practical for applications where size (as well as weight) are important.

In general, LCDs use *much less power* than their cathode-ray tube (CRT) counterparts. Many LCDs are reflective, meaning that *they use only ambient light to illuminate the display*. Even displays that do require an external light source (i.e. computer displays) consume much less power than Cathode ray tube(CRT) devices.

Liquid crystal displays have currently some drawbacks and could be solved with the tremendous research in-puts it is receiving. They are

(i) problems with viewing angle, (ii) contrast ratio, and (iii) response time

2. **The Role of Liquid Crystals in the digestion and assimilation :** The partially digested acidic slurry of food that leaves the stomach is called chyme. The chyme enters the first part of the small intestine which is called the duodenum. In the duodenum secretions from various secretion from digestive organs are added to the chyme. The pancreas produces various digestive enzymes and hormones like *insulin* and glycogen. Another secretion produced by the pancreas is sodium bicarbonate. This gas is the source of some of our burbs after a meal. Another secretion that enters the duodenum is Bile. Bile is classified as a amphiphilic molecule. This means that it has both a hydrophobic and a hydrophilic region. Amphiphilic molecules such as bile act as emulsifing agents. When these molecules are in solution they form liquid crystal aggregates known as micelles and vescicles. Although oil does not mix with either water or acid, its solution composed of water or acid and an amphiphile has no difficulty incorporating the oil. When the oil is added to the mixture it accumulates in the hydrophobic region of the micelles and forms an emulsion. When fats in the chyme enter the duodenum they are mixed with bile and form a special type of liquid crystal known as an emulsion. The function of the SDS is to simulate the addition of bile to the digestive tract. Thus this function of bile is facilitated by external addition of Sodium dodecyl (laurel) sulfate (SDS).

3. **Liquid Crystals as Lubricants :** Slippery Substances are called lubricants. They are very important in modern technology. Cars, trucks, airplanes, and machines, all have parts that rub against one another. These parts would heat up, wear down, and stop working if we did not have lubricants. Lubricants reduce the amount of friction between two surfaces that

move against each other. The fact that liquid crystals have lower viscosity than corresponding liquids is used in designing many lubricants, currently used in automobile engines and heavy machinery. For this purpose both the mesophases are suitable but smectic phase is preffered. Even in the biological systems, several cholesteric liquid crystals act as natural lubricants in the bone joints.

4. **Liquid Crystal Thermometers :** Chiral nematic (cholesteric) liquid crystals reflect light with a wavelength corresponding to the order, which inturn is dependent upon temperature. Liquid crystals make it possible to accurately gauge temperature just by looking at the color of the thermometer. By mixing different compounds, a device for practically any temperature range can be built. Physical problems, such as tumors, have a different temperature than the surrounding tissue and therefore diagnosed by administering an LC. Liquid crystal temperature sensors can also be used to find bad connections on a circuit board by detecting the characteristic higher temperature.

5. **Other Liquid Crystal Applications :** Liquid crystals have a multitude of other uses. They are used for nondestructive mechanical testing of materials under stress. This technique is also used for the visualization of RF (radio frequency) waves in waveguides. They are used in medical applications where, for example, transient pressure transmitted by a walking foot on the ground is measured. Low molar mass (LMM) liquid crystals have applications including erasable optical disks, full color "electronic slides" for computer-aided drawing (CAD), and light modulators for color electronic imaging.

Future of Liquid Crystals

The future of liquid crystals has just begun. There are still unsolved physical problems in this area, posing interesting challenges for the pure scientist. The need for liquid crystal applications grew, requiring engineers specialized in this field. Statistics show that cathode ray tube displays in TVs and computer monitors will be overtaken by liquid crystal displays in the near future. More recent applications involve switchable windows that can be changed from clear and opaque with the flip of a switch, liquid crystal displays with wider viewing angles, and displays that can remain in use for years without the need of power.

Knowledge in the field of liquid crystals is crucial for an understanding of biological membranes, thus biologists, medical researchers and pharmacists are also interested in liquid crystal research.

3.11 Questions

1. Distinguish Thermoset and Thermoplast polymers.
2. Important uses of Nylons.
3. What are the ratio of phenol and formaldehyde in novolac and resol.
4. Differentiate plasticized and unplasticized PVC.
5. Mention the conditions of HDPE manufacturing.
6. Explain the term plasticity.
7. Define co-polymerization

8. What are the advantages of polymers over metals?
9. Give the chemical structure of butyl rubber.
10. What are the important uses of Teflon?
11. What are the advantages of vulcanized rubber?
12. Define functionality of monomer and give the requirement for a molecule to act as monomer
13. What is vulcanization and its chemistry?
14. Mention any two synthetic rubbers and indicate how they are prepared. How are their properties compared to natural rubber?
15. What are the main differences between addition and condensation polymerization? Give an example in each case?
16. What is meant by thermosetting and thermoplastic polymers? Give at least two differences with examples.
17. Write the preparation and two properties each of
    (a) polythene (b) polyurethane
18. What is Buna-S. Mention its important uses.
19. Write short notes on the following
    (i) Phenol-formaldehyde resins (ii) Thiokol (iii) Teflon
20. What are the drawbacks of natural rubber?
21. Taking suitable example give the mechanism of free radical addition polymerization
22. Write informative notes on –
    (i) Fabrication of plastics (ii) Compounding of rubber
    (iii) Injection moulding (iv) Vulcanization
23. What are elastomers? Give the preparation and application of
    (i) Thiokol (ii) SBR
24. Write a note on compounding and fabrication of plastics.
25. Discuss the preparation, properties and uses of the following:-
    (a) Nylon (6,6) (b) Buna–S (c) PVC
26. (a) Explain the differences between thermoplastics and thermoset plastics.
    (b) Write a note on fabrication and moulding of plastics.
27. (a) Describe "Addition" and "Condensation" polymerizations with suitable examples.
    (b) What is vulcanization and why is it done?
    (c) Describe the preparation, properties and uses of Teflon
28. (a) Give an account of the compounding of natural rubber.
    (b) Outline the preparation, properties and uses of
    (i) Bakelite (ii) Polymethyl methacrylate.

29. (a) Differentiate between thermoplastics and thermoset plastics.
    (b) Write a note on compounding and fabrication of plastics.
30. (a) What is the constitution of natural rubber?
    (b) Describe the preparation, properties and uses of buna-N rubber.
    (c) Write briefly on the compounding of rubber and its applications as an engineering material.
31. Discuss the preparation, properties and applications of the following polymers:-
    (a) Nylon (b) Bakelite (c) Polythene
32. (a) What is the constitution of natural rubber?
    (b) What is Buna-S rubber? How is it manufactured? What are its applications?
    (c) Describe the preparation, properties and applications of PVC.
33. (a) Explain the differences between thermoplastics and thermo set plastics.
    (b) Discuss the preparation, properties and uses of different grades of polythenes such as LDPE, HDPE, and LLDPE.
34. Write a note on bakelite
35. (a) What is the constitution of natural rubber?
    (b) How is crude rubber obtained from latex?
    (c) Describe the uses of rubber as an engineering material.
36. (a) What is polymerization? Explain the different types of polymerizations with appropriate examples.
    (b) What is vulcanization and why it is done?
    (c) Write a note on preparation, properties and uses of Teflon.
37. Outline the preparation, properties and applications of the following polymers:-
    (a) Nylon (6,6) (b) Bakelite (c) PVC
38. Discuss the preparation, properties and applications of
    (i) Buna-S rubber (ii) Thiokol rubber
39. (a) Explain the difference between "addition polymerization" and "condensation polymerization" with suitable examples.
    (b) Write a note on fabrication and molding of plastics.
40. Write a short note on polythene
41. (a) What are high polymers? Describe their classification with suitable examples.
    (b) Describe the manufacture, properties and applications of PVC
42. (a) What is natural rubber?
    (b) How is raw rubber obtained from latex?
    (c) Explain the process of vulcanization and its uses.

3.12 Objective Questions

1. Which one among the following is a co-polymer
   (a) Nylon 6,6 (b) Teflon
   (c) PVC (d) Polybutadiene
2. A molecule is eliminated in
   (a) Addition polymerization (b) Condensation polymerization
   (c) Free radical polymerization (d) Ionic polymerization
3. Bakelite is prepared by the condensation of
   (a) Phenol-acetaldehyde (b) Cresol-formaldehyde
   (c) Phenol-formaldehyde (d) Urea-formaldehyde
4. Polyesters belong to the _____ type of polymer
   (a) Addition, thermoplastic (b) Addition, thermosetting
   (c) Condensation, thermoplastic (d) Condensation, thermosetting
5. Natural rubber is a _____ polymer of _____ monomer
   (a) Addition, chloroprene (b) Condensation, chloroprene
   (c) Addition, isoprene (d) Condensation, isoprene
6. _____ is the compound used in refrigeration instead of sulphurdioxide and ammonia
   (a) Teflon (b) sulphur hexafluoride
   (c) Freon-12 (d) Boron trifluoride
7. Polymer obtained by the condensation polymerization is
   (a) PVC (b) Teflon
   (c) Nitrite rubber (d) Bakelite
8. Which of the following is a copolymer
   (a) SBR (b) Buna-S
   (c) both a & b (d) none
9. Cellulose acetate is a
   (a) thermoplastic (b) thermosetting plastic
   (c) both (d) none
10. Coagulation of natural rubber is carried out by adding
   (a) CH_3COOH (b) HCl
   (c) alum (d) detergent

3.13 Fill in the Blanks

1. _____ is example for theromoset resin.
2. _____ plastic will become rigid on cooling.
3. _____ are permanent setting resins.
4. Thermoplastic resins are formed by _____ polymerization.
5. Thiokol rubber is made by reaction between _____ and _____.
6. Butyl rubber is a copolymer of isobutene and a small amount of _____.
7. _____ polymer is used for making of motor tyres.
8. Nylon is obtained by polymerization of _____.
9. Natural rubber is polymer of _____.
10. _____ can be used as vulcanizing agent.
11. Styrene rubber is a polymer of styrene and _____.
12. _____can be used for making gaskets.
13. _____ rubber is used for making tyre tubes.
14. Vulcanization of rubber improves_____.
15. _____ is used for making electric switches.
16. Nylon-6 is used for making _____.
17. In SBR-Rubber the percentage of styrene is _____ and percentage of 1,3 Butadiene is _____.
18. Phenol and Formaldehyde reacts in presence of Acidic catalyst and forms _____.
19. _____ Polymer is having excellent insulating properties.
20. _____ can be used as free radical initiator.

4 Nano-materials, Explosives, Propellants and Lubricants

4 Nano-materials, Explosives, Propellants and Lubricants

4.1 Chemistry of Nano-materials

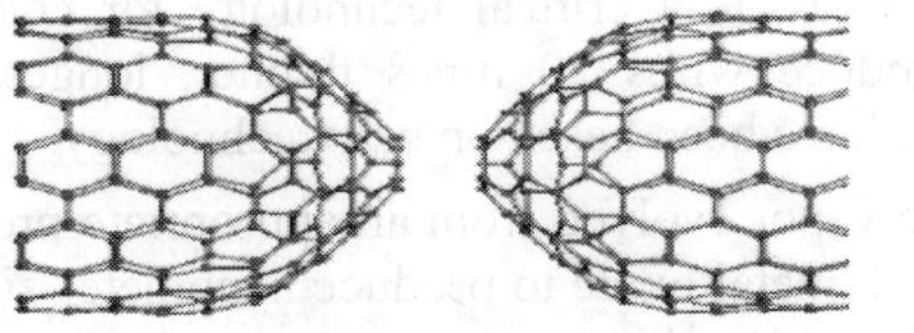

Nano tube welding

"NanoKids" consists of carbon, hydrogen and oxygen atoms

Introduction : Normal chemistry practices revolve around macroscopic particles and their mixing by mechanical means. So we could even call it as mechanochemistry. In mechanochemistry, individual chemical reactions at the molecular or atomic level are achieved by subjecting the reactants to high pressure conditions and pressing together to overcome the normal reaction barriers. This requires positioning of the reactants on a nanometer scale to precisely bring the reactants together. This possibility was first prophesied by Richard Feynman in 1959 in his talk entitled "*There Is Plenty Room At the Bottom*". He said: "The principles of physics, as far as I can see, do not speak against the possibility of maneuvering things atom by atom." His dream could not be realized for want of new class of miniaturized instruments that would allow manipulation and measurement of the properties of the small "nano" structures. In the 1980's with the invention of these instruments Feynman's dream started realizing. The scanning tunneling microscope, atomic force microscope and the near-field microscopes provided the eyes and fingers required for nanostructure measurement and manipulation.

In nanochemistry, the ***rules of bonding*** *are extended from atoms and molecules (with dimensions < 1 nm) to larger objects having the dimension of from 1 to 100 nanometers,* where one nanometer is one-billionth of a meter or 10^{-9} meter. In this regime, new chemical and physical properties appear that depend *on the size and shape of the particles.* Nanochemistry undertakes the synthesis of precisely defined nanoparticles to achieve novel materials for specific applications in such fields as advanced catalysis, ultra-thin films and membranes (separation and adsorption), storage devices, and optical and electronic devices. The nanometer regime is of particular interest because it is in this region where phenomena associated with atomic and *molecular interactions begin to be strongly influenced by the macroscopic properties of materials.*

The Nanorevolution is new but nanochemistry is not. Biological processes are essentially nanochemical processes that occur with low energy consumption and high efficiency. For instance, photosynthesis is a molecular-scale mechanical operation and enzymes are nanosize factories. The challenge is to learn how to control such processes. Once we can do this, there will be advances in everything from microelectronics to chemical engineering.

4.1.1 Applications of Nano-Materials

Nanochemistry has applications ranging from engineering to medicine. A summary of applications of the nano-materials produced is presented here.

1. **As Engineering Materials** : Applications of nanotechnology to materials and manufacturing include the following:
   - The fabrication of thin nanowires, is a critical technology for communications technology. Methods that produce wires 20 times thinner, longer, and better-defined than conventional ones have been based on nanotechnology.
   - In gas phase condensation, the vapor evolved from an appropriate precursor, say a metal, undergoes a reaction with metal oxide to produce nanometer-sized particles. *Indium tin oxide* particles made by this method are ***conductive***, and can be processed to form transparent thin films that are ***antistatic***. Similarly nano particles produced by combining yttria and some lanthanides phosphor (emit delayed emission) and therefore processed to form **light-emitting devices**.
   - Surface scientists have been able to exploit atomic self-organization processes for the controlled fabrication of nanostructures at surfaces having well defined size and shape. This allows forming nanostructure metals and **ceramics** in exact shapes *without machining*.
   - **Improved color printing** brought about by nanometer-scale particles that have the best properties of both dyes and pigments.
   - Nanoscale cemented and plated carbide materials and **nanocoating for cutting tools**, and other electronic, chemical and structural application.
   - **Nanofabrication on a chip** with high levels of complexity and functionality. Nanostructured chip devices prepared from nanoscale materials like complex oxides, carbides, borides and nitrides have been successfully *mass produced* in kilogram quantities.
   - In **nanocomposites**, one of the constituent phases has dimensions–length, width, or thickness, in the nanometer range. Because the building blocks of the nanocomposite have the dimensions of nanometers, they have *enormous surface areas* and large interfaces with other phases. The properties of the nanocomposite

result from interactions at these interfaces. In conventional composites where the building blocks have dimensions on the order of microns (one-millionth of a meter), there is less interfacial contact between phases, and consequently less effect by the interfaces on properties of the composite.

One naturally-occurring nanocomposite is bone. Bone consists to a large extent of nanoscale, platelike crystals of hydroxyapatite, $Ca_{10}(PO_4)_6(OH)_2$ dispersed in a matrix of collagen fibers. Conventional attempts to synthesize hydroxyapatite for artificial bone or bone implant materials have many difficulties, but scientists have found that these problems can be largely overcome if hydroxapatite is synthesized as a nanostructured material.

In organic and sol-inorganic nanocomposites, a nanoscale inorganic filler is typically dispersed in an organic polymer matrix. The filler carries the load, and its large surface area in contact with the matrix reduces the mobility of the polymer chains. Nylons are thermoplastic but *Nylon nanocomposites* containing small amounts of clay are capable of withstanding high temperature environments and have been designed into automobile air intake covers.

- Chemists have also been looking at *carbon* ***nanotubes*** (honeycomb **graphite** lattices rolled into cylinders having nanoscale thicknesses) as fillers in nanocomposites. The nanotubes, by virtue of their **electrical conductivity**, make the composite conductive as well. Nanotubes have been found by experiment to be stiffer than carbon fibers (which are thicker than 1 micron), but less brittle. Thus, nanocomposites containing carbon nanotubes are able to withstand much greater deformations before breaking than carbon fiber composites.

2. **In Chemical Industry :** In the field of **catalysis**, chemists have been searching for catalysts that are *stable at high temperatures and that can be used in smallest possible amounts*. The complex oxide barium hexaaluminate (BHA), or $BaO_6Al_2O_3$, has been of interest since it retains its catalytic activity at high temperatures. But conventional methods to synthesize BHA tend to reduce the material's surface area, and hence the activity. But when nanoparticles of BHA are prepared, a final grain size of 30 nanometers is achieved, and a *large surface area and catalytic activity* are achieved.

The new nano-catalysts designed and new nanochemical routes could pave way for environment friendly and atom economical **Green chemistry routes.** The nano chemical routes catalyze chemical reactions, sometimes at rates exceeding a million reactions per second, without requiring elevated temperature, pressure, or dangerous solvents.

3. **In Biotechnology :** Enzymes are known to catalyze more than 5000 diverse chemical reactions. The possibilities for using enzymes in small scale reactors are limitless. Protein catalysts or enzymes have precise control over specific chemical reactions, reducing unwanted side-reactions. Most enzymes work by lowering the energy barrier for a specific reaction; thus the energy needed for conversions is reduced to ambient temperatures and pressures. Recombinant DNA technology enables the production of novel proteins with enhanced properties not found in nature. *The main drawback with enzymes is fragility.* Combining nano-scale materials with enzymes could dramatically

improve the durability of enzymes, create localized high concentrations of protein and reduce costs by minimizing losses.

4. **In Medicine and Health :** Nanotechnology will provide new tools for medicine. It could radically change the way surgery is done. It will make it possible to do molecular-scale surgery to replace defective cells, repair and rearrange cells. Since disease is the result of physical disorder, misarranged molecules and cells, medicine at this level should be able to cure most diseases. Mutations in DNA could be repaired and cancer cells, toxic chemicals, and viruses could be destroyed through use of medical nanodevices. There are numerous other potential applications of nanoscience to biology :

   Rapid, efficient genome sequencing, revolutionizing diagnostics, and therapeutics

   Effective and less expensive healthcare using remote and in-vivo devices

   New formulations and routes for drug delivery that enormously broaden their therapeutic potential by effecting delivery of new types of medicine to a previously inaccessible sites in the body

   More durable, rejection-resistant artificial tissues and organs

   Sensor systems that detect emerging disease in the body, which will shift the focus of patient care from disease treatment to early detection and prevention

5. **Nanotechnology for Environment and Energy :** Nanotechnology has the potential to significantly impact energy efficiency, storage, and production. Solar cells are prohibitively expensive and nanometer-sized solar cells could be developed to provide much of the energy needed around the world at a cheap price. This would stop the usage of fossil and nuclear fuels. Thus this would provide a cleaner and safer method than any of our current methods for making power.

   Other new nanoscale capabilities that illustrate this potential are :

   The discovery of the ordered mesoporous material MCM-41 produced by Mobil Oil Co., with pore size in the range of 10 to 100 nm, is now widely applied in the removal of ultrafine contaminants.

   Nanoparticle-reinforced polymeric material discussed above could replace structural metallic components in the auto industry. Widespread use of those nanocomposites could lead to a reduction of 1.5 billion liters of gasoline consumption over the life of one year's fleet of vehicles and reduce carbon dioxide emission by more than five billion kilograms annually,

   The replacement of carbon black in tires by a nanometer-scale particles of inorganic clays and polymers is a new technology that is leading to the production of environmentally friendly, wear-resistant tires.

6. **Nanoelectronics and Computer Technology :** Nanotechnology will enable the construction of nanoscale circuits and computers. Smaller circuits run faster and this enables far greater computing speeds. Nanomaterials can help by providing nanocrystalline materials of ultra-high purity and with better thermal conductivity and longer life.

Other potential nanoelectronics and computer technology breakthroughs include :

Nanostructured microprocessor devices that continue the trend in declining energy use and cost per gate, thereby potentially improving the efficiency of computers by a factor of millions.

Higher transmission frequencies and more efficient utilization of the optical spectrum to provide at least ten times more bandwidth, with consequences in business education, entertainment, and defense.

Small mass storage devices with capacities at multi-terabit levels, a thousand times better than today.

Integrated nanosensor systems capable of collecting, processing and communicating massive amounts of data with minimal size, weight, and power consumption

7. **In Aeronautics and Space Exploration :** Birds have long inspired the development of aircraft, but our present man-made vehicles are primitive compared to Mother Nature's flying counterparts. To achieve the goal of flying like a bird, the challenges to be met are– material science advances in strength-to-weight ratios, reconfigurability, integration of sensing and actuation. Further in space crafts, the stringent fuel constraints of lifting payloads into earth orbit and beyond, and the desire to send spacecraft into other solar systems for extended missions, compel continued reduction in size, weight, and power consumption of payloads. Nano structured materials and devices promise solutions to these challenges. Further development of nanocomposites will be to make them tunable, adaptive, self-healing and stress-smart sensing systems. These materials will enable a revolution in air and space travel by increasing intelligence in aerospace systems and vehicles while reducing mass, size and power consumption.

To a certain extent explosives and rocket fuels are related materials. In both the cases, chemical reactions take place rapidly releasing a large amount of energy. In fact many materials were first used as explosives and later were used as rocket fuels. Due to this reason, explosives have been considered in this chapter before rocket fuels.

4.2 Explosives

4.2.1 Introduction

Explosives are compounds or mixtures which, on receiving thermal or mechanical shock, are rapidly converted into products of greatly increased volume. This is caused by the formation of gaseous products and the generation of a great deal of heat. Since the gases occupy a much larger volume, and the volume further increases because of the high temperature produced, extremely high pressure is exerted on the surroundings giving disruptive or propelling effect.

Most people associate explosives with their military uses. Explosives have enabled man to perform such engineering feats as would not have been possible without their aid. Large engineering constructions, like the various dams which have been built in a short span of time, would have taken decades or even hundreds of years and thousands of labourers, reminding one of medieval ages in which the great pyramids were built in Egypt, or huge and magnificent temples constructed by the Incas in South America. Much more common examples of the use of explosives are in the blasting of the ores of iron and other metals, coal mining, quarrying limestone, construction of tunnels and mountain roads and numerous similar heavy works. Less powerful explosives are increasingly being used for such odd jobs such as digging holes for tree plantations, making ditches for draining water and even for rivetting and welding. In this section the industrial explosives will be considered and military explosives will be referred to only in passing.

In India there are three large plants manufacturing industrial explosives with an installed capacity of 58,500 tonnes per annum. Of this 36,000 tonnes is for nitroglycerine based and the balance for slurry explosives. Units with additional capacity of 62,5000 tonnes (15,000 t nitroglycerine and 47,500 t slurry type) are being set up. About 50,000 t of industrial explosives were manufactured in 1981.

4.2.2 Classification

Explosive materials may be solid (e.g., trinitrotoluene, TNT) liquid (e.g., nitroglycerine) or gaseous (e.g., mixture of oxygen and acetylene). They are usually divided into three broad groups, low explosives, primary explosives and high explosives.

Low explosives, also called *propellants,* do not explode suddenly but burn. The chemical reactions take place comparatively slowly and burning proceeds from the surface inwards in layers at a rate of about 0.2 m per second. The evolved gases disperse readily, without building high pressures, and they can be controlled.

Primary explosives or initiating explosives are commonly known as *detonators*. These are highly sensitive explosives which explode on receiving a slight shock or fire. They are used in small amounts in cartridges and shells to start the explosion of the main explosive which is comparatively less sensitive. They should be handled with the utmost care.

High explosives have higher energy content than the primary explosives but are quite stable and insensitive to mechanical shock and fire. A small amount of a primary explosive is placed in contact with the high explosive so as to give it an explosive shock. This starts the rapid chemical reaction in the body of the high explosive which proceeds at the rate of 1,500 to 10,000 m per second. The large volume of the evolved gases, especially under the influence of the high temperature produced, has a shattering effect. The high explosives may be subdivided into military high explosives, and blasting or industrial explosives. The former consists of pure substances, and in the latter the pure explosive is mixed with non-explosive materials as in the dynamites.

4.2.3 Requirements

An explosive must fulfill the following basic requirements :

1. The explosive must be stable under ordinary conditions of storage. If it undergoes a rapid decomposition at 40°, it is considered thermally unstable and unsafe. If rapid decomposition does not take place up to 95°, it is considered very stable. It is noteworthy that the character of reaction on heating depends upon the rate of heating. In general, the higher the rate of heating the more violent is the chemical reaction.

2. The chemical reaction must be highly exothermic. Ammonium nitrate gives 346 cal/g whereas TNT gives 925 cal/g and nitroglycerine produces 1,486 cal/g of heat. In general, the energy released is of the order of 1000 cal/g.
3. The rate of reaction must be very rapid so that the heat of reaction accumulates faster than it is dissipated and the temperature rises sharply.
4. The products of reaction must be gaseous, preferably having small molecular weights so that the volume change is high. Under the influence of a sharp rise in temperature the gases expand rapidly. For example, 1 g of nitroglycerine yields about 40 ml of gases when cold, which occupy about 750 ml at the temperature of explosion. In general, the volume of the hot gases is about 10 to 15 times as much as at the room temperature (Water, which is a liquid at the room temperature, is in a gaseous state at the moment of explosion).

4.2.4 Properties and Tests

Since the explosives are required to do different kinds of work it is necessary to standardize, control and evaluate their properties so that the work is performed satisfactorily. In the military explosives this is all the more necessary so that the target may be hit accurately. Most of the tests are empirical and are useful in comparing one explosive with another.

Strength and velocity of detonation. These two properties determine the power or *brisance* of an explosive and are responsible for its shattering effect. Strength depends upon the gas pressure developed from a given weight or volume of the explosive. The velocity of the detonation wave decides the rate at which the effective strength is developed. A soft rock is blasted satisfactorily by an explosive of lower strength and lower velocity. A hard rock requires higher strength and higher velocity of detonation.

Brisance is usually tested by *Trauzl lead block* test in which a given weight of the explosive is exploded in a standard lead cylinder and the increase in its volume forms the basis of the test. Brisance is also evaluated by exploding the explosive in a closed cylinder in contact with standard sand. The amount of sand crushed is determined by the use of standard sieves and indicates the shattering power of the explosive.

Sensitivity to mechanical impact is of special importance to primary or initiating explosives. The minimum height from which a standard weight must be dropped to detonate the explosive is a measure of sensitivity.

The *heat* of explosion (determined calorimetrically) and the volume of the gaseous products are very important and indicate the work capacity of the explosive. This is different from the power of the explosive which is governed by the rate at which the heat energy is liberated.

It is extremely difficult to directly correlate the various properties of an explosive with its behaviour when it is used in a particular way. The different explosives do not differ very much from each other in the properties indicated above, but their actual performance varies considerably in different situations. Therefore, a number of practical field tests have been developed, especially for military explosives.

It is noticed that most of the explosives contain nitrogen in one form or another. The reason is that the strength of the bond linking the nitrogen atom to the adjoining atom is low with the result that the compound is relatively unstable and is sensitive to shock. In most compounds nitrogen is linked to oxygen which moves from nitrogen to carbon and hydrogen forming carbon monoxide, carbon dioxide and water. Nitrogen itself is liberated as a gas. In this process on an average, about 1000 cal/g of heat is liberated which helps the gases to expand and cause the destruction.

4.3 Manufacture and Properties of useful Explosives

4.3.1 Primary Explosives

As has already been explained, these are high explosives which are sensitive to heat, impact and friction and therefore explode easily. They have lower shattering power than the secondary explosives. Because they explode so easily and have small shattering effect, they are not used as the main explosive. The secondary explosives do not explode as readily, therefore, explosion is initiated in the primary explosive, which causes explosion in the secondary explosive. The most important primary or initiating explosives are lead azide, PbN_6, and mercury fulminate, $Hg(CNO)_2$.

Lead azide is very popular for military use because of the low cost, stability in storage, and excellent initiating action. It cannot initiate explosion in the less sensitive secondary explosive like cast TNT. As it reacts with brass, the caps of the containers loaded with it are made of aluminium. It is manufactured cheaply by the action of ammonia on sodium giving sodium amide which, with nitrous oxide, gives sodium azide. A soluble lead salt on treatment with sodium azide produces lead azide.

Mercury fulminate is more expensive and more sensitive than lead azide. It is slightly toxic and is used to a lesser extent than lead azide. For its manufacture, mercury is dissolved in excess of nitric acid and the solution is added to ethyl alcohol. The solution begins to boil spontaneously and mercury fulminate is precipitated.

Diazodinitro Phenol (DDNP) is widely used in commercial blasting caps. It is quite sensitive and has high brisance and therefore can initiate explosion in less sensitive explosive. DDNP and lead azide have replaced mercury fulminate for most uses.

Tetracene may also be mentioned as it is used in explosive rivets. Its initiating efficiency is low and it is not used by the military. It is easy to ignite, has high heat of explosion and produces a large volume of the gases.

4.3.2 Low Explosives or Propellants

Black powder and smokeless powder are the important low explosives.

Black powder or gun powder is cheap and is still used to a considerable extent in spite of a number of new explosives being available. It is an intimate mixture of potassium nitrate, charcoal and sulphur (75:15:10). The powdered mixture is moistened and pressed into a cake and then broken up into grains. A small spark is sufficient to explode the whole mass.

Although it has some desirable properties (e.g., non-shattering action) it is not good for coal mines because it produces a flame which can ignite gas and coal dust mixture and it also produces carbon monoxide, which is harmful to human beings. It is used for blasting where there is good ventilation, in fireworks and for igniting smokeless powders, etc.

Smokeless powder is nitrocellulose or more correctly cellulose nitrate. It is prepared by treating cellulose with nitric and sulphuric acids (The nitrocellulose plastics are not nitrated as fully as nitrocellulose explosives are nitrated). Nitrocellulose is dissolved in a mixture of alcohol and ether and the solvent is evaporated when a jelly-like solid is left behind. As it may react with traces of acid liberated during storage, a stabilizer (e.g., diphenylamine) is added to it. It is then pressed into cylindrical rods of smokeless powder.

It is called smokeless powder because on ignition it produces carbon monoxide, carbon dioxide, nitrogen and water vapour and almost no smoke.

4.3.3 Secondary High Explosives

These are quite insensitive and explode by detonation of a primary explosive. They release a large amount of energy required for various military purposes, demolition work and for certain recent methods of metal working operations. Some of them can be cast to give the required shape, some are pressed and some are available even as a plastic sheet–which makes them very useful for metal working operations. They are often classed into four groups: (1) Single compound explosives, (2) binary explosives, (3) plastic explosives and (4) dynamites.

4.3.4 Single Compound Explosives

Single Compound Explosives as the name suggests, contain only one chemical compound. Except ammonium nitrate, all other explosives are organic compounds–nitrates, nitro–compounds and nitramines of aliphatic, aromatic and heterocyclic series. The early explosives were obtained by nitration of glycerol and cellulose. As the coal tar industry developed, the nitro-compounds, like TNT came into prominence. The manufacture of nitramines started recently with the development of the synthetic organic chemicals industry.

Ammonium nitrate is cheap, has low brisance value, is quite insensitive, very stable and non-toxic. It cannot be used in contact with alloys of copper as it forms tetraminocupric nitrate which detonates easily. It is dangerous to store near any inflammable material. It is a high explosive though about half as powerful as TNT and is mostly used along with another explosive in making binary explosives. Its use is increasing.

Trinitrotoluene, TNT, is perhaps the best known of the high explosives and is widely used for firing shells. It is melted (m. p. 80.8^o) and poured into the containers. Its great importance is because: (1) it is not sensitive and therefore is safe to handle during manufacture, storage and transportation, (2) it does not absorb moisture, (3) it does not react with metals to form unstable compounds and (4) it has high brisance value. For all these reasons TNT is widely used for military as well as industrial purposes.

Pentaerythritol tetranitrate, PETN, is a standard, extremely powerful and sensitive military explosive. *Cylonite* or cyclotrimethylenetrinitramine, usually called RDX, came into prominence during World War II as a military as well as an industrial explosive. It is highly powerful. It is more sensitive but less toxic than TNT. *Picric acid*, trinitrophenol, was an important high explosive but has been largely replaced by TNT. *Tetryl* is a nitroaryl nitromine and its use as a single base explosive has been discontinued but it is used in binary explosives.

4.3.5 Binary Explosives

These consist of mixtures of various explosives, particularly of TNT with other explosives. For certain applications, they are superior to single compound explosives and are more convenient to make. TNT is an important ingredient of these explosives because it has a low melting temperature. Generally they are heated to liquid state and poured into the container and allowed to solidify. The widely used combinations are: *amatol* (ammonium nitrate and TNT, 50% of each), *pentolite* (PETN and TNT, 50% of each), *tetryol* (70% tetryl and 30% TNT), *torpex* (40% RDX, 40% TNT and 20% aluminium powder) and *tritonal* (80% TNT and 20% aluminium flakes). Each of these combinations has some special effect, hence is more suited for a particular use. Their chief advantages are higher energy output and lower cost of making the ammunition because they can be melted and poured.

4.3.6 Plastic Explosives

A combination of explosives which is in a plastic state and can be press-loaded, hand-moulded and made into various shapes without serious risk, is known as a plastic explosive. Plastic explosives are usually available in bulk, and are largely needed for military use, and in the form of plastic sheets which are mainly used for industrial applications. A simple combination consists, of a high explosive, which gives the destructive power, mixed with wax or oil, which makes it less sensitive and plastic. It may be made into a putty-like mass or a sheet.

On account of their engineering applications, special mention may be made of flexible plastic sheet explosives which are commercially available. These contain a high explosive like PETN with other materials, and are made into tough and flexible sheets of various sizes and thickness. The sheet can be cut into the desired shape and can be easily fixed to a metal part by an adhesive. Explosion is started by special caps. On explosion, due to the sudden impact, the metal can be made to take the desired shape under suitable conditions of metal forming.

4.3.7 Dynamites

Alfred Nobel, the founder of the Nobel prizes, gave the name dynamite to a solid mixture of nitroglycerine and kieselguhr (which is the mineral remains of unicellular algae). In a simple form it can be made by absorbing nitroglycerine in wood flour mixed with sodium nitrate. The mixture is encased in a cardboard cylinder to form sticks. By suitable compounding the dynamite can be used for blasting hard rocks or breaking soft minerals like coal into large lumps. It seems that detonation of dynamite is related to the porous, granular structure, because it is noticed that it often fails when highly compressed (as happens in wet bore holes when high water pressure acts on the dynamite cartridge).

Nitroglycerine is a high explosive and is used almost wholly for dynamite manufacturing. Nitroglycerine is manufactured by the action of strong nitric acid on high purity glycerine in the presence of strong sulphuric acid. The mixture is agitated and kept cold by circulating cold brine through closed coils. The temperature is not allowed to exceed 15°. The reaction may be represented as follows :

$$\begin{array}{lcl} CH_2OH & & CH_2ONO_2 \\ \vert & & \vert \\ CHOH + 3HNO_3 + (H_2SO_4) & \longrightarrow & CHONO_2 + 3H_2O + (H_2SO_4) \\ \vert & & \vert \\ CH_2OH & & CH_2ONO_2 \end{array}$$

It is thus seen that the compound is glyceryl trinitrate and the term nitroglycerine is a misnomer. Nitration is complete in about an hour and the mixture is run into a separating tank in which nitroglycerine rises to the top. It is separated and washed, first with water and then with dilute sodium carbonate solution. The operation is highly risky and must not be attempted by unskilled persons.

Nitroglycerine is a viscous white to yellowish liquid resembling glycerine. Its specific gravity is 1.6; it freezes at 13° and is very sensitive to shocks.

By varying the constituents, several types of dynamites are made. The important ones are the following:

Guhr dynamite is the name given to original dynamites made by absorbing nitroglycerine in porous material like charcoal or kieselguhr. This makes nitroglycerine safer to handle. It is dry and is granular in structure and relatively insensitive to shock. It is not used in the U. S. A. but is still used in Europe.

Straight dynamites of various strengths are on the market. A typical 40% dynamite used by the U. S. Bureau of Mines contains 40% nitroglycerine, 44% sodium nitrate, 15% wood pulp and 1% calcium carbonate (to act as an antacid). As nitroglycerine has relatively high freezing temperature, the dynamite is unsatisfactory for low temperature use. For low temperature use, nitroglycerine is partly replaced by nitration products of other related alcohols like glycol. Water resistance of straight dynamite is not good and should be fired within 24 hours if used under water.

In *ammonia dynamites* nitroglycerine and sodium nitrate are partly replaced by ammonium nitrate. This makes the dynamite cheaper. It has a great heaving (i.e., lifting or moving) ability and is less shattering.

Gelatin dynamites contain a small amount (usually < 6%) of nitrocellulose. Nitrocellulose is dissolved in nitroglycerine when a jelly-like mass is produced. It is mixed with other ingredients in the usual manner. The gelatin dynamites are highly water resistant and can be used for under-water operations. Another advantage of these over straight dynamites is that they are safer because the jelly-like mass has fewer tendencies to exude than the liquid nitroglycerine.

Ammonia gelatin dynamites have been developed to combine the advantages of ammonia dynamites with gelatin dynamites. They are similar to gelatin dynamites but a part of nitroglycerine is replaced by ammonium nitrate.

Military dynamites contain high explosive like TNT, RDX with binders. They do not contain nitroglycerine. They neither freeze in cold climate nor exude in hot weather. Their transporation, storage and handling are safer than those of the straight dynamites.

Permissible or permitted explosives have been developed for use in coal mines where there is the danger of violent explosion due to the frequent presence of methane and coal dust mixed with air. It is found that a hot and long lasting flame is more likely to cause explosion in such a surrounding than a cool flame of a short duration. The explosives for such uses contain about 80% ammonium nitrate and a small percentage of nitroglycerine as sensitizing agent soaked in a fibrous material. The black powder is certainly a better explosive in the sense that it produces large lumps of coal and not much fine dust but it is risky in the presence of methane and coal dust mixed with air.

The percentage *marking* on dynamites is for comparing strength on a weight for weight basis with that of straight dynamite with the percentage of nitroglycerine. For example, 60% ammonia dynamite was found to contain 20% nitroglycerine, 75% ammonium nitrate and 5% combustible material. It is marked 60% because its strength is comparable to that of straight dynamite which contains 60% nitroglycerine.

4.3.8 Some Recent Uses of Explosives

During the last few years explosive working of metals is being adopted on an increasingly large scare, especially in the aircraft and missiles industry, to solve difficult metal forming problems.

Generally the explosive is separated by air, water or oil from the work piece, which may be a metal plate or a metal sheet. Very high pressure, say of the order of a few thousand kg per cm^2, may be exerted on the work piece by the explosion and a variety of shapes of varying sizes may be produced.

Such operations have a number of advantages over the traditional methods of metal working. Some of these are:

(l) Capital investment of press is eliminated; (2) only female die is needed and a punch is not required; (3) the explosive pressure is uniformly distributed over the entire surface, thereby local stresses are absent. Details of the latest applications are worth the careful study (see bibliography) of the mechanical engineers (ASME, Detroit, has several publications on Advanced High Energy Rate Forming).

4.4 Propellants

4.4.1 Introduction

The term rocket is used to designate a wide variety of missiles, vehicles, thrust devices, etc. The fundamental principle governing the forward motion of the rockets was given three centuries ago by Newton in his third law of motion when he stated that for every action there is an equal and opposite reaction. The working of this law is noticed even by a small child when he releases an inflated rubber balloon and as air escapes from the small opening, the balloon moves forward against the force of gravity. The child has then "fired a rocket" just as the great scientists and engineers fire mighty rockets to the outer space.

The rocket engine or motor is unique in that both a fuel and an oxidizing agent are carried by it. Fuel and oxidizing agent are collectively referred to as the *propellant*. The oxidizing agent and the fuel component may be present in the same molecule, or the propellant may be a mixture. As the fuel does not depend upon atmospheric oxygen for combustion and the thrust produced is independent of the medium through which the rocket travels, the rocket engines are the only type of mechanism available for travel in the upper atmosphere (i. e. above about 30,000 meters) and the interplanetary space. The rockets are, however, being used for military purposes also for throwing explosives against targets thousands of miles away.

4.4.2 Principles of Rocket Propulsion

The rocket moves forward in accordance with Newton's third law of motion, viz. for every action there is an equal and opposite reaction. The thrust is not due to the jet of gases pushing against the surrounding air but is due to the 'equal and opposite reaction'. The propellant reacts chemically and produces gaseous products in the combustion chamber. The gases are at a high pressure and high temperature, of the order of 300 kg/cm^2 and 3,000°, and expand and emerge through a nozzle at supersonic velocity producing a thrust for propulsion.

The *propulsive force or thrust* is due to the momentum of the gas and can be calculated thus:

$$F = \frac{m}{g} \times v + (P_e - P_a)\, A_e$$

Where F= thrust (in lbs), m = mass flow rate of the propellant gases (in lbs/sec), g = the acceleration of gravity (in ft/sec), v = effective exhaust velocity (in ft/sec), P_e = exit gas pressure at the nozzle (in lbs/ft^2), P_a = ambient pressure (in Ibs/ft^2), A_e = Nozzle exit area (ft^2)

The equation shows that the rocket engine performance increases with increase in exhaust velocity. Since P_a decreases with altitude, and becomes zero in perfect vacuum, the performance increases with altitude. The exhaust velocity increases with increase in adiabatic (i. e. without heat change) flame temperature and with decrease in molecular weight of the combustion gases. The chemical fuels contain carbon, hydrogen, oxygen and nitrogen and give, in order of increasing molecular weights, H_2, H_2O, CO, N_2, O_2 and CO_2, Thus it is seen that lower molecular weight gases are produced by fuels rich in hydrogen and poor in carbon. For such materials the maximum temperature attainable is about 3500° as at higher temperatures highly endothermic reactions take place giving rise to atomic hydrogen, oxygen and nitrogen and the free hydroxyl group.

Another important term commonly used is *Specific impulse* (or *specific thrust* for liquid propellants). The term refers to thrust per pound per second of fuel burnt. In the simplest form specific impulse is Isp = F/m sec., where m is rate of propellant consumption (in lbs/sec) to produce the total thrust F (in Ibs). Detailed calculations show that specific impulse increases as the chamber pressure and temperature increases and the molecular weight of the exhaust gases decreases. In these relationships, the liquid propellants containing hydrogen and powerful oxidizing agents such as liquid oxygen, fluorine and ozone, show advantage over the solid propellants.

The *velocity* of a rocket, V, is approximately given by the following expression:

$$V = \nu \log e \frac{Mi}{Ms}$$

where, ν = effective exhaust velocity, Mi = initial mass of rocket, Mf = final mass of rocket

As the rocket moves, the mass of the propellant decreases, and the final mass of the rocket when all the propellant has been used up is less than the initial mass. In the expression given above the effects of air drag and gravity, which reduce the velocity have not been included.

In the design of an efficient rocket engine an attempt is made to keep the structural mass to a minimum and the mass of propellant to a maximum. In doing so several complicating factors have to be carefully considered. For example, the maximum theoretical specific impulse is for a propellant system of hydrogen and ozone. As the density of hydrogen is very low even under high pressures, the total mass of the complete rocket becomes high and a rocket using other propellants with lower specific impulse proves to be more efficient. Therefore, rockets are conveniently compared on the basis of their *over-all impulse ratio*, i.e., the ratio of the total impulse to total initial mass of the rocket.

In order to obtain very high velocities for rocket vehicles, as required for interplanetary flights, a series of rockets, one on top of the other, are built and fired successively and the earlier ones dropout. The advantage of such an arrangement is that each successive stage is smaller than the previous one and starts firing at a higher initial velocity. Theoretically the larger the number of such stages, the greater will be the final velocity but each additional stage means duplicating the entire mechanism with added costs. This puts a limit on the number of stages to be provided.

Most of the rocket engines burn chemical propellants. The engine has a combustion chamber. In the case of a liquid propellant it is injected in the chamber. There is a complicated arrangement of valves, regulators, igniters, etc. The propellant, either solid or liquid, burns in the combustion chamber and provides hot gas under high pressure, which escapes in a jet through a nozzle to the rear. This gives thrust for the rocket vehicle to move forward.

It has been reported that commencing at the launch, in the first 10 seconds the Saturn-5 moon rocket consumed 15 tonnes of fuel each second, producing a thrust of 3750 tonnes. More than 95% of the weight of the Staurn-5 rocket was made up of liquid fuels. The Apollo spacecraft was placed on top of the rocket. The rocket was detached successively and was not meant to return intact to the earth.

The above discussion helps one to understand the requirements and the working of the rocket fuel.

Although the nuclear reactor, plasma jet or hydrogen heaters have been used, most rockets at present use chemical propellants for providing the necessary power. The chemical propellants are classified into solid and liquid categories.

4.4.3 Solid Propellants

Depending upon their physical structure, the solid propellants are divided into two main groups: *homogeneous* and *composite* (or heterogeneous). Solid propellant mixtures in which the propellant or mixture of propellants are intimately mixed in a colloidal state are known as homogeneous solid propellants. Nitrocellulose and nitroglycerine combination is an example of this type. Composite propellants, like gun-powder or other combinations, in which an oxidizing agent is dispersed in a rubber-like fuel mass, have distinctly separate phases. Historically speaking, gun-powder was the only propellant used from the early 14th century to the late 19th century. In rocket engines it gives a specific impulse of about 60 sec., flame temperature of 800° to 1650° and the volume of the gases is about 400 times the volume of the charge.

Nitrocellulose, popularly known as smokeless powder or guncotton, replaced gunpowder. It is known as a *single base propellant*. Further improvement came by the discovery of the colloiding and solvent action of nitroglycerine (a powerful explosive in itself) on nitrocellulose. It is a powerful propellant and the name *ballistite* is given to such powerful propellants. The propellants which contain two materials, such as nitroglycerine and nitrocellulose, are called *double-base propellants*. The name *cordite* was given to a propellant using a plasticizer such as petroleum jelly. A typical composition is 65% nitrocellulose, 30% nitroglycerine and 5% jelly.

For smooth burning of a double-base propellant the operating chamber pressure of about 30 kg/cm^2 is required. The improvement over the earlier propellants can be seen from the fact that the flame temperature is about 2,750°, heat evolved is 850 to 1100 cal/g and volume of gas is about 1,500 times the original propellant volume.

The main constituents of a double-base propellant are nitrocellulose (50 to 55%) and nitroglycerine (40 to 45%). Smaller amounts of diethylphthalate (up to 5%) are also added, which along with nitroglycerine acts as solvent and plasticizer. They make the mixture into a homogeneous plastic mass which can be worked smoothly and is strong. As nitrocellulose decomposes in course of time, a stabilizer, like diphenylamine or ethylcentrallite is added (up to 1%). For military purposes, a flash suppressor (so that it may not be easily seen) like potassium nitrate (or sulphate) is added (about 1%).

The organic ingredients are mixed to form a paste and dried. Then the water soluble inorganic salts are added and the whole mass is made homogeneous by rolling in special mills and obtained in the form of flakes. The flakes are extruded in a press to give the required shape to the final grain.

4.4.4 Composite Propellants

Gun powder is the oldest composite propellant. During World War II 'Galcit series' of propellants, with about 75% potassium perchlorate and 25% asphalt oil, were developed.

Compositions of some other typical propellants of this type are; 80% ammonium perchlorate and 20% resin binder; 80% ammonium nitrate, 18% resin binder with 2% catalyst; 46% each of ammonium picrate and sodium nitrate with 8% plastic resin binder.

It will be noticed from the above data that oxidizing agents, commonly perchlorates and nitrates, have to be used in composite propellants. Obviously, the most important criterion in the selection of an oxidizing agent is the percentage of available oxygen. It should not be hygroscopic, should be stable in contact with the fuel at the atmospheric temperature, and should not form corrosive products. Generally it is desired that smoke is not formed. Metallic ingredients form smoky exhaust gases. For example, ammonium perchlorate leaves no solid residue but the products contain hydrogen chloride and water which form a corrosive and toxic fog. Potassium perchlorate leaves white particles of potassium chloride.

The composite propellants contain 50% to 90% solid oxidizer in the final product. The solid constituents of suitable particle size are combined with the blended liquid constituents. The whole mass is mixed at suitable temperature and pressure. The product now obtained is in a dough-like mass or a thick liquid and is cast into a rocket motor or into moulds.

One of the most important properties of the propellant, whether double-base or composite, is its burning characteristics. It should burn in layers parallel to the burning surface. The burning rate is the rate, expressed in cm per second, at which the burning surface recedes. Before putting a product into actual use, the effect of temperature and pressure of the gases produced on the burning rate needs to be studied. Once burning starts, the temperature and pressure on the remaining propellant increase greatly and may alter the burning characteristics. Irregular combustion is very dangerous and may give an unpredictable and erratic course to the projectile. To some extent the burning characteristics are controlled by adjusting the grain shape and size of the propellant.

The propellants should be mechanically strong and should be able to withstand high pressures developed during firing. They are essentially high energy, unstable substances and it should be ensured that the inner layers will not suddenly explode under the influence of the high pressures.

The combustion chamber is made or metal and expands at a rate different from that of the propellant. In the design, allowance should be made for this difference.

During storage, the propellant may be exposed to different conditions of temperature and humidity. Sometimes it shrinks during storage causing the formation of cracks which increase the surface area thereby increasing the combustion rate to a dangerous level. Hygroscopic materials may absorb moisture and retard combustion. All these factors have to be carefully considered in the actual formulation of the propellant.

4.4.5 Liquid Propellants

Liquid propellants have several advantages over solid propellants. They have higher specific impulse and are more versatile. Their engines can be checked and calibrated more precisely. The chief disadvantage is that the engine is more delicate and less suitable for rough handing than the engine for solid propellants.

Liquid propellants are classified into two groups: mono-propellants and hi-propellants. A mono-propellant may be a single compound, like ethyl nitrate, or it may be a solution of a fuel and an oxidizing agent, like alcohol and hydrogen peroxide.

Bi-propellants consist of two liquids, one a fuel and the other an oxidizer. They ignite spontaneously on contact and have to be handled, stored and injected separately into the combustion chamber. They are capable of developing more energy than the monopropellants and are commonly used for high energy rockets. On the other hand, mono-propellants are more convenient to use as they require a single storage tank and only one fuel injection and control system.

4.4.6 Mono-propellants

Very few available substances are satisfactory as monopropellants. The reason is that requirements of a mono-propellant are conflicting. The substance must be safe to store and at the same time it should easily ignite and burn smoothly.

Of the common monopropellants, a mixture of 21.4% methanol and 78.6% hydrogen peroxide (87% conc.) gives the maximum specific thrust. Next comes nitromethane. Pure hydrogen peroxide has also been used but it is not easy to store and handle because of its reactivity. Metal oxides catalyse its decomposition. Therefore the storage tanks are made of special materials and must be clean. Hydrogen peroxide ignites in contact with organic materials. One has to be careful in dealing with concentrated hydrogen peroxide as its decomposition is self-catalytic and may become violent. Other important mono-propellants are ethylene oxide, hydrazine and propyl nitrate. Each of these has certain advantages and certain disadvantages.

4.4.7 Bi-propellants

Bi-propellants are more widely used. The common *oxidizers* are liquid oxygen (LOX). hydrogen peroxide, ozone, fuming nitric acid and liquid fluorine. Liquid oxygen has to be stored under pressure in insulated containers made of special materials. It is safe, non-toxic and a good oxidizing agent. Ozone is a more powerful oxidizing agent but is toxic and may explode at high concentrations. Fuming nitric acid is insensitive to shock but attacks most metals and organic materials. White fuming nitric acid (WFNA) contains up to 5% water. Red fuming nitric acid (RFNA) contains up to 22% dissolved nitrogen peroxide.

Liquid fluorine is very reactive, corrosive, toxic and volatile. It is very good oxidizer but is difficult to store and handle.

Fuels. Liquid hydrogen with 100% ozone gives the maximum specific impulse but it has a very low boiling point and low specific gravity. Hydrazine, N_2H_4, (B. P. 1.7^o) is also a very good fuel but is expensive. 75% ethyl alcohol is good and readily available. Addition of water to alcohol reduces flame temperature but the molecular weight of combustion gases is also reduced which, to some extent, compensates for reduction in performance. Kerosene and aniline are the other important fuels. The specific impulse in all these cases (except hydrogen and ozone combination) is higher with liquid fluorine as oxidizer; next in order is liquid oxygen.

The Saturn rocket, which is perhaps the most powerful rocket in use, is reported to use kerosene oil and liquid hydrogen as fuel and liquid oxygen as oxidizer.

Of all the *properties* desirable in a liquid propellant, specific thrust and velocity considerations are the most important. These are obtained by high heat of reaction, smooth and rapid burning are next in importance. High specific gravity and low vapour pressure help to decrease the size and mass of the storage tanks. As long burning and large engines require cooling, at least one of the propellants must be satisfactory for use as a coolant by circulating it around the combustion chamber. This means it should have high specific heat, high thermal conductivity and high thermal stability. For facility in pumping, low viscosity is desired. It should be safe to store and should have low corrosivity and toxicity.

Although on the whole liquid bi-propellants are more satisfactory, the choice between a solid and liquid propellant depends upon the nature of launching and flight requirements and on the missile system. Liquid propellants give higher specific impulse but the arrangements for handling, storage and injection are more complicated. Solid propellants are easy and safer to handle, transport and store but have lower specific impulse. To combine the advantages of both, some of the large rockets use multi-stage systems employing solid and liquid propellants in different stages. Each stage has a separate engine and gets detached after the fuel has burnt thereby reducing the mass of the rocket that needs to be propelled further.

4.5 Lubricants

4.5.1 Introduction

When one solid surface is moved over another solid surface, there is a frictional resistance to motion. Even if the two surfaces were perfectly clean and smooth, this frictional resistance, which is primarily due to molecular attraction, would be great. These forces are much weaker than the atomic bonds and have been referred to in Chapter I under 'Secondary Forces of Attraction'. These are known as *van der Waals'* forces. Experiments have been made with clean and smooth surfaces made of different materials such as glass, copper, steel and various alloys. It has been found that on pressing the surfaces with a sliding motion, they stick so strongly that the lower surface may be lifted and even held in the vertical position without separating. Even the plane surfaces with which one deals in actual practice are not perfectly plane and smooth. These have minute projections which interlock each other, and also the surfaces are not perfectly clean. There may be even on the so-called 'clean' surface, some adsorbed moisture, dirt etc. Such impurities tend to reduce the friction. Thus, in practice, when one solid is moved over another, the projecting particles are torn off causing wear of the surface and a loss of energy which appears as heat. In general, both these effects are undesirable. Friction is desirable in cases like brakes, clutches and heat generation in friction-welding. Surface wear necessitates the replacement of the parts. As the equipment gets heated up; it is damaged, and even results in *seizure* i.e., welding of surfaces.

Any substance introduced between two moving surfaces with a view to reducing the frictional resistance is known as a lubricant. A thin film of lubricant, such as oil, is bound to the metal surface by the van der Waals' forces. In the fatty oils the glycerol end of the molecule is firmly held at the surface and the hydrocarbon chain sticks up normal to the metal surface. Thus, these oils have more oiliness and reduce metal-to-metal contact even under high load. Mineral oils are less polar but their behaviour is similar. Most of the lubricants commonly used are liquids, generally oils, but they may be solids or even gases. Graphite is a common solid lubricant. Air acts as a lubricant in high speed spindles. Quantitatively petroleum supplies the largest amount of material for making lubricants, but lubricants may be drawn from widely different sources, e.g., in cream separators milk acts as a lubricant; in certain very low temperature equipment, liquid air may be the lubricant.

Besides reducing energy losses due to friction and reducing wear, a lubricant may perform several other functions, such as providing a seal around the piston ring in an internal combustion engine so that the high pressure gases do not leak past the piston in the combustion chamber. In many instances, the lubricant helps in the dissipation of heat and, therefore, acts as a coolant. Yet another valuable function is that it prevents the entry of dirt and moisture between the moving parts.

4.5.2 Metallic Friction

The most important function of a lubricant is to reduce frictional resistance. The friction phenomena have been studied for many years. There are the classical works of Amontous (1699), Coulomb (1785), Towers (1885), Reynolds (1886), Erving (1892) and many others. Later, valuable contributions were made by Blok, Hardy, Bowden and Tabor. Such studies have led to the development of specialized branches of study, like Physics and Chemistry of Surfaces and Tribology (which deals with the motion of surfaces).

The well-known laws of friction are:

1. Frictional force is proportional to the normal load; and
2. Frictional force is independent of the apparent area of contact. (The real area of contact is often less than one ten-thousandth of the apparent area.)

   To these laws a third law can be added:
3. Frictional force is nearly independent of the sliding velocity.

These laws are fairly well obeyed over a wide range of experimental conditions. Very hard materials like diamond and very soft materials like Teflon show deviations from the first law. Surfaces with a very good finish and polish exhibit deviations from the second law. Deviations from the third law are quite common. The force required to start sliding is greater than to keep the surfaces in motion. The former is known as static friction and the latter as kinetic friction. The kinetic friction remains constant over changes in speeds which are commonly come across. The coefficient of friction almost doubles over speed range of a million times.

According to the modern adhesion theory, developed chiefly by Bowden and Tabor at Cambridge, even the best polished flat surface of a metal on examination under a high magnification (e.g., with an electron microscope) shows the surface to be like a hilly area.

Peaks and valleys of different height, and depth, are observed. This is known as surface roughness and the highest peaks are called surface asperities (Fig. 4.1).

When two flat surfaces are placed over one another, contact is established between the asperities i.e., the peaks of the upper surface rest over the peaks of the lower surface (Fig. 4.2a).

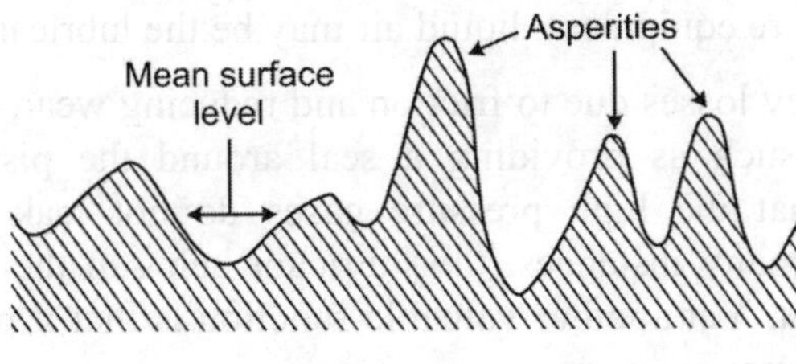

Fig. 4.1 Surface roughness.

Thus the real area of contact is much smaller than the apparent area of contact. A load of a few kilograms per square centimeter develops very high pressures (of the order of a few tonnes per sq. cm) at the real area of contact. Under such high loads the asperities are crushed and deform plastically (i.e., permanently) until the area of contact is large enough to support the load. This real area of contact is independent of the size and geometry of the surface but is proportional to the applied load (Fig. 8.4b). If the real area of contact = A; applied load = W and mean pressure $=P_m$, then $A = W / P_m$.

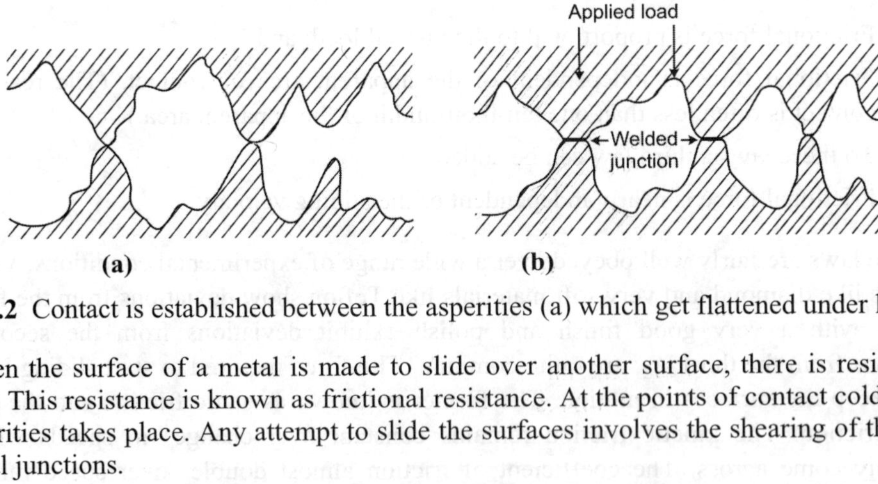

Fig. 4.2 Contact is established between the asperities (a) which get flattened under load (b)

When the surface of a metal is made to slide over another surface, there is resistance to motion. This resistance is known as frictional resistance. At the points of contact cold welding of asperities takes place. Any attempt to slide the surfaces involves the shearing of the metal-to-metal junctions.

The mean shear stress S_m at which sliding takes place depends on the strength of adhesion of the junction. Hence the force required to cause motion, $F = A \times S_m$ = friction force.

$$\text{Coefficient of friction} = \mu = \frac{F}{W} = \frac{A \times S_m}{A \times P_m} = \frac{S_m}{P_m}$$

Thus it is seen that coefficient of friction is independent of the apparent area of contact. In this process of shear, the peaks of the softer metal get broken more easily than the peaks of the harder metal. This fact is also supported by experimental evidence. This accounts for the surface wear and is diagrammatically illustrated in figure 4.3.

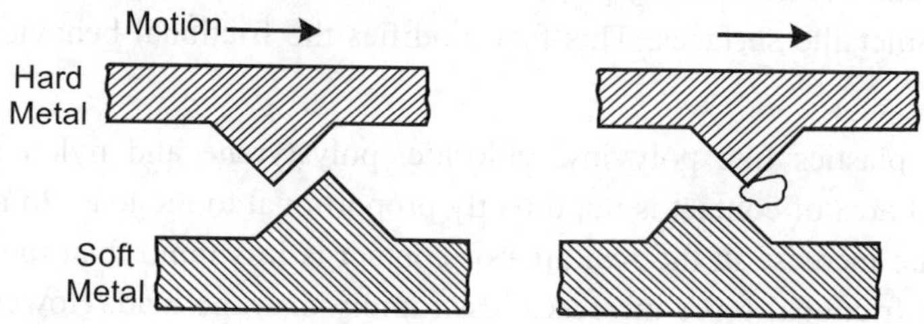

Fig. 4.3 Surface wear is due to shearing of asperities.

From the foregoing it is clear that metallic friction is not purely a surface effect. When solids slide over one another, deformation and distortion take place to a relatively greater depth below the sliding surfaces. Frictional force is, therefore, the sum of ploughing force P and shearing force S. Therefore, F = P + S.

Ploughing is due to hard metal asperities ploughing or cutting grooves in the softer metal. This becomes a significant factor when a hard metal slides over a relatively soft metal. In other cases its effect is not significant.

As it is the softer metal whose particles are broken off, the mean shear stress S_m is approximately equal to the shear strength S of the softer metal. In the same way since plastic deformation of the softer metal takes place more readily, the factor P_m can be assumed to be the yield pressure P of the softer metal. Therefore, the coefficient of friction can be expressed as follows:

$$\mu = \frac{S_m}{P_m} = \frac{\text{shear strength of softer metal}}{\text{yield pressure of softer metal}} = \frac{S}{P}$$

Shear strength and yield pressure of a metal vary together and their ratio is approximately the same for all metals. This explains the experimental observation that the coefficients of friction of different clean metals do not vary greatly. Since small changes in temperature affect both shear strength and yield pressure of a metal to the same extent, the coefficient of friction of clean metals does change with small changes in temperature.

4.5.3 Friction of Non-metals

The non-metals can be classed in two broad groups: (a) brittle materials, and (b) plastics.

The frictional characteristics of metals are primarily due to the fact that under load plastic deformation and welding take place. The resistance to motion is due to the shearing of the welds. Such a mechanism is exhibited by brittle non-metals like glass and common salt. These break easily. However, when they are subjected, static load some plastic deformation similar to that in metals takes place. Under such circumstances the nature of friction is similar to that met within the metals. A noteworthy point is that the metals have higher surface reactivity and energy than non-metallic surfaces. This fact modifies the frictional behaviour of metals to a great extent.

Some of the plastics like polyvinyl chloride, polystyrene and nylon are visco-elastic; therefore, the real area of contact is not directly proportional to the load. In most cases, as the load increases, the increase in the real area of contact is less rapid than the increase in load, consequently the frictional force increases less rapidly than the load. However, the frictional behaviour of even widely varying classes of non-metal is quite uniform.

4.6 Types of Lubrication and Mechanisms

Science of lubrication deals with the problems met within the movement of solid surfaces in contact with each other. Lubrication is broadly classified as hydrodynamic or fluid film lubrication, boundary or thin film lubrication and extreme pressure lubrication.

4.6.1 Hydrodynamic Lubrication

In the case of a shaft running at a fair speed in a well-lubricated bearing, with not too high a load; a film of the liquid lubricant covers the shaft as well as the bearing surfaces. In such a case the two metal surfaces do not come in contact with each other and the coefficients of friction for unlubricated surfaces range from 0.5 to 1.5 and for well-lubricated surfaces from 0.001 to 0.03. These figures show how great an advantage can be achieved because of reduction in friction and also reduction in wear and heating. Figure 8.6 shows formation of a fluid film of lubricant in a journal bearing with shaft under vertical load. When the journal is at rest, Fig. 4.4 (a), almost all the lubricant is squeezed out of the line of contact between the journal and bearing. The clearance between journal and bearing is full of the lubricating oil and its layers adhere to the journal and bearing surfaces. As the journal begins to rotate on the application of a torque to it, it tends to roll up the bearing surface. Rotation of the journal drags the oil between the journal and bearing.

Besides other factors, the quantity of the oil so dragged depends upon its viscosity. As the speed of rotation increases, more and more oil is pumped between the journal and bearing. The pressure exerted by the oil further increases as the clearance between the journal and bearing decreases (due to the rolling action of the journal on the bearing surface). This causes the formation of a wedge-shaped oil film between the journal and bearing surfaces which is essential for hydrodynamic lubrication. A stage comes when the oil pressure exceeds the load and the journal floats, and the journal and bearing surfaces separate completely with an oil film between them as shown in Figure 4.4 (b).

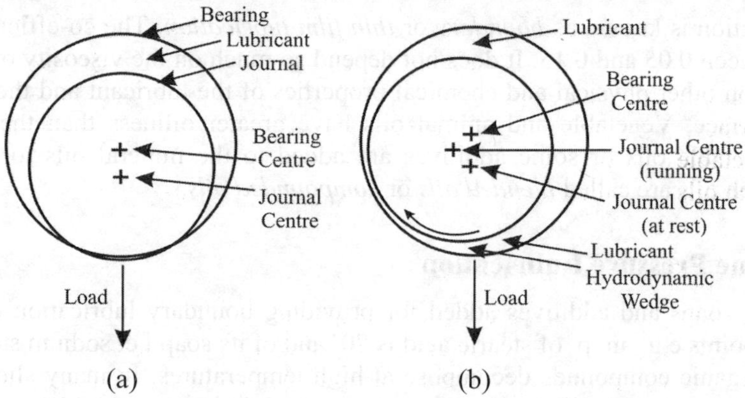

Fig. 4.4 Fluid film formation in a bearing (a) at rest (b) running.

If an oil of low viscosity is used, the bearing force, due to the rotation of the journal, will not be sufficient to drag the oil between the journal and the bearing. Therefore, for hydrodynamic lubrication viscosity is an important consideration. Right clearance space between the moving surfaces, filled with fluid of correct viscosity and such a design of the equipment that a wedge-shaped film is formed between the moving surfaces are the essentials for hydrodynamic lubrication. Theoretically in such a system, friction is independent of the load but depends on the viscosity and the thickness of the film of the lubricant, and relative velocity and area of the moving surfaces. Only the intermediate planes of the lubricant shear over one another. This type of lubrication is known as *hydrodynamic* or *thick film* or *fluid-film lubrication*. In hydrodynamic lubrication, the oil film is sufficiently thick to cover the irregularities of the surfaces and the metal surfaces do not come in contact with each other. Therefore, there is no wear and the resistance to movement is only due to the resistance between the particles of the lubricant moving over each other. This internal resistance of the lubricant is known as its *viscosity*. A liquid which flows easily has low viscosity and the one which flows sluggishly has high viscosity. Therefore, in hydrodynamic lubrication, the lubricant chosen should have the minimum viscosity under working conditions, and yet should remain in place and separate the surfaces.

4.6.2 Boundary Lubrication

The second kind of lubrication takes place when a continuous fluid film cannot be maintained. This happens (i) When a shafting comes into action from rest or (ii) the speed is very low or (iii) the load is very high or (iv) the viscosity of the oil is too low. In such cases it is necessary that a layer of oil be adsorbed by physical, chemical or both physical and chemical forces on the metal surfaces. The property which is responsible for this kind of adsorption is known as *oiliness*. An oil which has more oiliness than another oil of the same viscosity, gives a lower coefficient of friction at low speeds or high loads. Under such conditions when the fluid film cannot be maintained but the lubricant is absorbed on the metal surfaces, the metal surfaces come very close to each other, yet they do not touch each other and the load is carried by layers of the lubricant which have been adsorbed on the metal surface or which chemically reacts with the metal surface forming a thin film of metal soap which acts as lubricant. This kind of lubrication is known as *boundary* or *thin film lubrication*. The co-efficient of friction is usually between 0.05 and 0.15. It does not depend so much on the viscosity of the lubricant as it depends on other physical and chemical properties of the lubricant and the nature of the underlying surface. Vegetable and animal oils have greater oiliness than the mineral oils. Therefore vegetable oils or some additives are added to the mineral oils to improve their properties. Such oils are called *blended oils* or *compounded oils*.

4.6.3 Extreme Pressure Lubrication

The fatty oils, soaps and additives added for providing boundary lubrication does not have high melting points e.g., m. p. of stearic acid is 70° and of its soap i.e. sodium stearate is 280°. Further, the organic compounds decompose at high temperatures. In many sliding processes where load and speed are high, the local temperatures attained are much higher with the result that the boundary lubricants fail. For such extreme conditions certain compounds are added to mineral oils. These are called extreme pressure additives. They have active chlorine (as in chlorinated esters), sulphur (as in sulphurized oils) and phosphorus (as in tricresyl phosphate) group. At high temperatures the active groups react with the metal forming films of metallic chlorides, sulphides and possibly phosphides. These compounds have much higher melting points (e.g., m. p. of iron chloride is 650°, of iron sulphide is 1100° and of iron phosphide is 1290°) and low shear strengths. Thus they provide lubrication under extreme pressures and temperatures. This type of lubrication is called *extreme pressure lubrication*. In a sense it is really extreme temperature lubrication.

4.7 Criteria, Properties and Testing of Lubricants

The suitability of a lubricating oil for a particular use can be decided finally by its actual behaviour in the machine. Laboratory tests help to classify the lubricants and to specify them when obtaining a fresh supply. Comparison and evaluation of different products available in the market is possible by correct specifications and interpretations of the tests.

Correct assessment of the lubricant saves risk of destruction of expensive machinery. It is necessary that standard equipment and procedure, say, as laid down by Indian Standards Institute, be followed so that the results are comparable and accurate

The important criteria of lubricants and their testing by suitable methods are detailed below:

1. Viscosity, 2. Flash and the fire points, 3. Cold tests – cloud and pour points, 4. Emulsification, 5. Volatility, 6. Gravity, 7. Colour and Fluorescence, 8. Carbon residue, 9. Oxidation stability, 10. Aniline point, 11. Neutralization number, 12. Saponification number and 13. Mechanical tests.

4.7.1 Viscosity

Viscosity is the most important single property of a lubricating oil. It may be defined as 'a measure of the internal resistance to motion of a fluid by reason of the forces of cohesion between molecules'. In simpler words, viscosity is resistance to flow or 'thickness' or 'body' of the oil. The formation of a fluid lubricating film, the viscous frictional losses in the bearings and heat generation under particular conditions of load, bearing speed, and lubricant supply, depend on viscosity and to some extent on the oiliness of the lubricant. 'Therefore it is necessary that the lubricant selected has viscosity sufficient to maintain a lubricating film under the working conditions but does not have excessive viscosity which would cause undue friction.

Absolute viscosity may be defined as the force required to move a plane surface of one square centimeter in area at a speed of one cm per second over another fixed parallel plane separated by one cm thick layer of the liquid. Poise is the unit of viscosity and is equal to one dyne per second per square centimeter. This unit being very large, viscosity is often expressed in terms of centipoise (0.01 poise). The viscosity of water at 20° is about 1 centipoise.

In commercial viscometers, a fixed volume of the liquid is allowed to flow through a standard capillary tube and the time of flow is noted. There are three makes of commercial viscometers in wide use. The Redwood viscometer is used in the Common wealth countries, Saybolt in the U.S.A. and Engler in Europe. The results are sometimes expressed in terms of time taken by a given quantity of the oil to flow through the particular instrument, e.g., 'viscosity is 100 Redwood seconds at 20°'.

The instruments are of standard dimensions; therefore, viscosity in terms of poise can be calculated from the time taken by the oil to flow.

It is noteworthy that in most cases the viscosity of oil decreases with rise in temperature. Therefore viscosity at the working temperature needs to be considered. The change in viscosity with change in temperature is expressed by the *viscosity index*. The viscosity index is an empirical number indicating the rate of change in viscosity of an oil within a given temperature range. Low viscosity index means a large change in viscosity with change in temperature. A high viscosity index shows a relatively small change in viscosity with change in temperature. The naphthenic oils have higher rates of change of viscosity with temperature than the paraffinic oils. An oil of high viscosity index does not become very thick when cooled or very thin when heated. Therefore in equipment if the change in temperature in service is large, as in hydraulic system in aircraft the viscosity index of the oil should be high.

It may be noted that viscosities of gases *increase* with rise in temperature. The viscosity of a liquid decreases as temperature rises because intermolecular attraction is reduced due to expansion. In the case of a gas, as the temperature rises, the velocity of the molecules is increased with the result that the collisions of the molecules is increased, thereby the collisions of the molecules become more frequent, therefore there is increased resistance to motion.

The viscosity of an oil may be significantly higher at high pressures; hence this change also needs to be taken into account. The viscosity of a gas also increases with increase in pressure.

Redwood Viscometer : The Redwood viscometer is made in two sizes, No. 1 and No. 2. Viscometer No. 1 is commonly used for determining viscosities of lubricating oils and is shown in the Fig. 4.5. Viscometer No. 2 is similar to it but the jet for the outflow of the oil is of larger diameter, and therefore it is more satisfactory for oils of higher viscosities, e.g., the fuel oils. The bottom of the oil cylinder is also heated by the liquid in viscometer No. 2.

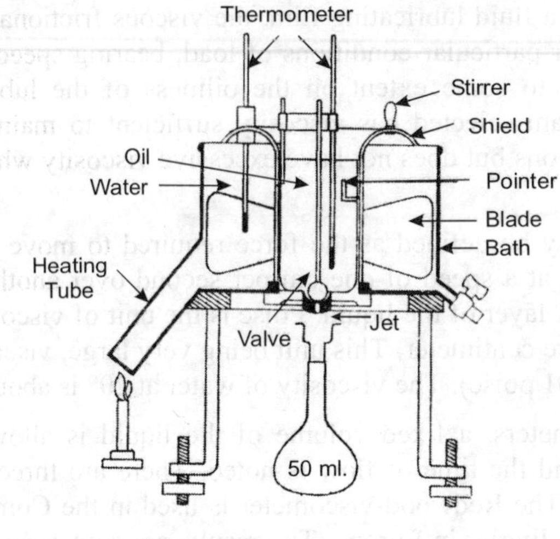

Fig. 4.5 Redwood viscometer.

The viscometer may be heated by gas, as shown in the figure, or electrically heated. The oil is taken in the inner cylinder which is of silver-plated brass, 90 mm in height and 46.5 mm in diameter. At the bottom of the cylinder a jet made of agate is fixed. It is 10 mm long and its bore is 1.62 mm in diameter. It is opened or closed by a valve which is a small silver-plated brass ball fixed to a stout wire. There is a pointer which indicates the level up to which the oil is to be taken. A thermometer indicates the oil temperature. Outside the oil cylinder is a stirrer carrying blades for stirring water for maintaining the desired temperature uniformly. It has a shield which prevents water splashing into the oil cylinder. A thermometer indicates the water temperature.

The oil cylinder is cleaned and the valve is closed. Oil, free from suspensions or emulsions, is taken in the cylinder up to the pointer level. Water is taken in the bath and the temperature is carefully regulated. A 50 ml measuring flask is kept just below the jet. When the oil is at the desired temperature and the water is as nearly as possible at the same temperature, the ball valve is lifted and suspended from the thermometer bracket. The time taken for 50 ml of the oil to collect is noted and the ball valve is closed to prevent overflow of the oil. The result may be expressed in Redwood seconds or converted into absolute units at the particular temperature. The Saybolt Universal viscometer is worked in a similar manner.

Properties and use: Viscosity is the most important single property of lubricating oils. Under conditions of hydrodynamic lubrication, the bearing friction, heat generation, the rate of flow of the lubricant under given conditions of load and speed, and the design of the bearing are determined by the viscosity of the lubricant at the operating temperature.

There is no simple method of deciding the optimum viscosity for a particular use. Experience is the guiding factor. In some cases much higher viscosity oils have been used because the gear boxes are difficult to make tight, thus leakages are minimised. High viscosity oils reduce noise and help in making a good fit.

4.7.2 Flash and Fire Points

The flash point is the lowest temperature at which the oil gives off vapours that will ignite when a flame is passed over the surface of the oil. The fire point of an oil is the lowest temperature at which it ignites and continues to burn for at least five seconds.

The two tests are carried out in the same apparatus on the same sample. The apparatus used may be the 'open' or 'closed' cup. In the open cup apparatus the oil is heated with its upper surface exposed to the atmosphere. The closed cup gives more reproducible results. The Pensky Marten's closed tester is commonly used and is shown in Fig. 8.8. It consists of an oil cup about 5 cm in diameter and 5.5 cm deep. The level up to which oil is to be filled in is marked inside the cup. The cup is supported by its flange over a heating vessel in such a manner that there is clearance between the cup and the heating vessel. The oil is dried because the presence of water makes it difficult to determine the flash point. After taking oil in the cup to the level indicated, it is closed by a cover. Through the centre of the cover passes a stirrer which stirs the oil as well as the vapours as shown in the Fig. 8.8. The cover has four openings of standard dimensions. Through one of these passes a thermometer. Through the second opening shown in the Figure, passes the test flame and the other two openings are for the admission of air. The shutter opens the openings for the test flame and air, and dips the burner and brings the flame over the oil Surface. As the test flame gets extinguished it is automatically lighted by the pilot burner.

Heat is applied to raise the oil temperature at the rate of about 5° per minute. The stirrer is worked between tests at the rate of about 1 to 2 revolutions per second. The flame is applied to the oil every 1° up to 105° and afterwards at increments of 3°. The temperature at which a distinct flash is seen inside the cup is noted as the flash point.

The heating of the oil is continued and the test flame applied as before. When the oil ignites and continues to burn for at least 5 seconds, the temperature reading is recorded as the fire point. In most cases the fire points are 5° to 40° higher than the flash points.

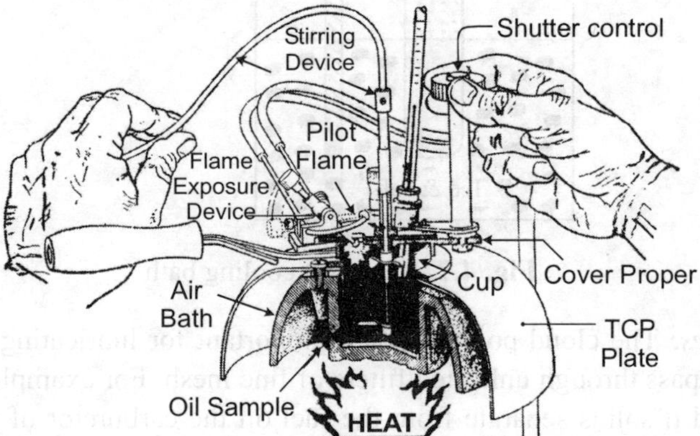

Fig. 4.6 A section through Pensky Marten's flash point apparatus
(Lubrication Engineering Manual, Courtesy, U.S. Steel Corporation)

Properties and uses: The flash point and fire point do not indicate directly the lubricating value of the oil but these are important when the oil is exposed to high-temperature service. Thus these tests provide safeguards against risks of decomposition and fire during storage and use. Flash and fire points are now being omitted from many specifications although some insurance companies insist on it from the point of view of safety during storage.

Formerly an oil of high flash point was supposed to offer more resistance to spontaneous combustion and explosion in the lubrication of air compressors. Flash point does not indicate tendency to undergo spontaneous combustion. Spontaneous ignition temperature is the temperature at which ignition can take place in contact with a hot body. It is reported that a hot bearing disintegrated and a portion of it fell into the crank case of large Diesel engine with the result that violent explosion took place.

4.7.3 Cloud Point and Pour Point

When an oil is cooled at a specified rate, the temperature at which it becomes cloudy or hazy in appearance is called the cloud point of the oil. This is due to the separation of tiny crystals of wax. The temperature at which the oil just ceases to pour or flow is called the pour point.

The oil under test is taken up to a fixed level in a glass test jar of standard dimensions. The jar is kept in a jacket which is surrounded by a freezing mixture. A thermometer is introduced in the oil. The test jar is withdrawn from the jacket with every degree of fall in the oil temperature and examined. It is replaced in the jacket within three seconds. The temperature at which cloudiness is noticed is taken as the cloud point.

Cooling is continued and the test jar is withdrawn with every 3° fall in temperature and tilted to observe the flow of oil. The temperature at which it does not flow even when kept horizontal for five seconds is taken as the pour point (Fig. 4.7).

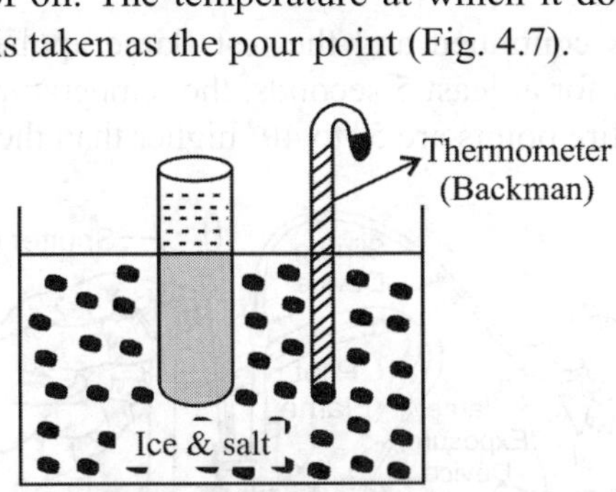

Fig. 4.7 Ice & salt cooling bath

Properties and uses: The cloud point is not so important for lubricating oils as it is for fuel oils which have to pass through unheated filters of fine mesh. For example, a jet plane may be exposed to 60° and if solids separate from the fuel oil the carburetor of jets may be blocked up. The pour point is of greater significance for lubricating oils exposed to low temperature. It indicates the temperature below which the oil cannot be poured out from its container or the equipment cannot be lubricated by it by gravity flow.

4.7.4 Emulsification

There are several tests to indicate the emulsification tendencies of the oils and a particular test is chosen according to the conditions met with by an oil in actual use. In a test, oil and water are stirred and the time taken for the emulsion to break is noted. In a more widely used test, steam is passed into the oil under specified conditions, and the time in minutes taken for the separation of the oil from the emulsion is recorded as the steam emulsion number.

The time taken by the oil for separation from the emulsion depends upon its refining and viscosity. Good refinement and lower viscosities are conducive to quicker separation of the oil from the emulsion.

The emulsification tendency is undesirable in most applications of the oil. Formation of a little emulsion in a circulating oiling system may not be harmful but sticky emulsions and sludges interfere with oil circulation and are certainly undesirable. The formation of stable emulsions with water may be desirable in the case of certain steam-cylinder oils and water-soluble cutting lubricants. The general principle is that formation of emulsion is desirable when an oil is to be used in the presence of water. In the case of steam turbines the oil should separate quickly from the emulsion so that it can be recirculated for use.

4.7.5 Volatility

The volatility test has not been standardized. Volatility may be found by exposing a weighed quantity of the lubricant to a current of hot air, under specified conditions and determining the loss in weight. Its application is limited to those conditions in which the oil is exposed to a temperature at which a fraction of it volatalises leaving a residual oil which may have different lubricating properties.

4.7.6 Gravity

Besides specific gravity, A.P.I. (American Petroleum Institute) gravity is in use. A.P.I. gravity is an arbitrary scale in which the specific gravity of water is taken as 10. Liquids lighter than water are assigned values higher than 10 and the heavier liquids, values lower than 10. Naphthene base oils have lower A P I gravity than those of paraffin base oils.

Gravity need not be mentioned in the specifications of lubricating oils because it gives little indication of the quality or usefulness of the lubricant. It is used for inter-conversion of weight and volume measurements.

The terms 'density', 'specific gravity' and 'gravity' are often loosely used. Density is the mass of unit volume of a substance under specified conditions. It has a unit e.g. grams per milli litre, pounds per cubic foot, etc. Specific gravity is the ratio of the weight of a substance to the weight of an equal volume of water. Gravity is a reading of an arbitrary scale on a hydrometer. e.g., degree Baume, A. P. I. gravity, etc.

4.7.7 Colour and Fluorescence

Colour is determined by the usual colorimetric method of comparison of the colour of the oil with that of the standard colour discs. Colour indicates the degree of refinement. A pale colour does not indicate lower viscosity, as many think. Distillates are lighter in colour than the residual oils. A pale-coloured lubricant is desirable in such applications as in the textile industry where staining of fabrics has to be avoided. In other cases it is unnecessary to insist on colour requirements.

The mineral oils show fluorescence in reflected light. Fixed oils do not show fluorescence; hence fluorescence helps to detect adulteration.

4.7.8 Carbon Residue

The carbon residue test determines the percentage of carbon residue left in an oil after subjecting it to an extreme heating condition out of contact with air. In the process, evaporation and pyrolysis of the oil take place. The Conradson method has been widely used but recent recommendations are in favour of the Ramsbottom method. The Conradson apparatus is shown in Fig. 4.8. In the Conradson method, a weighed quantity of the lubricant is taken in a crucible which is placed in another iron crucible having a cover with a small opening. These are then placed in a third crucible, and covered with a hood and strongly heated at a specified rate and the residue is cooled and weighed. The result is expressed as percentage of the residue on the original weight taken.

In the Ramsbottom method about 4 g of the oil are accurately weighed in a coking bulb made of hard glass, 57 mm in height and 25 mm in diameter. It is then placed for 20 minutes in a furnace which has been previously heated to 550°. The bulb is then removed from the furnace, cooled in a desiccator and weighed, and the result expressed in percentage.

Efforts have been made to correlate the results obtained by the two methods but not much success has been achieved. The Ramsbottom method gives more reproducible results.

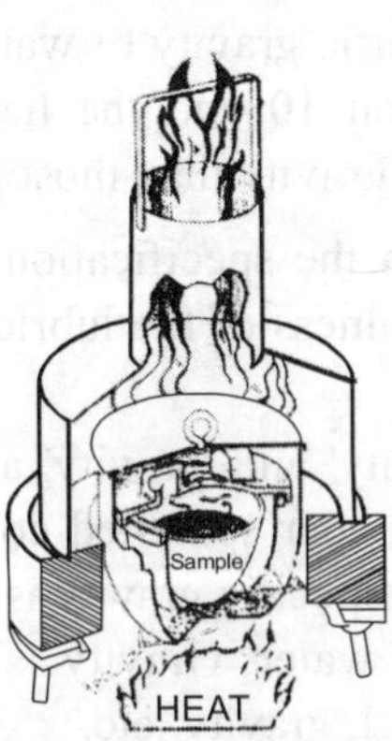

Fig. 4.8 Conradson method for carbon residue determinations *(Lubrication Engineers' Manual, Courtesy, U.S. Steel Corporation)*

The carbon residue test indicates the carbon-forming tendency of lubricating oils in internal combustion engines, especially as regard piston-crown deposits around the ring, and on valve seals in the combustion chambers. Under very good conditions some indication of carbon deposits in the cylinder may be obtained. It has been suggested that carbon residue varies directly with the viscosity. In good lubricating oils, stabilizing additives are added to the oils to enable them to resist chemical decomposition at the high temperature developed in internal combustion engines. It is noteworthy that the black deposits in the engine do not entirely consist of carbon, though it is the main constituent. Dust from the air may also be present. The carbon residue test is important for lubricants used for internal combustion engines and air compressors.

4.7.9 Oxidation Stability

Lubricating oils tend to oxidize in contact with air, slowly at room temperature but rapidly as the temperature increases above 100°. The reaction is further accelerated in the presence of metal particles which act as catalysts. These conditions are common in internal combustion engines, steam turbines, and other machinery which have oil circulating systems. Oxidation products are undesirable. Some of these are insoluble, gummy and sticky materials which clog oil holes, filters, pipe lines and other parts of the oiling systems. The soluble products are acidic and corrode the surface of the bearings and also form varnish like deposits. Therefore, resistance to oxidation is a desirable property of a lubricating oil. Oxidation inhibitors are added to oils to improve their resistance to oxidation. The inhibitors get oxidized first and the oil is protected.

A common test is the Sligh oxidation test. In this test a 10 g sample of the oil is taken in a special flask and air is displaced by oxygen. The flask is kept in an oil bath maintained at 200°. After 2½ hours the flask is removed, cooled, and the contents are diluted with petroleum naphtha and allowed to stand for an hour. The precipitate is filtered, washed, dried and weighed. The weight is expressed as percentage of the weight of the sample taken to give Sligh oxidation number.

4.7.10 Aniline point

The aniline point is defined as 'the minimum equilibrium solution temperature for equal volumes of aniline and sample'. For its determination equal volumes of the sample and aniline are taken in a tube and mechanically mixed. The mixture is heated until homogeneous and then cooled at a controlled rate. The temperature at which the two phases separate is recorded as the aniline point.

The aniline point of an oil gives an indication of the possible deterioration of rubber sealing, packing, etc. in contact with the oil. The aromatic hydrocarbons have a tendency to dissolve natural rubber and certain types of synthetic rubber. Therefore, in such cases a low aromatic content in the lubricant is desirable. As the aniline point is a measure of 'paraffinicity' of an oil, the higher the aniline point the lower is the percentage of aromatic hydrocarbons and the higher is the content of paraffins. Since paraffins give easy starting and a minimum of knock in a Diesel engine, a high aniline point usually indicates a high cetane number (See Diesel fuels).

4.7.11 Neutralization Number

The neutralization number refers to the determination of acidic or basic constituents. Determination of acidic constituents is more common and is referred to as acid value (or number). It is the number of milligrams of potassium hydroxide required to neutralize the free acid in one gram of the oil.

About 10 g of the oil is weighed in a conical flask and 50 ml of neutral alcohol (or alcohol toluene mixture) is added and boiled for a minute. The acidity is titrated against N/l0 standard potassium hydroxide solution using phenolphthalein as the indicator. The results are expressed in terms of the number of milligrams of potassium hydroxide required per gram of the sample. Basic constituents are determined in a similar manner by titrating against acid solution using the appropriate indicator.

Generally, mineral acids are not present in the lubricants unless refined in a faulty manner. Organic acids are not as corrosive as the mineral acids and in good samples the acid value should be very low, say, less than 0.1. Increase in acid value should be watched because it indicates oxidation of the oil which will lead to gum and sludge formation, besides corrosion. For certain uses, storage of the oil in galvanised iron tanks is discouraged because the free acidity produced attacks the zinc forming zinc soaps.

4.7.12 Saponification Number

The saponification number of an oil is the number of milligrams of potassium hydroxide required to saponify the fatty material present in 1 g of the sample.

About 5 g of the sample is weighed in a flask and 50 ml of semi-normal potassium hydroxide solution in alcohol is added to it and boiled for an hour under a condenser. The unreacted potassium hydroxide is titrated back and the result calculated.

Mineral oils do not form soaps but vegetable and animal oils do. Thus this test helps to determine approximately the proportions of these in a mixture.

4.7.13 Mechanical Testing

Physical and chemical tests do not give a completely satisfactory basis upon which to select a lubricant for a particular purpose. Several machines are in use for determining frictional resistance. Recently, there has been the problem of lubrication under conditions of very high pressure. To judge the suitability of a lubricant under extreme conditions several machines have been designed which determine the load-carrying capacity of a lubricant, torque, friction and wear. One such machine is a Shell Four Ball Extreme Pressure Lubricant Tester which is shown in the Fig. 4.9.

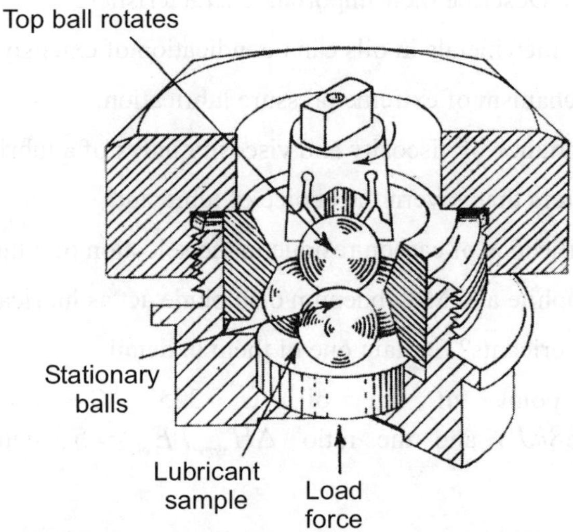

Fig. 4.9 Shell four ball extreme pressure lubricant tester

(adopted from Lubrication Engineer's Manual, Courtesy, U.S. Steel Corporation)

The Shell Four Ball machine has four steel balls which are immersed in the lubricant under test. The lower three balls remain stationary while the top ball is rotated. Any desired load can be applied. If the lubricant is satisfactory under the test conditions the balls after the test come out undamaged. The load is gradually increased and the balls withdrawn and examined at specified intervals. In the earlier stages formation of scars on the balls is observed. When the load is further increased, at a certain value the lubricant fails completely and so much heat develops that the balls weld together. The test enables the determination of the load that can be safely carried by a lubricant. Similarly torque, friction and wear test can be performed. The results can be approximately correlated to conditions met within actual use. Comparison between two lubricants can also be made.

4.8 Short and Essay Type Questions

1. Classify the explosives and what are the requirements for an explosive.
2. Write a note an primary and binary explosives.
3. Write an essay on Dynamites and their uses.
4. What is propellant? Flow many types are there? Mention them. Describe the solid propellants used as Rocket Fuels?

5. Explain the theories of lubrication. Write about the applications of lubricants in industry.
6. Describe the important properties of lubricants.
7. What are lubricants and what are their important properties?
8. Define lubricants. Describe their important characteristics.
9. "The analysis of trace metals in oils can be indication of extensive wear". Explain.
10. Describe the mechanism of extreme pressure lubrication.
11. What is the importance of viscosity and viscosity index of a lubricant?
12. What are the factors that govern the choice of lubricant?
13. What is the significance of carbon residue in the selection of a lubricant?
14. Explain how Graphite and Molybdenum disulphide act as lubricants.
15. What are solid lubricants? Explain one of them in detail.
16. The boiling point of an oil is 325 ^{0}C. Using Troutons Rule ($\Delta H_{vap} / T_{bp} = 88kJ$) and the ratio $\Delta H_{vap} / E_{vis} = 5$, determine E_{vis} for the lubricant
17. What lubricants would you use for operating in a space craft intended for a journey of 6 months
18. Distinguish between hydrodynamic lubrication, boundary lubrication and extreme pressure lubrication.
18. How can a viscous lubricant be converted into a grease.
19. What are synthetic lubricants? How are they superior to petroleum based lubricants?
20. Explain viscosity index and aniline point as applied to a lubricating oil.
21. What is meant by oiliness of a lubricant? How can this be improved?
22. Explain the following properties of lubricants giving their significance.

    (i) viscosity & viscosity index (ii) flash & fire points

    (iii) cloud & pour points (iv) saponification number

    (v) Neutralisation number (vi) carbon residue
23. Explain different additives employed for improvement of lubricants
24. Write a note on primary and Binary explosives.
25. Write an essay on Dynamites and their uses.

4.9 Self Test Questions

1. The smokeless powder explosive is
   (A) Lead Azide (B) TNT
   (C) Nitrocellulose (D) Ethylnitrate
2. Cyclotonimethylenetinitramine is known as
   (A) TNT (B) PETN
   (C) RDX (D) TNP
3. Dynamine main constituent is
   (A) Ammonium nitrate (B) Sodium nitrate
   (C) Nitroglycerine (D) nitrocellulose
4. Military dynamics do not contain
   (A) TNT (B) RDX
   (C) PETN (D) Nitroglycerine
5. Vegetable oils are
   (A) suitable at high T (B) stable
   (C) thickened in use (D) oxidation resistant
6. Oiliness is less for
   (A) Vegetable oils (B) Mineral oils
   (C) Greases (D) Waxes
7. Greases are not suitable for
   (A) engine parts (B) gear boxes
   (C) watches (D) instruments with bearings
8. Lubricants are aimed to reduce
   (A) abrasion (B) corrosion
   (C) wearing (D) all
9. Lubricant used in machines working at low temperature should have
   (A) high pour point (B) low pour point
   (C) high flash point (D) low flash point
10. Graphite dispersed in oil is called
    (A) dispersion lubricant (B) grease
    (C) oildag (D) graoil

11. Generally for liquid lubricants

    (A) flash and fire points are identical

    (B) flash point > fire point

    (C) flash point < fire point

    (D) flash point > pour point

12. Silicones are known for

    (A) high viscosity index, high oxidation resistance and low volatility

    (B) high viscosity index, high oxidation resistance and high volatility

    (C) high viscosity index, low oxidation resistance and low volatility

    (D) low viscosity index, high oxidation resistance and low volatility

5 Electrochemistry, Cells and Insulators

5 Electrochemistry, Cells and Insulators

5.1 Introduction

All matter is electrical in nature. An atom, the basic particle is made up of subatomic particles, viz., electrons, protons, neutrons, etc., where the positively charged nuclei is surrounded by the revolving cloud of electrons. Any chemical change may be considered as rearrangement of atoms and their electron clouds. Thus the formation of HI by the combination H_2 and I_2 may be represented as

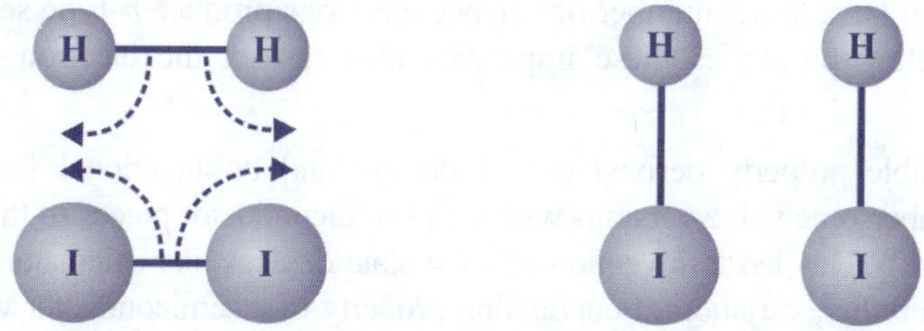

Here the atoms rearrange themselves by changing partners with simultaneous rearrangement of covalent bonds. This suggests that a chemical change and an electrical change are two faces of the same coin.

Electrochemistry is the branch of physical chemistry that is concerned with the inter-relationship of electricity and chemistry. It has a two pronged coverage, firstly, with the nature and properties of substances containing ions and charged particles and, secondly, with the relationship between electric currents and chemical reaction. The former is sometimes called **ionics** and the latter **electrodies**.

5.2 Electrical Conduction

The substances of this universe can fall into four different categories depending upon their capability of flow of electrons, viz., electrical conductors, insulators, semiconductor and super-conductors.

A. Electrical Conductors (or Simply Conductors)

With few exceptions like mica, generally all the good thermal conductors are good electrical conductors. They allow electrons to pass through them with minimum resistance, and show no preference of direction of flow. Some examples of this class are metals, metal sulphides, acids, bases, ionizable aqueous and nonaqueous salt solutions and fused salts. The electrical conduction of a solid depends on the spacing and state of occupancy of the energy bands. Their valence and conduction bands are placed closely and side by side.

B. Insulators or Non-Conductors

Substances of this category do not allow the electricity to pass through them. Some examples of this class are pure water, majority of the substances of plant origin (dry wood, rubber, paper, etc.), many synthetic organic compounds (benzene, CCl_4, ether, etc.), mica and all non-metals except carbon. In an insulator, the gap between the valence band and conduction band is considerably large making the flow of electrons difficult.

C. Semiconductors

Elements of 4A group, especially Si and Ge have properties that are intermediate between those of metals and nonmetals and therefore are called semiconductors. Their conducting properties are considerably enhanced by the addition of certain impurities, called doping elements. Addition of impurities like phosphorus (P) that have more valence electrons than Si generally called donor impurities give ***n*-type semiconductors**. Similarly acceptor impurities like boron (B) Aluminium (Al) that have lesser number of valence electrons produce ***p*-type semiconductors**. In addition to conducting properties these impurities also restrict the electron movement in the crystal.

The most valuable property derived out of doping and unidirectional flow of electrons is replacement of cumbersome valve versions with small semiconductor pieces of thickness of a pencil eraser called **"Chips"**. This led to invention of solid state devices like calculators, mini computers, pocket radios, etc. Further, variation of conducting property of a semiconductor with temperature is exploited by making temperature measuring and controlling devices.

D. Superconductors

Electrical conductivity decreases with decrease in temperature. Lead was found to conduct electricity with zero resistance below 7.3 *K*. In 1987, physicists found that certain mixtures of copper, barium and rare earth metal oxides exhibit this zero resistance to as high as 130 *K*. The property of certain metals, alloys and compounds by virtue of which they conduct electricity with zero or negligible resistance is known as **super conductivity**. The characteristics temperature at which a metal becomes superconducting is called transition temperature T_c.

Because of zero resistance, super conductors attain the capacity to store infinite amount of current. This allows transfer of power from the place of generation to place of utility, few hundred kilometers away without any loss of power during transmission. Further superconductivity is expected to create very high field strength electromagnets that can be used for many purposes like lifting super heavy weights.

E. Solution State Equilibrium

Salt and ions are solvated (aquated)

$$NaCl\ (aq) \rightleftharpoons Na^+_{(aq)} + Cl^-_{(aq)}$$

The negatively charged ions, called anions, move towards **anode** or positive electrode. The electrode connected to negative terminal of the battery is called **cathode** or negative electrode and the cations move towards it. This assembly of cathode and anode in an electrolytic medium is referred as **cell**. Electricity or flow of electrons from a battery enters the cell through cathode and leaves through anode. Regardless of the kind of cell or battery, the cathode is defined as an electrode at which reduction occurs i.e., the species around it gain electrons and the anode is defined as an electrode at which oxidation occurs i.e., the species around it lose electrons. In the fused sodium chloride, the sodium ions are reduced at cathode to liberate sodium metal and at anode, the chloride ions are oxidized to chlorine gas. Thus the mechanism of electrolytic conduction may be summarized as follows.

1. Electrolytic conduction mainly occurs through movement of cations and anions, and therefore depends on their concentration.
2. Electrolytic conduction is always accompanied by electrochemical changes.
3. When there are no secondary reactions due to solvent or any other foreign species, there is a progressive decrease in conduction. This fall is attributed to the fall in the concentration of electrolyte on account of deposition or development of non-conducting atmosphere around electrode.
4. A rise in temperature decreases the viscosity of the medium and increases the kinetic motion of the ion. Both these factors act cumulatively and the electrolytic conduction invariably increases with increase in temperature.
5. The electrodes may participate in the electrochemical secondary reactions, in which case they are corroded and sometimes may even contribute for the rise in electrolytic conduction.

5.3 Electrical Resistance and Ohms' Law

We already discussed that the metallic conduction increases with fall in temperature. With the approach of absolute zero the resistance will be almost equal to zero and therefore *all substances behave like perfect conductors at absolute zero*. G. S. Ohm in 1827 found that resistance (R) is related to the current passed (I) and the potential difference across the conductor (E)

$$I = \frac{E}{R} \qquad .....(5.1)$$

Ohm's law may thus be defined that the current strength flowing through a conductor at uniform temperature is directly proportional to the potential difference applied across the conductor and inversely proportional to the resistance of the material.

When current is measured in amperes and potential difference in volts the resistance offered is a fixed value called ohm (represented by symbol omega 'Ω')

$$\text{Ohm} = \frac{\text{Volt}}{\text{Ampere}}$$

Thus, Ohm is the resistance offered by a body when a current of 1 ampere flows across it under an application of a potential difference of 1 volt. Ohm's law is strictly obeyed by all metallic conductors. Electrolytic conductors obey the law, when the current passed is small enough not to bring any significant concentration changes of the electrolyte.

In addition Ohm also found that the resistance of a conductor of uniform cross section is directly proportional to its length and inversely proportional to the area of cross section.

$$R \propto \frac{1}{a} \quad \text{or} \quad R = \rho \frac{1}{a} \qquad .....(5.2)$$

where ρ (rho) is a proportionality constant called **specific resistance** or **resistivity**. Specific resistance is nothing but the resistance offered by a material of unit length and unit area of cross section. The units of specific resistance are

$$\rho = \text{Ohm} \times \frac{\text{cm}^2}{\text{cm}} = \text{Ohm cm}$$

5.3.1 Conductance and Specific Conductance

Resistance and specific resistance are commonly used for metallic conductors, where the atoms are static. In case of electrolytic solutions the electricity is virtually conducted by constantly moving ions. So it was thought more meaningful to define another quantity called **conductance** (this term should not be equated to conductivity, as done in older literature, as the term conductivity is reserved by IUPAC to represent specific conductance introduced below). It gives the ease with which electricity flows through an electrolytic conductor. The conductance of an electrolytic solution is defined as the reciprocal of its resistance.

$$\textbf{Conductance } (L) = \frac{1}{R} \textbf{ Ohm}^{-1} \textbf{ or mho or } \Omega^{-1} \textbf{ or Siemens (S)} \qquad .....(5.3)$$

The conventional unit of conductance is ohm^{-1} or mhos or Ω^{-1}. In SI system the same unit is called Siemens and is represented by symbol S.

The reciprocal of specific resistance is defined as **conductivity** or **electrolyte conductivity** or **specific conductance.** Of these three terms only conductivity and electrolytic conductivity are recognised by IUPAC and therefore specific conductance and specific resistance are no longer in use. This quantity, designated by κ (kappa), may be defined as the conductance of unit volume of the solution. In other words electrolytic conductivity is the conductivity of a solution taken in between electrodes of unit area of cross section and separated by a unit distance, Fig. 5.1.

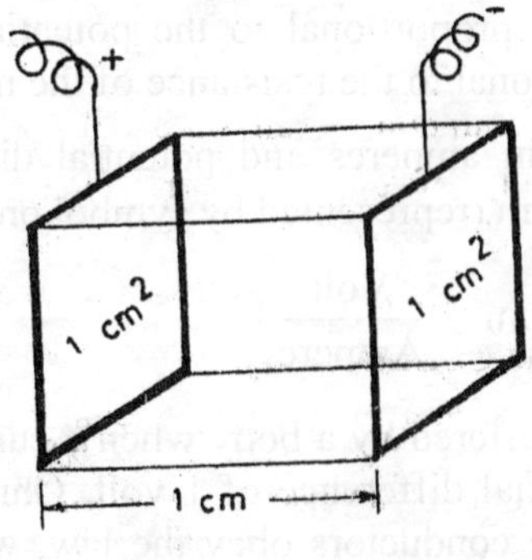

Fig. 5.1 Electrolyte conductivity.

In CGS system the conductivity is expressed in S cm^{-1} or mho cm^{-1}

Conductivity $$\kappa = \frac{1}{\rho} = \frac{1}{\text{Ohm cm}} = \text{mho cm}^{-1} = \text{S cm}^{-1} \qquad .....(5.4)$$

In SI system, meter being the fundamental unit of distance, the conductivity is expressed as S m^{-1}. The conversion factor of both CGS & SI systems is,

$$1 \text{ S cm}^{-1} = 100 \text{ S m}^{-1} \quad \text{or} \quad 1 \text{ S m}^{-1} = 0.01 \text{ S cm}^{-1}$$

The electrolyte conductivity is dependent upon the concentration of electrolyte. The conductivity increases with increase in the concentration of electrolyte at a given temperature, Table 5.1. A relation between conductivity and conductance could be obtained starting from our definition represented by eqn. (5.4).

Conductivity $$\kappa = \frac{1}{\rho}$$

But from eq. (5.6) $$\frac{1}{\rho} = \frac{1}{R} \times \frac{l}{a}$$

i.e., $$\kappa = \frac{1}{R} \times \frac{l}{a}$$

Substitute L for $1/R$ (from eqn. 5.3)

$$\kappa = L \frac{1}{a} \qquad .....(5.5)$$

$$L = \kappa \frac{a}{l}$$

Because the concentration of all pure solids is taken as unity, the electrolytic conductivity of electrolytes remains constant at a constant temperature.

5.3.2 Equivalent Conductivity

Since conductance and electrolyte conductivity are concentration dependent terms, to allow a comparison between different electrolytes, F. W. Kohlrausch (1876) suggested the use of **equivalent conductivity/ equivalent conductance**, represented by Λ_v (Greek : Capital Lambda).

It is defined as the *conductance of all the ions produced by the dissociation of one gram equivalent of an electrolyte, dissolved in a certain volume, V, of the solvent at a constant temperature.*

Mathematically it is expressed as;

$$\Lambda_v = 1000 \frac{\kappa}{C} \text{ ohm}^{-1} \text{ cm}^2 \text{ equivalent}^{-1} \qquad .....(5.6)$$

where C is the concentration of the solution in gram equivalents per litre. Since V is the volume of solution containing 1 equivalent of electrolyte

$$V = \frac{1000}{C}$$

Therefore eqn. (5.10) becomes

$$\Lambda_v = \kappa V \text{ Ohm}^{-1} \text{ cm}^2 \text{ eq}^{-1} \quad .....(5.7)$$

The equivalent conductivity of a solution can thus be readily obtained from a knowledge of concentration (or volume of solution containing one equivalent of electrolyte) and electrolyte conductivity. Eq. (5.7) also suggests that equivalent conductivity of a solution of unit volume containing one gram equivalent of a substance is equal to the electrolyte conductivity.

In SI system equivalent conductance is expressed as;

$$\Lambda_V = \frac{1000\ \kappa\left(\text{S m}^{-1}\right)}{C\left(\text{eq m}^{-3}\right)} \text{ S m}^2 \text{ eq}^{-1}$$

Conversion of SI units into CGS units and vice-versa is done using the conversion factor;

$$1 \text{ S m}^2 \text{ eq}^{-1} = 10^4 \text{ S cm}^2 \text{ eq}^{-1}$$

5.3.3 Molar Conductivity

In recent times it has become a practice to express the concentration in moles per litre rather than equivalents per litre. Keeping in pace with this new practice the term **molar conductivity/molar conductance** was introduced. This quantity is also sometimes referred to as **molecular conductivity**. Molar conductivity is defined as the conductance of all the ions produced by the dissociation of one gram molecular weight of an electrolyte dissolved in a certain volume, V of the solvent at constant temperature. This quantity is generally denoted by the symbols : and Λ_m. Similar to equivalent conductance, molar conductance is given by the expressions,

$$\mu = 1000 \frac{\kappa}{C}$$

and $$\mu = \kappa V \quad .....(5.8)$$

where V is the volume of the liquid containing 1 mole of the electrolyte.

The units of : in CGS system are $\text{Ohm}^{-1} \text{ cm}^2 \text{ mol}^{-1}$ or $\text{S cm}^2 \text{ mol}^{-1}$ and in SI system it is $\text{S m}^2 \text{ mol}^{-1}$. The conversion factors between these systems is

$$1 \text{ S m}^2 \text{ mol}^{-1} = 10^4 \text{ S cm}^2 \text{ mol}^{-1}$$

5.3.4 Relation between Molar Conductivity and Equivalent Conductivity

The relationship between molar conductivity and equivalent conductivity can be obtained by comparing eqn. (5.7) and (5.8).

$$\frac{\Lambda_v}{\mu} = \frac{\dfrac{1000\kappa}{C(\text{in equivalents per litre})}}{\dfrac{1000\kappa}{C(\text{in moles per litre})}}, \quad \frac{\Lambda_v}{\mu} = \frac{C(\text{in moles})}{C(\text{in equivalents})}$$

But $\quad C(\text{mol/L}) = \dfrac{W}{M}, \; C(\text{eq/L}) = \dfrac{W}{E}$

and $E = \dfrac{W}{z}$ where z is the total charge carried by cations (or anions) liberated out of dissociation of the electrolyte.

$$C(\text{eq/L}) = \frac{W}{(m/z)} = Z.\, C(\text{mol/L})$$

Therefore

$$\frac{\Lambda_v}{\mu} = \frac{C(\text{mol/L})}{z\, C(\text{mol/L})} = \frac{1}{z} \qquad \text{.....(5.9)}$$

This can be well demonstrated by considering the following examples.

For	KCl	$\longrightarrow$	$K^+ + Cl^-$	$z = 1$ Therefore $\Lambda_v = \mu$
For	$ZnCl_2$	$\longrightarrow$	$Zn^{2+} + 2Cl^-$	$z = 2$ Therefore $2\Lambda_v = \mu$
For	Na_2HPO_4	$\longrightarrow$	$2Na^+ + HPO_4^{2-}$	$z = 2$ Therefore $2\Lambda_v = \mu$
For	$AlK(SO_4)_2$	$\longrightarrow$	$Al^{3+} + K^+ + 2SO_4^{2-}$	$z = 4$ Therefore $4\Lambda_v = \mu$

Problem 1.

The equivalent conductivity of 0.1 N barium chloride was found to be 105.2 S cm^2 eq^{-1} at room temperature. What would be its molar conductivity at the same temperature?

Barium chloride dissociated through following equation,

$$BaCl_2 \longrightarrow B^{2+} + 2Cl^-$$

The total charge carried by individual ions $z = 2$

We know from eqn. (5.13) $\dfrac{\Lambda_v}{\mu} = \dfrac{1}{z}$

given $\Lambda_v = 105.2$ S cm^2 eq^{-1} and $z = 2$ (sum of change on cation/anion)

$$\frac{105.2}{\mu} = \frac{1}{2}$$

or Molar conductivity $\mu = 105.2 \times 2$

$$= 210.45 \text{ cm}^2 \text{ mol}^{-1}$$

5.3.5 Measurement of Conductivity of Electrolytes

Experimental determination of conductivity of electrolytic solutions is important in covering both the basic and applied aspects of electrochemistry. The basic aspects such as nature and quantification of intermediate ionic attractions, ion-ion and ion-solvent pairings, etc., can be done through conductometric measurements. On the applied front the study of corrosion and passivity occupies the top priority. In addition, these measurements are widely used for testing the purity of water obtained from an ion-exchanger to determine quantities such as solubility of sparingly soluble salts, dissociation constants of weak electrolytes, ionic product of water, to compare the strengths of acids and bases and to carry out conductometric titrations of pure and mixture of electrolytes.

Basically determination of conductance, *L*, involves estimation of resistance, *R*, of the electrolytic solution. This is usually done by means of **Wheatstone bridge circuit** which involves a comparison of unknown resistance with a standard resistance. The Wheatstone bridge circuit is shown in Fig. 1.2.

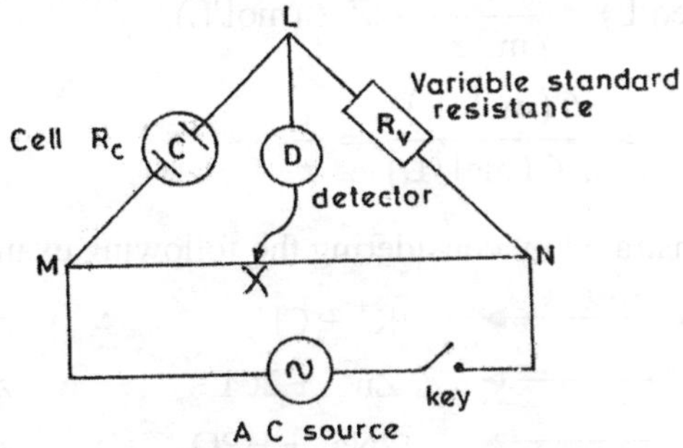

Fig. 5.2 Wheatstone bridge circuit.

The electrolyte of known concentration is taken in a container called **conductivity cell**. It consists of two platinum plates of area of cross section *a* cm^2 and separated by a distance 1 cm. These plates are generally coated with platinum black to decrease the polarization effects. This forms one arm of the circuit. The other arm of the circuit is fitted with a variable standard resistance. These two arms are attached to both ends of a meter bridge. A source of alternating current is also attached to both extremes of the meter bridge. The preference of alternating current over direct current is for the following reasons.

(i) Direct current changes the concentration of electrolyte due to electrolysis,

(ii) Gases liberated at the electrodes undergo polarization and thereby affect the conductance, and

(iii) By employing alternating current the direction of current is reversed few thousand times and pulses produced in opposite direction mutually cancel the polarization effects.

The current balance detector, D, fixed between R_c and R_v is generally a head piece or magic eye for less accurate measurements and a cathode ray oscillator for precession measurements.

Now the sliding contact, often called jockey, is moved over the long and uniform meter bridge wire MN, the point of least current through the detector, X, indicated by least buzzing noise of a head phone or smallest signal of an oscilloscope is noted. Under these conditions Wheatstone bridge principle suggests that the ratio of the resistances in the meter bridge arms, i.e., MX to NX is equalto the ratio of LM to LN.

$$\frac{\text{MX}}{\text{NX}} = \frac{\text{LM}}{\text{LN}}$$

But the resistance in LM is that of conductivity cell, i.e., R_c, and the resistance in arm LN corresponds the variable resistance, R_v.

Therefore,

$$\frac{\text{MX}}{\text{NX}} = \frac{R_c}{R_v}$$

Since R_v is known and MX and NX are determined through experiment, the resistance of the cell R_c can be calculated. Inverse of R_c gives the conductance of the experimental solution.

To calculate the electrolyte conductivity of the solution we take help of the eqn. (5.5).

$$\kappa = L\,\frac{l}{a} \qquad \text{.....(5.10)}$$

A knowledge of l and a allow us to calculate the electrolytic conductivity of the solution. But in the normally available cells the electrodes are far from parallel to obtain an exact value for l and far from perfect geometrical shape to obtain an exact value for a. So, instead of attempting to find the values of l and a of the conductivity cell, it is a matter of great simplification to use the literature electrolytic conductivity values of standard electrolytes like KCl Table 5.2. With these values one can determine its conductivity and use these two values to find the ratio l/a. This ratio is a characteristic of a conductivity cell and is known as cell constant, X. The eqn. (5.10) may thus be written as

or

$$\boxed{\kappa = L \times X} \qquad \text{.....(5.11)}$$

Electrolyte conductivity = Conductance × Cell constant.

5.3.6 Determination of Cell Constant

In finding the cell constant, first and foremost step is to remove organic and oily impurities sticking to the walls of the cell and electrodes by treating with dilute chromic acid. It is then thoroughly rinsed with conductivity water. The electrodes of the cell are then dipped in a standard KCl solution, say N/10 KCl, thermostated to required temperature and its conductance is measured. The electrolytic conductivity of N/10 KCl, as determined accurately by Kohlrausch and verified by other scientists at 25 °C is equal to 0.012856 S cm^{-1}. If our experimental temperature is also 25 °C, then cell constant is obtained from the following expression.

$$\text{Cell constant} = \frac{0.012856}{\text{Observed conductance}}$$

The cell constant so obtained can be used in finding the electrolytic conductivity of experimental electrolyte solution.

Various conductivity cells are designed to meet the requirements of different conductance measurements. For solutions of high conductance measurements the electrodes are widely spaced, Fig. 5.3(a), while for low conducting liquids they are closely spaced Fig. 5.3(b).

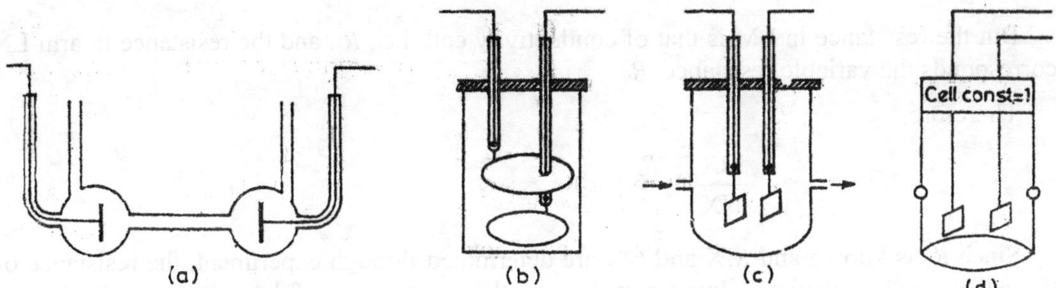

Fig. 5.3 Types of conductance cells.

Fig. 5.3(c) gives the arrangement of a conductivity cell used for on line conductance measurements and Fig. 5.3(d) is a commercial dipping type cell.

For liquids having large conductance values normal distilled water is satisfactory. For low conducting substances, the electrolyte is dissolved in a special grade water called conductivity water. It is prepared in all quartz still. First the water is refluxed, for sufficiently long time, with alkaline permanganate to oxidize the volatile organic impurities. Then it is distilled and collected into an air tight quartz container. Purity of the water can be established by comparing the conductivity with the literature value.

Conductivity of water after single distillation in glass at 25 °C

$$= 2.0 \times 10^{-6}\ \text{S cm}^{-1}$$

Conductivity of triple distilled water at 25 °C

$$= 0.8 \times 10^{-6} \text{ to } 1 \times 10^{-6}\ \text{S cm}^{-1}$$

Conductivity of conductivity grade water at 25 °C

$$= 0.056 \times 10^{-6}\ \text{S cm}^{-1}$$

Conductivity of purest water available at 25 °C

$$= 0.038 \times 10^{-6}\ \text{S cm}^{-1}$$

If the measured conductivity is low, the real conductivity of an electrolytic solution is obtained by subtracting conductivity contribution (due to dissociation) of water used for making the solution.

Problem 2.

The resistance of an 0.1 N solution of an electrolyte is 40 ohm. If the distance between the electrodes is 1.2 cm and area of cross section is 2.4 cm^2, calculate the equivalent conductivity.

Given $l = 1.2$ cm, $a = 2.4$ cm^2, $C = 0.1$ N, $R = 40$ ohm

$$\text{Cell constant} = \frac{l}{a} = \frac{1.2}{2.4} = 0.5\ \text{cm}^{-1}$$

$$\text{Observed conductivity} = \frac{1}{40} = 0.025\ \text{S}$$

Electrolyte conductivity $\kappa = L \times X$

$$= 0.5 \times 0.025$$

$$= 0.0125\ \text{S cm}^{-1}$$

$$\text{Equivalent conductivity } \Lambda = \frac{1000\kappa}{C} = \frac{1000 \times 0.0125}{0.1}$$

$$= 125\ \text{S cm}^2\ \text{eq}^{-1}.$$

5.3.7 Effect of Dilution on Conductance

We already learnt that conductance depends on three factors, viz.,

(i) on the number of ions,

(ii) on the charge of the ions, and

(iii) on the mobility (mass) of ions.

Addition of an electrolyte, till saturation limit, increases the population of free ions and it is reasonable to expect a simultaneous increase of conductance. But the number of ions of an electrolyte are decided by the ionic character of a bond. Further, with the increase in concentration of a solute, the ions get closer amounting to an increase in the electrostatic forces of attraction and viscosity of the medium. These factors tend to reduce the conductance of the solution. A cumulative effect of all the above factors is:

(i) The conductance increases with increase in the concentration of the electrolyte to a certain maximum level, and

(ii) Any further increase will only result in a fall in the conductance value.

This trend is well demonstrated in case of highly soluble electrolytic solutions of KOH and acetic acid, Fig. 5.4. However, in majority of the cases, saturation limit precedes the segment representing a fall in the conductance. Therefore we often notice only an increase in conductance with the increase in concentration followed by a simple plateau representing the arrival of saturation limit. Because electrolytic conductivity is defined as the conductance of a unit cube of solution, it adopts the trends followed by conductance.

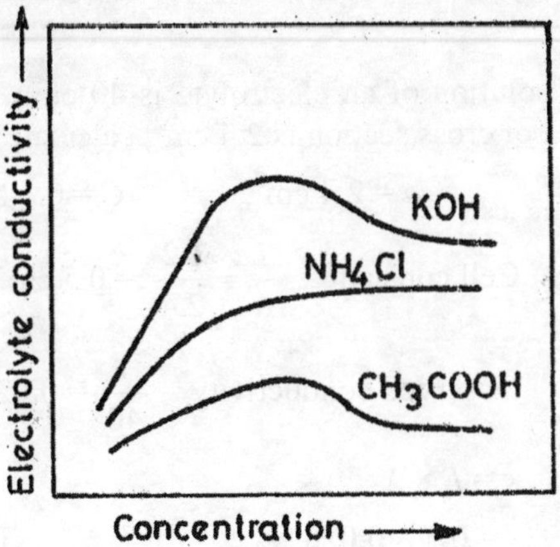

Fig. 5.4 Variation of conductivities of electrolytes with concentration.

On the other hand equivalent conductivity and molar conductivity are inversely proportional to the concentration and therefore, increase with increase in dilution (decrease in concentration), Fig. 5.5.

Equivalent conductivity

$$\Lambda_v = 1000\ \frac{\kappa}{C} = \kappa V$$

Molar conductivity

$$\mu = 1000\ \frac{\kappa}{C} = \kappa V$$

The trends of decrease in electrolytic conductivity and increase in equivalent and molar conductivity values with dilution is shown in Table 5.1.

Table 5.1 Electrolytic and molar conductivities of aqueous solutions of KCl at different concentrations (25 °C).

Concentration C mol L^{-1}	Volume containing 1 mole, V mL	Electrolytic conductivity κ S cm^{-1}	Molar* conductivity : S cm^2 mol^{-1}
1.0	1,000	0.11173	111.73
0.1	10,000	0.012856	128.56
0.05	20,000	0.00667	133.4
0.02	50,000	0.002766	138.3
0.01	1,00,000	0.001413	141.3
0.005	2,00,000	0.0007175	143.5
0.001	10,00,000	0.0001469	146.9
0.0005	20,00,000	0.0000739	147.8

* KCl being uni-univalent electrolyte its equivalent and molar conductances are same.

Comparative information on the variation of equivalent conductivities with concentration of various electrolytes is given in Table 5.3. A close observation of data given in Table 5.2 & 5.3 and the curves (in Fig. 5.5) obtained basing on the data, prompt us to draw the following conclusions.

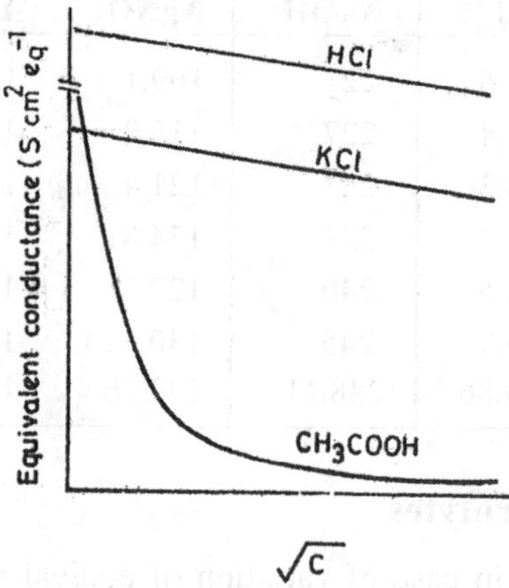

Fig. 5.5 Variation of equivalent conductivity of HCl, KCl and CH_3COOH with square root of concentration.

(i) The electrolytic conductivity increases with increase in concentration of electrolyte.

(ii) After reaching a particular concentration level, the conductance of some electrolytes may marginally drop and ultimately reach a constant value.

(iii) In case of less soluble or sparingly soluble salts the conductance may reach the constancy without showing any fall in its value.

(iv) Equivalent conductivity invariably falls with the increase in the concentration or increase with increase in dilution.

(v) A gradual, linear change in Λ or : with square root of concentration is found in case of strong electrolytes like KCl and HCl, known for their complete dissociation in aqueous media.

(vi) Electrolytes like acetic acid, known for their low ionization, show insignificant change in conductance at higher concentrations but abruptly rise at higher dilutions (lower concentrations).

(vii) The maximum, value of equivalent conductivity obtained by the extrapolation to zero concentration or infinite dilution of an electrolyte is a characteristic value and is called equivalent conductivity at infinite dilution, Λ_4.

Table 5.2 Variation of equivalent conductivity (S cm^2 eq^{-1}) with concentration of various electrolytes at 25 °C.

Concentration	Strong electrolytes					Weak Electrolyte
C mol L^{-1}	HCl	KCl	NaOH	$AgNO_3$	½$BaCl_2$	CH_3COOH
0.10	391.3	128.6	221	109.1	105.2	5.2
0.05	399.1	133.4	227	115.2	111.5	7.4
0.02	407.2	138.3	233	121.4	119.1	11.6
0.01	412.0	141.3	237	124.8	123.9	16.2
0.005	415.8	143.5	240	127.2	128.0	22.8
0.001	421.4	146.9	245	130.5	134.3	48.6
0.0005	429.16	149.86	248.11	133.36	139.98	390.75

5.3.8 Classification of Electrolytes

On the basis of curves obtained in case of variation of equivalent conductivity with square root of concentration the electrolytes are grouped into two categories.

A. Strong Electrolytes

Strong electrolytes are the substances whose equivalent conductivity do not change appreciably with the changes in concentration. The equivalent conductances at any dilution are of the same order as the equivalent conductance at infinite dilution. The solutions of mineral acids, bases and many of their simple salts fall in this category of compounds.

B. Weak Electrolytes

Weak electrolytes are the substances which possess low and constant equivalent conductivity values at high concentrations, and these values increase very rapidly in extremely dilute solutions. The carboxylic acids, amines and their salts fall in this category of compounds.

The linear variation of equivalent conductivity with $\sqrt{C}$ of different strong electrolytes, led to the formulation of an empirical relation called **Kohlrausch equation,**

$$\Lambda_v = \Lambda_4 - b\sqrt{C} \qquad .....(5.12)$$

where b is a constant.

Kohlrausch equation suggests that for strong electrolytes the intercept of Λ_v vs $\sqrt{C}$ plots is equal to Λ_4. Weak electrolytes do not obey Kohlrausch relation and therefore this method of calculation of Λ_4 is not suitable for weak electrolytes. This is achieved through indirect calculation based on **Kohlrausch's law of independent migration of ions** discussed later.

Problem 3.

The resistance of 0.2 N solution of an electrolyte was found to be 250 ohms at 25 °C. Calculate the equivalent conductivity if the cell constant is 0.75 cm^{-1}.

Observed conductance

$$L = \frac{1}{R} = \frac{1}{250}$$

but Electrolytic conductivity

$$\kappa = X \times L$$
$$= 0.75 - (1/250)$$
$$= 0.003 \text{ mho cm}^{-1}$$

Equivalent conductivity of 0.2 N electrolyte

$$\Lambda_V = \frac{1000\kappa}{C} = \frac{1000 \times 0.003}{0.2}$$

$$\Lambda_V = 15 \text{ mho cm}^2 \text{ eq}^{-1}$$

Problem 4.

The equalent conductivity of HCl of concentrations 0.1, 0.01 and 0.001 N solutions are 391.3, 412.0 and 421.4 S cm^2 $equiv^{-1}$ respectively. Calculate Kohlrausch constant and Λ_4. (HCl.)

According to Kohlrausch law $\Lambda_v = \Lambda_4 - b\sqrt{C}$

Substituting the given data, we have;

$$391.3 = \Lambda_4 - b\sqrt{0.1} \qquad \text{I}$$
$$412.0 = \Lambda_4 - b\sqrt{0.01} \qquad \text{II}$$
$$421.4 = \Lambda_4 - b\sqrt{0.001} \qquad \text{III}$$

II-I gives,

$$(412 - 391.3) = 0 - b\left[\sqrt{0.01} - \sqrt{0.1}\right]$$
$$20.7 = -b\,(0.1 - 0.316)$$
$$b = \frac{20.7}{0.216} = 95.8$$

Substituting the value of b is equation II

$$412 = \Lambda_4 - (95.80)\sqrt{0.01}$$
$$\Lambda_4 = 412 + 95.8\,(0.1)$$
$$\Lambda_4 = 421.6$$

Similar result can be obtained by taking any other pair of equations given above.

5.4 Kohlrausch Law of Independent Migration of Ions

In a state of infinite dilution all electrolytes ionize or dissociate cent percent, and the ions can be expected to be wide apart not to have any electrostatic or any other type of attraction or repulsion.

Under these conditions, we achieve an ideal situation where all the charge is expected to be carried by the ions independent of each other. During our discussion on the variation of equivalent conductivity with square root of concentration we noticed that equivalent conductivity increases with dilution to reach a limiting value characteristic of the electrolyte and named it as equivalent conductivity at infinite dilution, Λ_4. In 1875 Kohlrausch made a series of measurements, involving electrolytes with common cations and common anions and calculated their differences. He obtained a constant difference in Λ_4 values of an ion pair, irrespective of the nature of the common counter ion employed, as exemplified by following results obtained at 25 °C.

Difference between Λ_4 values of some pairs of electrolytes having K^+ and Na^+ cations and common anions :

KCl	149.85	KBr	151.64	KOH	272.10	KNO_3	144.96
NaCl	126.45	NaBr	128.24	NaOH	248.70	$NaNO_3$	121.56
Diff	23.40		23.40		23.40		23.40

Difference between Λ_4 values of some pairs of electrolytes having OH^- and Cl^- ions and common cations :

NaOH	248.70	NH_4OH	272.15	LiOH	237.28	½$Ba(OH)_2$	262.23
NaCl	126.45	NH_4Cl	149.9	LiCl	115.03	½$BaCl_2$	139.98
Diff	122.25		122.25		122.25		122.25

Basing on similar observations, Kohlrausch advanced a hypothesis called **Law of independent migration of ions**. This law states that *at infinite dilution each ion makes a definite contribution towards the equivalent conductivity of an electrolyte, irrespective of nature of the co-ion with which it is associated in the solution.* Therefore equivalent conductivity at infinite dilution is made up of two independent terms, called **ion conductivities**, characteristic of each ionic constituent of an electrolyte in solution. We can express the same conclusion mathematically as

$$\Lambda_4 = \lambda_+ + \lambda_- \qquad \text{.....(5.13)}$$

where λ_+ and λ_- are the ion conductances of cation and anion respectively. The ion conductance of any ion is a fixed characteristic value in a given solvent. (In the absence of any specific mention, these values are always for aqueous solutions), and are dependent on temperature. The ion conductivity values of some common ions at 25 °C is presented in Table 5.3.

Table 5.3 Ion conductivities at infinite dilution at 25 °C in S cm^2 equiv.

Cation	λ_+	Anion	λ_-
H^+	349.81	OH^-	197.6
Li^+	38.68	Cl^-	76.35
Na^+	50.10	Br^-	78.14
K^+	73.50	I^-	76.84
Ag^+	61.92	NO_3^-	71.46
NH_4^+	73.55	ClO_4^-	67.36
½Mg^{2+}	53.05	IO_4^-	54.38
½Ca^{2+}	59.50	HCO_3^-	44.50
½Sr^{2+}	59.46	½SO_4^{2-}	80.02
½B_a^{2+}	63.63	Acetate$^-$	40.90
½Cu^{2+}	54.00	Benzoate$^-$	32.30
½Zn^{2+}	53.00	½Oxalate^{2-}	24.00

5.4.1 Ionic Conductivity and Ionic Mobilities

Equivalent conductivity of any electrolyte at any dilution is directly proportional to the charge carried by the ions and their velocities. The conductivity is thus given by the products of charge and velocity of individual ions. At infinite dilution the ionization is complete and solution containing one equivalent of various electrolytes contain equivalent number of ions. Hence at infinite dilution total charge carried by all ions is same in every case. Because the total charge is constant at infinite dilution, the $\wedge_4$ must depend exclusively on ionic velocities.

Defining the **ionic velocity or mobility** as the speed with which a charged particle at infinite dilution moves under a potential gradient of one volt per cm, we have,

$$\lambda_+ \propto u_+ \qquad \lambda_- \propto u_-$$

or
$$\lambda_+ = ku_+ \qquad \lambda = ku_-$$

where k is a proportionality constant and u_+ and u_- represent the ionic velocities at infinite dilution. Since 1 equivalent of an ion under unit potential gradient carries a charge of 1 Faraday per sec., the proportionality constant $k = 96500$ coulombs.

Therefore,

$$u_+ = \frac{\lambda_+}{k} \ \& \ u_- = \frac{\lambda_-}{k} \qquad .....(5.14)$$

$$\text{Ionic mobility} = \frac{\text{Ionic conductivity}}{96500} \qquad .....(5.15)$$

Ionic conductivity is expressed in S cm^2, while ionic mobility is expressed in cm s^{-1}. Calculated ionic mobilities of few common ions at 25 °C are given in Table 5.4.

Table 5.4 Calculated ionic mobilities of few common ions at 25 °C

Cations	u_+ (cm s^{-1})	Anions	u_- (cm s^{-1})
H^+	36.2×10^{-4}	OH^-	20.5×10^{-4}
K^+	7.61×10^{-4}	SO_4^{2-}	8.27×10^{-4}
Ba^{++}	6.60×10^{-4}	Cl^-	7.91×10^{-4}
Na^+	5.19×10^{-4}	NO_3^-	7.40×10^{-4}

Exceptionally high mobilities of H^+ and OH^- ions is attributed great affinity of these ions for water molecules, which can initiate a chain of transfers from one to other as shown below;

Mechanism of H^+ ion transfer :

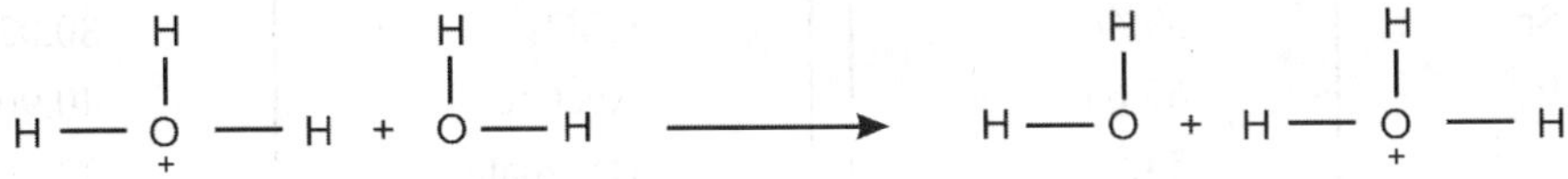

Mechanism of OH^- ion transfer:

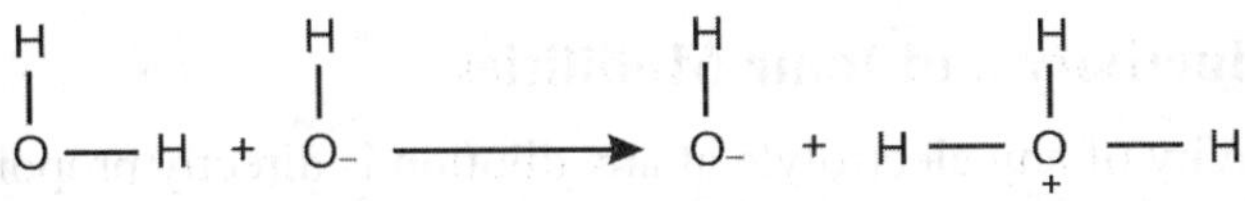

This additional mobility/conductivity is referred to as **Grotthus conductivity**.

5.4.2 Applications of Kohlrausch's law

The introduction of concept of ionic conductivity opened the vistas towards numerous applications. A few of them are discussed below.

I. Estimation of equivalent conductivity at infinite dilution of weak electrolytes

Method 1: The extrapolation technique used for the determination of λ_4 of strong electrolytes is not useful for weak electrolytes. This is due to the steepness of the curve (Fig. 5.7) at the approach of zero concentration or infinite dilution. Λ_4 of weak electrolytes is calculated using Kohlrausch's law. Using acetic acid as an example, we have at infinite dilution

$$\Lambda_{\infty(CH_3COOH)} = \lambda_{(CH_3COO-)} + \lambda_{(H^+)}$$

Adding Λ_4 of the salt of a strong acid and strong base, e.g., NaCl

$$\Lambda_{\infty(CH_3COOH)} = \Lambda_{\infty(CH_3COO^-)} + \lambda_{H^+} + \lambda_{Na^+} \; \lambda_{Cl^-}$$

on rearranging the ions on R.H.S, we may write

$$\Lambda_{\infty(CH_3COOH)} = \Lambda_{\infty(NaCl)} + \lambda_{(CH_3COONa)} + \lambda_{\infty(HCl)}$$

or

$$\Lambda_{\infty(CH_3COOH)} = \Lambda_{\infty(CH_3COONa)} + \Lambda_{\infty(HCl)} - \Lambda_{\infty(NaCl)}$$

Because all the electrolytes on the right hand side are strong electrolytes, Λ_∞ values of these electrolytes can be obtained by the extrapolation of a Λ_v versus $\sqrt{C}$ plots and there by Λ_4 value of acetic acid.

II. Calculation of degree of dissociation and dissociation constant of a weak electrolyte

The ratio of equivalent conductivity at any dilution (Λ_v) to that at infinite dilution (Λ_4) is called **conductance ratio.**

$$\alpha = \frac{\Lambda_v}{\Lambda_\infty} \quad .....(5.16)$$

In dilute solutions of strong electrolyte, the conductance ratio is almost independent of the nature of the salt and is a constant value close to unity at any given concentration. For weak electrolytes this ratio is quite small and is a characteristic quantity called **degree of dissociation**. This in fact represents the fraction of molecules dissociated out of unit concentration of the electrolyte.

	$CH_3COOH \rightleftharpoons$	$CH_3COO^- +$	H^+
Initial conc.	1	0	0
Equilibrium conc.	$1-\alpha$	α	α
When C moles are taken initially	$(1-\alpha)C$	αC	αC

Dissociation constant

$$K_D = \frac{[CH_3COO^-][H^+]}{[CH_3COOH]} = \frac{(\alpha C)(\alpha C)}{(1-\alpha)C}$$

K_D of uni-uni valent electrolyte

$$= \frac{\alpha^2 C}{(1-\alpha)} \quad \text{where} \quad \alpha = \frac{\Lambda_v}{\Lambda_\infty}$$

where C is the concentration of the weak electrolyte and Λ_v is the corresponding equivalent conductivity.

III. Calculation of Ionic Product of Water

Water is a weak electrolyte, which dissociates partially to give H^+ and OH^- ions.

	$H_2O \rightleftharpoons$	$H^+ +$	OH^-
For 1 mole at equilibrium	$1-\alpha$	α	α
For C moles	$C(1-\alpha)$	$C\alpha$	$C\alpha$

Ionic product of water $K_w = [H^+]\,[OH]^- = (C\alpha)\,.\,(C\forall) = C^2\alpha^2$

Weight of 1000 mL of water = 1000 × 0.997 = 997 g

18 g (1 mole) of water dissolved in 1000 mL of water gives 1 M solution.

So 997 g of water present in 1000 mL of water has a concentration

qual to $= \frac{997}{18} = 55.4 \text{ M} = 55.4 \text{ N}$

Equivalent conductivity of water

$$\Lambda_v = \frac{1000\kappa}{C} = \frac{1000 \times 5.54 \times 10^{-8}}{55.4}$$

$$= 1 \times 10^{-6} \text{ S cm}^2 \text{ equiv}^{-1}$$

Limiting equivalent conductivity $\Lambda_\infty = 349.8 + 197.6 = 547.4$

Degree of dissociation

$$\alpha = \frac{\Lambda_v}{\Lambda_\infty} = \frac{1 \times 10^{-6}}{547.4} = 1.827 \times 10^{-9}$$

$$[OH^-] = [H^+] = \alpha C$$

$$= 1.827 \times 10^{-9} \times 55.4 = 1.01 \times 10^{-7}$$

Therefore, Ionic product of water

$$= [H^+][OH^-] = (1.01 \times 10^{-7})^2 = 1.02 \times 10^{-14}$$

IV. Determination of Solubility Products

Another application of conductance measurements is the determination of solubility products of sparingly soluble salts. Salts like AgCl, $PbSO_4$, AgI, $BaSO_4$, PbI_2, etc., are *soluble to a negligible extent* and whatever that gets into solution dissociate completely. If ***S*** is the solubility, in gram equiv. per litre, of a given salt and κ is the specific conductance of the saturated solution, then Λ_v of the solution is given by;

$$\Lambda_v = \frac{1000\kappa}{S}$$

These solutions are usually so dilute that their equivalent conductivity, Λ_v, is almost equal to Λ_4. Therefore,

$$\Lambda_4 = \frac{1000\kappa}{S}$$

$$S = \frac{1000\kappa}{\Lambda_\infty} \quad \text{.....(5.17)}$$

Since the solutions are extremely dilute, the conductance contribution due to pure water is appreciable part of the total conductance and needs to be subtracted from the total value to get the contribution of dissolved salt.

Therefore,

$$S = \frac{1000(\kappa - \kappa_w)}{\Lambda_\infty} \quad .....(5.18)$$

where κ_w is the specific conductance of pure water. In case of uni-univalent electrolyte like AgCl.

The solubility product $K_{sp} = (a_{Ag^+})(a_{Cl^-})$

The solution is dilute enough to regard the activity coefficient are unity. Therefore, $a_{Ag^+} = a_{Cl^-} = S$

Therefore $\quad K_{sp} = S^2$

In case of AgCl

$$\Lambda_{4(AgCl)} = \lambda_{Ag^+} + \lambda_{Cl^-}$$
$$= 61.90 + 76.35$$
$$= 138.25 \text{ S cm}^2$$

k of AgCl solution $= 3.41 \times 10^{-6}$ S cm^{-1}

k of pure water $= 1.60 \times 10^{-6}$ S Cm^{-1}

k of dissolved AgCl $= (3.41 - 1.60) \times 10^{-6}$ S cm^{-1}

$= 1.81 \times 10^{-6}$ S cm^{-1}

Substituting the above results in equation (5.17)

$$\text{Solubility of AgCl} = \frac{1000 \times 1.81 \times 10^{-6}}{138.25}$$

$$S = 1.31 \times 10^{-5} \text{ eq L}^{-1}$$

$$\text{Solubility product} = K_{sp} = S^2 = (1.31 \times 10^{-5})^2$$
$$= 1.72 \times 10^{-10} \text{ eq}^2 \text{ L}^{-2}$$

Problem 5.

The equivalent conductivities of sodium benzoate and HCl at 25 °C are 82.4 and 426.16 S cm^2 equiv^{-1} respectively. If the transport numbers of benzoate and H$^+$ ions are 0.392 and 0.821 in the above mentioned compounds, find the equivalent conductivity of benzoic acid at 25 °C.

Given, $\quad \Lambda_4$ (Sodium benzoate) = 82.4 S cm^2 equiv^{-1}

Λ_4 (HCl) = 426.16 S cm^2 equiv^{-1}

t_+ (in HCl) = 0.821 & t_- (in Sodium benzoate) = 0.392

$\lambda_+ = t_+ \Lambda_4$ & $\lambda_- = t_- \Lambda_4$

$\lambda_{H^+} = 0.821 \times 426.16 = 349.9$ S cm^2 equiv^{-1}

$\lambda_{benzoate} = 0.392 \times 82.4 = 32.3$ S cm^2 equiv^{-1}

$\lambda_{4\,(Benzoic\ acid)} = \lambda_{benzoate} + \lambda_{H^+}$

$\lambda_{4\ Benzoic\ acid} = 32.3 + 349.9 = 382.2$ S cm^2 equiv^{-1}

Problem 6.

The equivalent conductance values of 0.0005M (at 25 °C) NaCl, NH_4Cl, NaOH and NH_4OH are 124.50, 149.1, 245.6 and 9.38 mho cm^2 $equiv^{-1}$ respectively. Calculate the dissociation constant of NH_4OH.

$$\Lambda_{\infty(NH_4OH)} = \Lambda_{\infty(NH_4Cl)} + \Lambda_{\infty(NaOH)} - \Lambda_{\infty(NaCl)}$$

$$= 149.1 + 245.6 - 124.50$$

$$= 270.2 \text{ S cm}^2 \text{ equiv}^{-1}$$

Degree of dissociation

$$\alpha = \frac{\Lambda_v}{\Lambda_\infty} = \frac{9.38}{270.2} = 0.0347$$

Dissociation constant $K = \alpha^2 C$

$$= (0.0347)^2 \times (1/100)$$

$$= 1.2 \times 10^{-5}$$

Problem 7.

Calculate the conductance of pure water at 25 °C, when measured employing a cell with a cell constant 1.0. The ion conductivities of H^+ and OH^- are 349.8 and 197.6 S cm^2 $equiv^{-1}$ respectively. Density of water at 25 °C is 0.997 g cm^{-3}.

$$\Lambda_4 = \lambda_{H+} + \lambda_{OH^-} = 349.8 + 197.6 = 547.4$$

Ionic product of water $K_w = 1 \times 10^{-14}$

$$H_2O \rightleftharpoons H^+ + OH^-$$

Initial	1	0	0
At equilibrium	$C(1-\alpha)$	$C\alpha$	$C\alpha$

$$[H^+] = [OH^-] = C\alpha = 1 \times 10^{-7}$$

Therefore,

Degree of dissociation $= \alpha = \dfrac{1\times10^{-7}}{C}$

Concentration of water C

$$= \frac{997}{18} = 55.4 \text{ M}$$

Degree of dissociation

$$\alpha = \frac{1\times10^{-7}}{55.4} = 1.8 \times 10^{-9}$$

But $\alpha = \dfrac{\Lambda_v}{\Lambda_\infty}$

Therefore,

$$\Lambda_v = \alpha\,\Lambda_4 \quad = 547.4 \times 1.8 \times 10^{-9} \quad = 9.85 \times 10^{-7}$$

$$\Lambda_v = \frac{1000\kappa}{C}$$

$$\kappa = \frac{9.85 \times 10^{-7} \times 55.4}{1000} = 5.46 \times 10^{-8} \text{ S Cm}^{-1}$$

$$\text{Conductance} = \frac{\text{Conductivity}}{\text{Cell constant}}$$

$$\text{Conductance} = \frac{5.46 \times 10^{-8}}{1} = 5.46 \times 10^{-8} \text{ mho}$$

Problem 8.

The conductivity of a saturated solution at 298 K of a sparingly soluble salt MX is 2.5×10^{-6} S cm^{-1}, while that of pure water is 1.61×10^{-6} S cm^{-1}. If the equivalent conductivity of MX at infinite dilution is 156 S cm^2 equiv^{-1}, calculate the solubility product of the salt.

Conductivity of a saturated solution $= 2.5 \times 10^{-6}$ S cm^{-1}

Conductivity of pure water $= 1.61 \times 10^{-6}$ S cm^{-1}

Conductivity contribution of only salt (k) $= (2.5 - 1.61) \times 10^{-6}$

$$= 0.89 \times 10^{-6} \text{ S cm}^{-1}$$

$$\Lambda_4 \text{ of the salt} = 156 \text{ S cm}^2 \text{ equiv}^{-1}$$

Solubility $S = \dfrac{1000\kappa}{\Lambda_\infty} = \dfrac{1000 \times 0.89 \times 10^{-6}}{156} = 5.7 \times 10^{-6}$ mol L^{-1}

MX being a uni-uni valent salt

	MX	$\rightleftharpoons$	M^+	$+ X^-$
Initial	1 mole		0	0
at equilibrium	$1-x$		x	x

Solubility product $(K_{sp}) = [M^+]\,[X^-] = x^2$

$$= (5.7 \times 10^{-6})^2 = 32.5 \times 10^{-12} \text{ mol}^2 \text{ L}^{-2}$$

Solubility product of MX $= 32.5 \times 10^{-12}$ mol^{-2} L^{-2}.

5.5 Conductometric Titrations

This is another interesting application which exploits the variation of conductance of a solution during a titration. This technique can be used advantageously with,

(a) colored solutions where selection of proper indicator is difficult;

(b) acids and bases and their mixtures;

(c) ions and their mixtures that are readily precipitated by the addition of titrant.

A grave disadvantage of conductivity based titrations is their non-applicability with solutions contaminated with high concentrations of electrolyte impurities. Presence of excess of electrolyte (ions) and hence high electrolytic conductivity masks the minor changes in conductivity, taking place as a result of addition of titration. Few examples of acid-base and precipitation titrations are discussed below.

5.5.1 Titration of a strong acid (HCl) with a strong base (NaOH)

Strong acids dissociate completely to give hydrogen ions. The ionic conductance data, Table 5.6 suggest that protons (H^+) are almost two times more conducting than hydroxy ions (OH^-), which themselves are more conducting than all the other ions. When NaOH is added to a solution of HCl we have the reaction.

$$(H^+ + Cl^-) + (Na^+ + OH^-) \longrightarrow (Na^+ + Cl^-) + H_2O$$

During the course of neutralization initially high conducting H^+ ions present in the acid are replaced by sodium ions having a much lower conductance. Therefore, addition of alkali is accompanied by a fall (curve AB) in the conductance till all the H^+ ions are totally neutralized. Further addition of alkali may be interpreted as only an addition of hydroxyl ions which are no longer used up in the acid-base reaction. Hence, after neutralization conductance of the reaction solution increases linearly (curve BC) with the volume of the alkali added. The titration graph will have the form shown in Fig. 5.6. Interpolation of both the linear segments AB and BC to meet at B gives the end point of the titration.

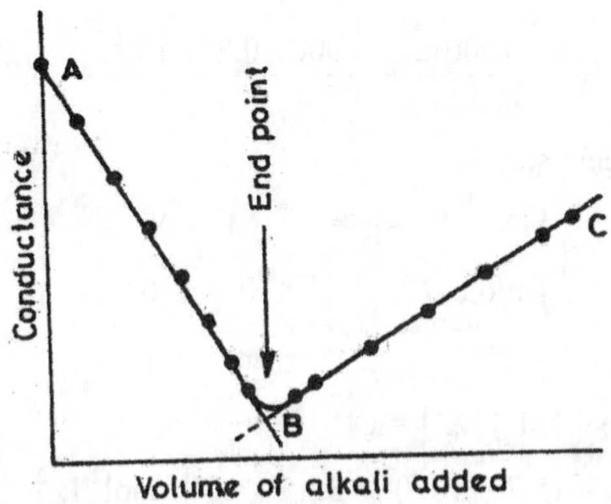

Fig. 5.6 Conductometric titration of a strong acid Vs strong base.

Precise measurements leave a small rounding near equivalent point B and is attributed to water dissociation.

5.5.2 Titration of Weak Acid (CH_3COOH) by Strong Base (NaOH)

A weak acid, like acetic acid, is characterized through its insignificant dissociation. Therefore, at the beginning of titration the acid solution has a low conductance. Addition of alkali to this moderately weak acid solution, initially result in a small fall of conductance owing to the replacement of free H^+ ions by acetate ions. If the acid is very weak, e.g., boric acid and phenol, the initial decrease of conductance upon neutralization is not observed due to insignificant dissociation. After the addition of a few drops of the alkali, the conductance increases in a regular fashion. This is due to suppression of ionization equilibrium of acetic acid by the acetate ions formed (buffering action) and increase in ionic population (sodium acetate is a strong electrolyte) in the solution.

$$(CH_3COOH) + (Na^+ + OH^-) \rightleftharpoons (CH_3COO^- + Na^+) + H_2O$$

Beyond neutralization point, addition of alkali result in a steep raise in conductance, the strong base being a better conductor than the salt. This trend in indicated by the segment B, C, D in Fig. 50.7.

5.5.3 Conductometric titration of a mixture of a strong (HCl) and a weak acid (CH_3COOH) with a strong base (NaOH)

One of the valuable features of conductometric titrations is the titration of a mixture of a strong acid and a weak acid in one titration without employing any indicator. In the presence of a strong acid, the dissociation of a weak acid is completely suppressed due to common ion effect. Therefore, addition of a base to such a mixture will first result in the neutralization of the strong acid. The weak acid starts reacting only after neutralization of the strong acid. In the titration graph, Fig. 5.7, the line segment AB represents the neutralization of strong acid and the line segment BC represents the neutralization of weak acid.

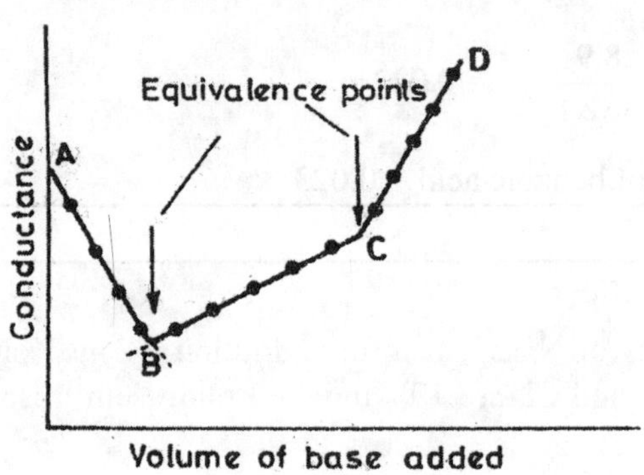

Fig. 5.7 Conductometric titration of a mixture of strong and weak acid *Vs* strong base.

Problem 9.

At 18 °C the conductivities at infinite dilution of NH_4Cl, NaOH and NaCl are 129.8, 217.4 and 108.9 mhos respectively. If the equivalent conductivity of N/100 solution of NH_4OH is 9.33 mhos, calculate the degree of dissociation of NH_4OH

$$\Lambda_{\infty(NH_4Cl)} = \lambda_{NH^{4+}} + \lambda_{Cl^-} = 129.8 \text{ S cm}^2 \qquad \text{I}$$

$$\Lambda_{4(NaOH)} = \lambda_{Na^+} + \lambda_{OH^-} = 217.4 \text{ S cm}^2 \qquad \text{II}$$

$$\Lambda_{4(NaCl)} = \lambda_{Na^+} + \lambda_{Cl^-} = 108.9 \text{ S cm}^2 \qquad \text{III}$$

I + II – III

$$\Lambda_{\infty(NH_4OH)} = (129.8 + 217.4) - 108.9 = 238.3$$

Degree of dissociation

$$\alpha = \frac{\Lambda_v}{\Lambda_\infty} = \frac{9.33}{238.3} = 0.039$$

Dissociation constant

$$K = \frac{\alpha^2 C}{1-\alpha} = 1.58 \times 10^{-5}$$

Problem 10.

At 25 °C the equivalent conductivity of 0.1M benzoic acid is 8.9 ohm^{-1} cm^2 equiv^{-1}. The ionic conductivities at infinite dilution, of the hydrogen ion and benzoate ions are 349.8 and 32.3 ohm^{-1} cm^2 respectively. Calculate the degree of dissociation of 0.1 M benzoic acid.

$$\Lambda_v = 8.9 \text{ ohm}^{-1} \text{ cm}^2 \text{ eq}^{-1}$$

$$\Lambda_4 = \lambda_{H^+} + \lambda_{benzonate} = 349.8 + 32.3 = 382.1$$

$$\alpha = \frac{\Lambda_v}{\Lambda_\infty} = \frac{8.9}{382.1} = 0.023$$

Degree of dissociation of benzoic acid = 0.023

Problem 11.

The equivalent conductivity of NH_4Cl at infinite dilution is 149.7 ohm^{-1} cm^2 equiv^{-1} and the ionic conductances of OH^- and Cl^- are 198 and 79.3 ohm^{-1} cm^2 respectively. Calculate Λ_4 of NH_4OH.

$$\Lambda_{\infty(NH_4Cl)} = \lambda_{NH_4^+} + \lambda_{Cl^-} = 149.7 \text{ ohm}^{-1} \text{ cm}^2 \text{ eq}^{-1} \qquad \text{I}$$

$\lambda_{OH^-} = 198$ II

$\lambda_{Cl^-} = 79.3$ III

$I + II - III = 268.4$

Λ_4 of NH_4OH is 268.4 $ohm^{-1}\ cm^2\ eq^{-1}$.

Problem 12.

Calculate the cell constant of a cell where the area of the electrode is 2.5 sq cm and distance between them is 10 cm

$$\text{Cell constant} = X = \frac{l}{a} = \frac{10}{2.5} = 4\ cm^{-1}$$

Problem 13.

Specific conductance of a 0.02 M KCl solution at 25 °C is 0.002768 mhos m^{-1}. If the resistance of a cell containing this solution is 400 ohms, what is the cell constant?

$$\text{Cell constant} = \text{Specific conductance x Resistance}$$
$$= 0.002768 \times 400 = 1.1$$

Problem 14.

When a certain conductance cell was filled with 0.02 M KCl, it had a resistance of 82.4 ohms at 25°C. When it is filled with 0.005 N K_2SO_4, it had a resistance of 326 ohms. Find the equivalent conductivity of the given K_2SO_4 solution. The specific conductivity of 0.02 M KCl is 0.002768 $ohm^{-1}\ cm^{-1}$ at 25°C.

$$X = \kappa R$$
$$X = 0.002768 \times 82.4 = 0.2281\ cm^{-1}$$
$$\kappa_{K_2SO_4} = \frac{0.2281}{326} = 0.0006996\ mho\ cm^2\ eq^{-1}$$
$$\kappa_{K_2SO_4} = \frac{1000\kappa}{C} = \frac{1000 \times 0.0006996}{0.005}$$
$$= 139.9\ mho\ cm^2\ eq^{-1}$$

Problem 15.

The equivalent conductivity of 0.05 N solution of propionic acid at 25 °C is 6.32 $equiv^{-1}\ ohm^{-1}\ cm^2$ and the value increases to 385.6 $cm^2\ equiv^{-1}\ ohm^{-1}$ at infinite dilution. Calculate the degree of dissociation of the acid at that temperature.

Degree of dissociation of propionic acid

$$\alpha = \frac{\Lambda_v}{\Lambda_\infty} = \frac{6.32}{385.6} = 0.0164$$

Problem 16.

In a conductivity cell 0.01 N KCl solution gave a resistance of 225 ohms, while a 0.01 N solution of HCl gave a resistance of 77.1 ohms. Specific conductance of 0.01N KCl is 0.00141 ohm^{-1} cm^{-1}. Calculate the cell constant and equivalent conductance of HCl.

Cell constant = sp. conductance × resistance

$$= 0.00141 \times 225 = 0.31725\ cm^{-1}$$

$$\kappa_{HCl} = \frac{0.31725}{77.1} = 0.0041\ mho\ cm^{-1}$$

$$\Lambda_{(HCl)} = \frac{1000\kappa}{C} = \frac{1000 \times 0.0041}{0.01}$$

$$= 410\ mho\ cm^2\ eq^{-1}$$

5.6 Introduction to EMF

Chemical changes are always accompanied by energy changes, where the difference appears generally in the form of heat (either evolution or absorption). Thus, when a zinc rod is placed in chlorine water, electron exchange takes place between Zn atoms and Cl_2 molecules at the point of inter-phase and Zn dissolves accompanied by evolution of heat.

$$Zn + Cl_2 \longrightarrow ZnCl_2 + \Delta H$$

This is an oxidation - reduction reaction for which partial ionic equations are

$$Zn - 2e^- \longrightarrow Zn^{2+}$$

$$Cl_2 + 2e^- \longrightarrow 2Cl^-$$

$$Zn^{2+} + 2Cl^- \longrightarrow ZnCl_2$$

Since electrical energy is another form of energy, we could devise a reaction system in such away that all electron movement takes place through an external electrical conductor and there by tap electrical energy. Conversely, a chemical reaction could be initiated by passing electricity through a reaction system. *The chemical and electrical changes are carried in a physical entity called cell. A cell may be defined as a single arrangement consisting of two electrodes and capable of producing electricity due of chemical reaction and vice versa.* A battery is a combination of two or more cells arranged either in series or parallel.

There are two types of cells in which the above electrochemical processes are carried out.

(i) **A galvanic or voltaic cell** uses the energy released from a spontaneous chemical reaction to generate electricity.

(ii) **An electrolytic cell** uses the electrical energy to cause a non-spontaneous chemical reaction to occur in spontaneous manner, and the process is commonly called **electrolysis**.

Modern society depends on the galvanic cells for running cars, radios, calculators, medical instruments, computers, on the field analytical devices, emergency flash lights and so on. The electrolytic cells are used in the chemical and metallurgical industries for the extraction, synthesis and purification of various chemicals. In this chapter our interest lies mainly with the galvanic cells.

5.7 Galvanic Cells

Under normal circumstances a redox reaction occurs when the oxidizing agent comes in contact with the reducing agent as in the reaction between Zn and Cl_2 described above. The electrons are transferred directly from the reducing agent to the oxidizing agent. In order to tap the energy of the reaction in the form of electrical energy, the reactants must be physically separated so as to prevent direct transfer of electrons. Electrons are then allowed to go from Zn to Cl_2 through an external circuit, made out of a wire or metallic conductor. In the current example both Zn rod and Cl_2 taken in two separate compartments which are separated either through a porous partition or a salt bridge. The porous partition or salt bridge prevents direct mixing of the reactants and at the same time are permeable to ions, Fig. 5.8.

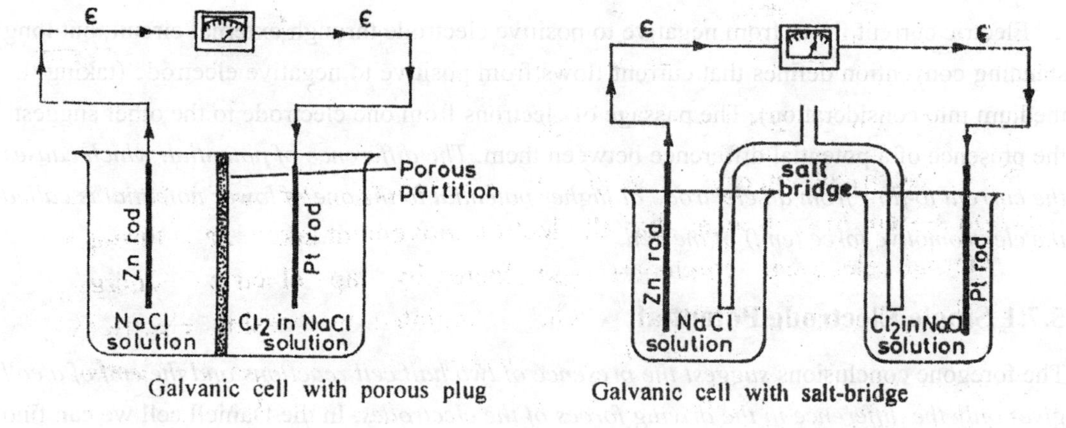

Fig. 5.8 A Galvanic cell.

Since the Cl_2 itself is a non-conductor unlike Zn rod, to conduct electricity in the Cl_2 compartment an inert metal wire like Pt is placed. The Zn rod and Pt wire are called electrodes. By *definition the electrode at which oxidation occurs is called the anode* (*Zn electrode*); *the electrode at which reduction occurs is known as cathode* (electrode is also called chlorine gas electrode).

For this system the *individual oxidation* and *reduction reactions at the electrodes are called* **half-cell reactions**.

Half-cell reaction of Zn electrode (anode) :

$$Zn(s) \longrightarrow Zn^{2+}(aq) + 2e^-$$

Half-cell reaction of Pt electrode (cathode) :

$$Cl_2(g) + 2e^- \longrightarrow 2Cl^-(aq)$$

Total cell reaction :

$$Zn(s) + Cl_2(g) \longrightarrow Zn^{2+}(aq) + 2Cl^-(aq)$$

In order to maintain electrical neutrality of both the compartments the Cl^- and Zn^{2+} move across the porous partition. For the movement of each Zn^{2+} ion from anodic compartment to cathodic compartment there is a concurrent migration of two Cl^- ions from cathodic compartment to anodic compartment. In the present example addition of NaCl is only to facilitate easy movement of ions before sufficient concentration of ions is built up with the course of time. A cell operates only so long as the chemical reaction, on which the cell is based, proceed. The electricity ceases as soon as one of the reactant is exhausted or the reaction reaches equilibrium.

Electric current flows from negative to positive electrode through external circuit, but long standing convention defines that current flows from positive to negative electrode (taking the medium into consideration). The passage of electrons from one electrode to the other suggests the presence of a potential difference between them. *The difference of potential, which causes the current to flow from an electrode at higher potential to the one of lower potential is called the electromotive force (emf) of the cell.*

5.7.1 Single Electrode Potential

The foregone conclusions *suggest the presence of two half cell reactions and the emf of a cell gives only the difference in the driving forces of the electrodes*. In the Daniell cell we can find more rational answer for the question of why copper does not displace zinc ion from the solution by understanding more about this driving force. When the cell reaction takes place, what we actually observe as **emf** or **cell potential** or **cell voltage** is a kind of tug-of-war between these driving forces. The species with greater tendency to reduce acquires an electron supplied by the less reducing species that gets oxidized. In the zinc-copper cell, therefore, copper ion must have a larger reducing nature or **reduction potential** than zinc. Thus;

$$E_{cell} = \textbf{Reduction potential of reduced species} - \textbf{Reduction potential of oxidized species}$$

or E_{cell} = **Reduction potential of reduced species** + **Oxidation potential of oxidized species**

or in short

$$E_{cell} = E_{cathode} - E_{anode} \quad .....(5.19)$$

(Remember that when no specific mention is made the electrode potential is always the reduction potential)

or

$$E_{cell} = E_{right} - E_{left} \quad .....(5.20)$$

In the Daniell cell, therefore;

$$E^0_{cell} = E^0_{Cu^{2+}|Cu} - E^0_{Zn^{2+}|Zn} \quad .....(5.21)$$

where $E^0_{Cu^{2+}|Cu}$ is the standard reduction potential for copper ion and $E^0_{Zn^{2+}|Zn}$ is the standard reduction potential for zinc ion.

Since $E^0_{Cu^{2+}|Cu}$ is larger than $E^0_{Zn^{2+}|Zn}$, E^0_{cell} should be positive for a spontaneous reaction. Here the prefix standard implies experimental conditions where the **half-cell potential** or **single electrode potential** are measured at 25°C, maintaining unit concentration of reversible ions. The **standard cell potential,** E^0_{cell}, represents the magnitude of difference between reduction potentials of two half cells of a redox reaction or sum of the standard oxidation potential and standard reduction potential of respective electrodes. *The emf is an intensive property like temperature and therefore 'independent' of the volume of the solution and size of the electrodes.*

Knowing the standard electrode potential of one of the electrode, it is possible to determine electrode potentials of all the other electrodes by suitably coupling and measuring the cell voltage. Unfortunately, even to start with, it is impossible to determine the absolute values of a single electrode potential. So arbitrarily the emf of a standard hydrogen electrode is chosen as zero and relative to this standard electrode, electrode potentials of various electrodes were determined. A standard hydrogen electrode consists of a platinum foil encased in a glass sleeve through which hydrogen gas is passed at a pressure of 1atm and is placed in one molar acid solution, Fig. 5.3. Platinum foil is coated with a layer of platinum black. In addition to providing a conducting surface, the platinum electrode adsorbs molecular hydrogen and helps in maintaining the following equilibrium.

$$2H^+(aq) + 2e^- \rightleftharpoons H_2(g)$$

Table 5.5 Standard Reduction Potentials at 25 °C*

Oxidising power increases ↓

Electrode	Electrode Reaction	E^0(V)
$Li^+ \mid Li$	$Li^+(aq) + e^- \longrightarrow Li(s)$	– 3.045
$K^+ \mid K$	$K^+(aq) + e^- \longrightarrow K(s)$	– 2.925
$Ba^{2+} \mid Ba$	$Ba^{2+}(aq) + 2e^- \longrightarrow Ba(S)$	– 2.90
$Sr^{2+} \mid Sr$	$Sr^{2+}(aq) + 2e^- \longrightarrow Sr(s)$	– 2.89
$Ca^{2+} \mid Ca$	$Ca^{2+}(aq) + 2e^- \longrightarrow Ca(s)$	– 2.87
$Na^{2+} \mid Na$	$Na^{2+}(aq) + e^- \longrightarrow Na(s)$	– 2.714
$Mg^{2+} \mid Mg$	$Mg^{2+}(aq) + 2e^- \longrightarrow Mg(s)$	– 2.37
$Be^{2+} \mid Be$	$Be^{2+}(aq) + 2e \longrightarrow Be(s)$	– 1.85
$Al^{3+} \mid Al$	$Al^{3+}(aq) + 3e^- \longrightarrow Al(s)$	– 1.66
$Mn^{2+} \mid Mn$	$Mn^{2+}(aq) + 2e^- \longrightarrow Mn(s)$	– 1.18
$OH^- \mid H_2(g) \mid Pt$	$H_2O + e^- \longrightarrow H_2(s) + OH^-(aq)$	– 0.828
$Zn^{2+} \mid Zn$	$Zn^{2+}(aq) + 2e^- \longrightarrow Zn(s)$	– 0.763
$Cr^{3+} \mid Cr$	$Cr^{3+}(aq) + 3e^- \longrightarrow Cr(s)$	– 0.74
$Fe^{2+} \mid Fe$	$Fe^{2+}(aq) + 2e \longrightarrow Fe(s)$	– 0.440
$Co^{2+} \mid Co$	$Co^{3+}(aq) + 2e^- \longrightarrow Cd(s)$	– 0.408
$SO_4^{2-} \mid PbSO_4(s) \mid Pb$	$PbSO_4 + 2e^- \longrightarrow Pb(s) + SO_4^{2-}(aq)$	– 0.31
$Co^{2+} \mid Co$	$Co^{2+}(aq) + 2e^- \longrightarrow Co(s)$	– 0.28
$Ni^{2+} \mid Ni$	$Ni^{2+}(aq) + 2e^- \longrightarrow Ni(s)$	– 0.25
$Sn^{2+} \mid Sn$	$Sn^{2+}(aq) + 2e^- \longrightarrow Sn(s)$	– 0.14
$Pb^{2+} \mid Pb$	$Pb^{2+}(aq) + 2e^- \longrightarrow Pb(s)$	– 0.126
$H^+ \mid H_2(g) \mid Pt$	$2H^+(aq) + 2e^- \longrightarrow H_2(s)$	0.000
$Sn^{2+}, Sn^{4+} \mid Pt$	$Sn^{4+}(aq) + 2e^- \longrightarrow Sn^2(s)$	+ 0.13

Table 5.5 Contd...

Oxidising power increases ↓

Electrode	Electrode Reaction	E^0(V)
$Cu^{2+} \mid Cu$	$Cu^{2+}(aq) + 2e^- \longrightarrow Cu^{+}(s)$	+ 0.15
$Cl^- \mid AgCl(s) \mid Ag$	$AgCl(s) + 2e^- \longrightarrow Ag(s) + Cl^-(aq)$	+ 0.2223
$Cl^- \mid Hg_2Cl_2(s) \mid Hg$ $	$Hg_2Cl_2 + 2e^- \longrightarrow 2Hg(s) + 2Cl^-(aq)$	+ 0.268
$Cu^{2+} \mid Cu$	$Cu^{2+}(aq) + 2e^- \longrightarrow Cu(s)$	+ 0.337
$OH^- \mid O_2(g)\ Pt$	$O_2(g) + 2H_2O + 4e^- \longrightarrow 4OH^-(aq)$	+ 0.401
$Cu^+ \mid Cu$	$Cu^+(aq) + 2e^- \longrightarrow Cu(s)$	+ 0.521
$I^-, I_2 \mid Pt$	$I_2(s) + 2e^- \longrightarrow 2I^-(aq)$	+ 0.5355
$MnO_4, OH^- \mid MnO_2 \mid Pt$	$MnO_4(aq) + 2H_2O + 3e^- \longrightarrow MnO_2(s) + 4OH^-(aq)$	+ 0.59
$Fe^{2+}, Fe^{3+} \mid Pt$	$Fe^{3+}(aq) + e^- \longrightarrow Fe^{2+}(aq)$	+ 0.771
$Ag^+ \mid Ag$	$Ag^+(aq) + e^- \longrightarrow Ag(s)$	+ 0.7991
$Hg_2^{2+} \mid Hg \mid Pt$	$Hg_2^{2+}(aq) + 2e^- \longrightarrow 2Hg(l)$	+ 0.85
$Hg_2^{2+} \mid Hg^{2+} \mid Pt$	$2Hg^{2+}(aq) + 2e^- \longrightarrow Hg_2^{2+}(aq)$	+ 0.920
$Br^-, Br_2 \mid Pt$	$Br_2(l) + 2e^- \longrightarrow 2Br^-(aq)$	+ 1.0652
$H^+ \mid O_2 \mid Pt$	$O_2(g) + 4H^+(aq) + 4e^- \longrightarrow 2H_2O$	+ 1.229
$Cr_2O_7^{2-}, H^+, Cr^{3+} \mid Pt$	$Cr_2O_2(aq) + 14H^+(aq) + 6e^- \longrightarrow 2Cr^{1+}(aq) + 7H_2O$	+ 1.33
$Cl^- \mid Cl_2 \mid Pt$	$Cl_2(g) + 2e^- \longrightarrow Cl^-(aq)$	+ 1.3595
$Au^{3+} \mid Au$	$Au^{3+}(aq) + 3e^- \longrightarrow Au(s)$	+ 1.50
$MnO_4^-, Mn^{2+} \mid H^+ \mid Pt$	$MnO_4^-(aq) + 8H^+(aq) + 5e^- \longrightarrow Mn^{2+}(aq) + 4H_2O$	+ 1.51
$Ce^{3+}, Ce^{4+} \mid Pt$	$Ce^{4+}(aq) + e^- \longrightarrow Ce^{3+}(aq)$	+ 1.61
$H^+ \mid H_2O_2 \mid Pt$	$Ce^{3+}(aq) + 2H^-(aq) + 2e^- \longrightarrow 2H_2O$	+ 1.77
$Co^{2+}, Co^{3+} \mid Pt$	$H_2O_2(aq) + e^- \longrightarrow Co^{2+}(aq)$	+ 1.82
$H^+ \mid O_3, O_2 \mid Pt$	$O_3(g) + 2H^+(aq) + 2e^- \longrightarrow O_2(g) + H_2O(l)$	+ 2.07
$F^- \mid F_2 \mid Pt$	$F_2(g) + 2e^- \longrightarrow 2F^-(aq)$	+ 2.87

* All the ions are at unit activity and gases at 1 atm pressure.

$ *EMF* of the calomel electrode containing saturated KCl is 0.2415 V and with 1 N and 0.1 N KCl solutions are 0.280 and 0.334 V respectively.

When hydrogen electrode is paired with another electrode (half-cell) in a galvanic cell, it can undergo either oxidation or reduction depending on the magnitude of potential of coupled electrode relative to hydrogen electrode. The cell *emf* so determined, directly gives the single electrode potential of the coupled electrode. Some of the half cell reactions and corresponding standard reduction potentials are listed in Table 5.5.

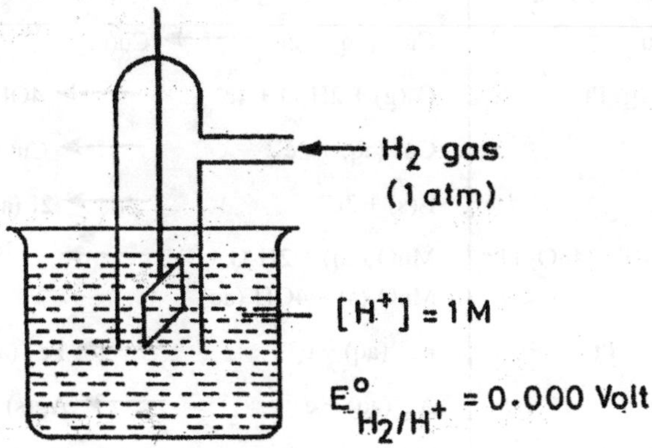

Fig. 5.9 Hydrogen gas electrode.

The standard reduction potential table serves many useful purposes.

1. From these tables, at a glance, *we can make out the good reducing agents and good oxidizing agents*. Thus fluorine tops the list of oxidizing agents as it has highest standard reduction potential. From a higher positive reduction potential we can understand that H^+ is a much powerful oxidizing than Zn^{2+}. Similarly lithium metal may be considered as best reducing agent.

2. A knowledge of reduction potentials also helps in deciding whether a given redox reaction will proceed spontaneously in the forward direction or not. Suppose if we consider the Zn-Cu cell reaction.

$$Zn + Cu^{2+} \rightleftharpoons Cu + Zn^{2+}$$

We have the two half cell reactions

Oxidation (anode) $\quad Zn \longrightarrow Zn^{2+} + 2e^- \quad E^0_{Zn^{2+}|Zn} = 0.76\ V$

Reduction (cathode) $\quad Cu^{2+} + 2e^- \longrightarrow Cu \quad E^0_{Cu^{2+}|Cu} = 0.34\ V$

$$E^0_{cell} = E^0_{Zn^{2+}|Zn} - E^0_{Cu^{2+}|Cu}$$

or

$$E^0_{cell} = E^0_{oxidation} + E^0_{reduction}$$

$$E^0 = 0.76 + 0.34$$

$$= 1.10\ V$$

The positive E^0_{cell} value suggest that the redox reaction is spontaneous in the forward direction.

Let us now test the spontaneity of the reaction

$$Fe^{2+}(aq) + Ni(s) \longrightarrow Fe(s) + Ni^{2+}(aq)$$

The two half cell reactions are

$$Fe^{2+} + 2e \longrightarrow Fe \qquad E^0_{Fe^{2+}|Fe} = -0.44V$$

$$Ni \longrightarrow Ni^{2+} + 2^{e-} \qquad E^0_{Ni^{2+}|Ni} = -0.25\ V$$

$$E^0_{cell} = E^0_{Fe^{2+}|Fe} - E^0_{Ni^{2+}|Ni}$$

$$= -0.44 - (-0.25)$$

$$= -0.19V$$

Because E^0_{cell} computed is negative, we draw the conclusion that this reaction is not spontaneous under standard conditions.

3. *The standard electrode potential data is highly helpful in deciding the course of electrolysis reactions*. From the data given we see that the reduction potential for copper is 0.39V, for H_2O -0.83V and for aluminium –1.66V. During electrolysis of a Cu^{2+} solution, copper ions are more readily reduced (potential being more positive) than water and will plate on the electrode according to the half reaction.

$$Cu^{2+}(aq) + 2e^- \longrightarrow Cu(s)$$

In case of aluminium, due to more positive reduction potential (or less negative potential) of H_2O than aluminium, H_2O is preferentially reduced and no deposition of aluminium is seen.

$$2H_2O + 2e^- \longrightarrow H_2 + 2OH^-$$

$$2Al^{3+} + 6OH^- \longrightarrow 2Al(OH)_3$$

5.7.2 Representation of a Cell

Sketching an electrochemical cell as in Fig. 5.8 is a tedious process. A symbolic representation is often used to describe a cell. This representation is called a **cell diagram**. Following rules are adopted in these representations. For example, a Daniell cell is represented as;

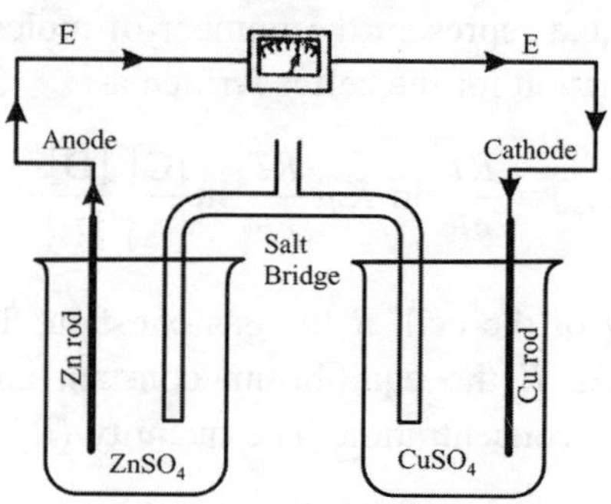

Anode   Salt bridge   Cathode

$$Zn(s)\ |\ Zn^{2+}(0.1M)\ ||\ Cu^{2+}(0.1M)\ \ |\ Cu(s)$$

Left Half cell   Right Half cell

(i) The anode, where oxidation occurs is represented at the left and at cathode, where reduction takes place, is represented on the right of the cell diagram.

This order can be easily remembered by recollecting the Hindi phrase "**LAO SaR CaR**" or simply remembering the south asian country **LAOS**. Here the letters LAOS, stands for **L**eft **A**node **O**xidation **S**alt bridge.

In the Daniell cell Zn | Zn^{2+} half cell is an oxidizing electrode and is written on the left.

(ii) A single vertical line represents the boundary between different phases that constitute an electrode.

(iii) A double vertical line indicates that the liquid junction potential has been decreased to minimum by suitable means like using a salt bridge.

(iv) The concentration of the dissolved substances are indicated in the parenthesis.

(v) In case of gases at equilibrium it is necessary to represent the pressure of the gas.

$$\text{Pt} \mid H_2(1\text{ atm}) \mid H^+ (0.1\text{ M}) \parallel Cu^{2+} (0.1\text{ M}) \mid Cu(s)$$

(vi) When two oxidation states of the same species are present in a homogenous medium, they are usually separated by a comma.

$$\text{Pt} \mid H_2 (1\text{ atm}) \mid H^+ (0.1\text{ M}) \parallel Fe^{3+} (0.2\text{ M}), Fe^{2+} (0.1\text{ M}) \mid \text{Pt}$$

5.8 Nernst Equation Relationship between Cell reaction and EMF

Nernst (1864-1941) studied the theoretical relationship between the electrochemical reaction and the corresponding cell emf. This relationship is generally known as **Nernst equation**.

Let us consider a galvanic cell whose cell reaction is written as

$$a\text{A} + b\text{B} \longrightarrow c\,\text{C} + d\,\text{D}$$

where lower case letter a,b,c,d represent the number of moles of A, B, C, D respectively at equilibrium. The Nernst equation for the cell is written as;

$$E_{cell} = \frac{RT}{nF}\ln K - \frac{RT}{nF}\ln\frac{[\text{C}]^c[\text{D}]^d}{[\text{A}]^a[\text{B}]^b}$$

where E_{cell} denotes the *emf* of the cell, *R* the gas constant, T the temperature, n number of faraday of current *F* passed, *K* the equilibrium constant and the quantities in the square brackets refer the respective concentration. The quantity (*RT/nF*) ln *K* is known as standard emf of the cell $E^{\text{o}}_{\text{cell}}$.

$$E_{\text{cell}} = E^0_{\text{cell}} - \frac{RT}{nF}\ln\frac{[\text{C}]^c[\text{D}]^d}{[\text{A}]^a[\text{B}]^b} \quad \text{.....(5.22)}$$

or $$E_{cell} = E^0_{cell} - \frac{2.303\,RT}{nF} \log \frac{[C]^c [D]^d}{[A]^a [B]^b} \quad .....(5.23)$$

At room temperature (T = 298 K), taking R = 8.314J K^{-1}, F= 96457 C, the above equation simplifies to

$$E_{cell} = E^o_{cell} - \frac{0.05916}{n} \log \frac{[C]^c [D]^d}{[A]^a [B]^b} \text{ at } 25\ ^oC \quad .....(5.24)$$

From the equation (5.24) we notice that the standard cell *emf* is equal to the cell *emf* when the activities of both reactants and products is equal to unity.

5.9 Types of Electrodes and Nernst Equation

Nernst equation is also applicable to half cell reactions. Few half cell reactions and corresponding Nernst's expressions are discussed below.

5.9.1 Metal/Metal ion electrode – Zn | Zn^{2+} electrode

This a typical representative example of *Metal/Metal ion electrode.* This electrode can be constructed by bringing a zinc rod in contact with zinc ions. Oxidation of zinc atoms to give zinc ions is expressed by the following equation;

$$Zn \longrightarrow Zn^{2+} + 2e^-$$

In the reaction two electrons are exchanged between both the zinc species in the overall redox reaction making n = 2. Therefore,

$$E_{Zn|Zn^{2+}} = E^0_{Zn|Zn^{2+}} - \frac{0.05916}{2} \log \frac{a_{Zn^{2+}}}{a_{Zn}}$$

Since the activity of all the pure solids is unity (a = 1)
Therefore,

$$E_{Zn|Zn^{2+}} = E^0_{Zn|Zn^{2+}} - \frac{0.05916}{2} \log a_{Zn^{2+}}$$

This expression suggests that the oxidation electrode potential, E, decreases with increase 1n $\left[a_{Zn^{2+}}\right]$ and equals to standard electrode potential, E^o, at unit concentration of the electrolyte.

5.9.2 Gas Electrode-Hydrogen electrode

The electrode reaction of this electrode is written as;

$$\tfrac{1}{2} H_2 \longrightarrow H^+ + e^-$$

Applying Nernst equation

$$E_{H_2|H^+} = E^0_{H_2|H^+} - \frac{0.0592}{1} \log \frac{a_{H^+}}{p^{1/2}_{H_2}}$$

at 1 atm pressure

$$E_{H_2|H^+} = E^0_{H_2|H^+} - 0.0592 \log a_{H^+}$$

$$E_{H_2|H^+} = E^0_{H_2|H^+} + 0.0592 \text{ pH} \qquad \text{.....(5.23)}$$

Advantages of Hydrogen Gas Electrode : It gives satisfactory results in the entire pH range of 0-14. The potential and pH are free from salt errors.

Limitations : The hydrogen gas used must be sufficiently pure. The electrode can not be used in the presence of gas impurities like air and oxygen and electrolyte impurities like oxidizing and reducing agents. It is very susceptible to poisoning by cyanides, arsenic, antimony compounds and alkaloids. It is cumbersome to handle in a normal laboratory.

5.9.3 Redox electrode - Quinhydrone Electrode

Platinum rod dipped in a mixture of two different oxidation states of a metal or two redox states of any compound is called a **redox electrode**. Platinum wire immersed in a mixture of ferrous and ferric ion mixture is a typical example of this class.

Electrode reaction $\quad Fe^{2+} \longrightarrow Fe^{3+} + e^-$

$$a_{Fe^{2+}} \qquad\qquad a_{Fe^{3+}}$$

The electrode potential at any $[Fe^{2+}]/[Fe^{3+}]$ ratio is given by the expression,

$$E_{Pt|Fe^{2+}, Fe^{3+}} = E^0_{Pt|Fe^{2+}, Fe^{3+}} - \frac{0.0592}{1} \log \frac{a_{Fe^{3+}}}{a_{Fe^{2+}}}$$

Quinhydrone electrode : This is a redox electrode reversible to protons and often replaces the hydrogen electrode. Quinhydrone is a 1:1 molar mixture of quinone and hydroquinone. The electrode consists of a shiny platinum electrode dipped in a acid / base test solution, which is saturated with quinhydrone. The electrode reaction is given by

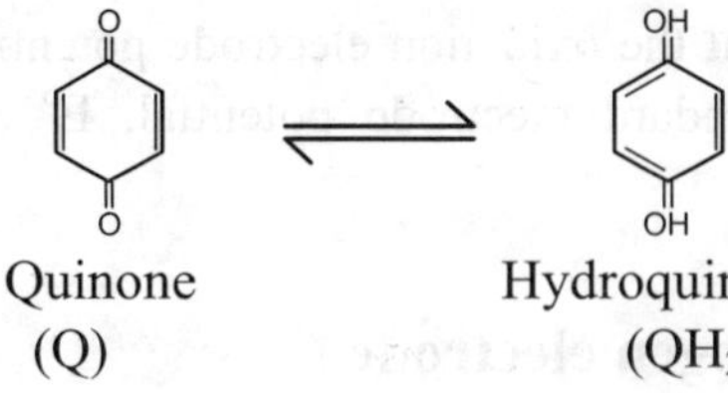

Quinone (Q) $\qquad$ Hydroquinone (QH_2)

The electrode potential at 25°C is given by,

$$E_{Pt|Q,H^+, QH_2} = E^0_{Pt|Q,H^+, QH_2} - \frac{0.0592}{2} \log \frac{a_{QH_2}}{a_Q\, a^2_{H^+}}$$

Since quinone and hydroquinone are taken in equimolar amounts, i.e., $a_Q = a_{QH_2}$

We have

$$E = E^0 - \frac{0.0592}{2} \log \frac{1}{a^2_{H^+}}$$

$$E = E^0 + 2 \times \frac{0.0592}{2} \log a_{H^+}$$

$$E = E^0 + 0.0592 \log a_{H^+}$$

$$E_{Q,QH_2} = E^0_{Q,QH_2} - 0.0592 \text{ pH}$$

Quinhsydrone electrode can thus be used to measure pH of a solution. Due to instability of quinone in strong alkaline medium, this hydrogen ion indicator electrode is suitable only upto a pH of 8.

Advantages : The quinhydrone electrode is simple to set up and needs no removal of air. The reversibility equilibrium is achieved faster than hydrogen gas electrode thereby allowing a quicker measurement. pH values of solutions containing reducible substances like Cu^{2+}, Cd^{2+}, unsaturated acids, NO_3^-, etc., and catalytic poisons can be measured using quinhydrone electrode, where hydrogen electrode can not be used.

Limitations : The electrode can not be used at pH values greater than 8. Even this electrode fails in the presence of strong oxidizing and reducing agents.

4. **Metal / Metal insoluble Salt electrode – Calomel Electrode:** This electrode has Hg in contact with its insoluble salt Hg_2Cl_2 and chloride ions. Construction, Electrode reaction and Nernst equation of calomel electrode is discussed below.

5.10 Reference Electrodes

Although all the electrode potentials are measured relative to hydrogen electrode, whose potential is arbitrarily taken as zero, it is not always convenient to use this gas electrode in day to day potentiometric measurements. The main difficulties with this **primary reference electrode** is;

(a) Inconvenience in handling gases

(b) Maintenance of accurate pressures throughout the measurements

(c) Availability of sufficiently pure hydrogen gas.

(d) Necessity to platinize the platinum electrode with a solution of chloro platinic acid quite often.

For these reasons some **secondary reference electrodes** like silver chloride electrode and calomel electrode whose standard potentials are very accurately determined against a hydrogen electrode are widely used.

5.10.1 Standard Calomel Electrode - SCE

Calomel electrode is particularly very simple to construct, free from surface sensitivity and accurate to use even in a very normal laboratory.

A compact form of calomel electrode, as supplied by many commercial manufacturers is shown in Fig. 5.10.

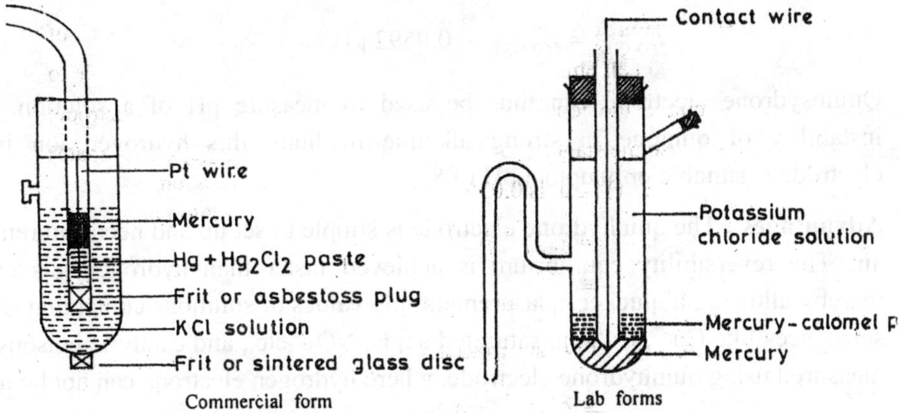

Fig. 5.10 The calomel electrode.

The calomel electrode consists of an inner glass tube and an outer jacket. In the inner glass tube a platinum wire is dipped into mercury which rests on a paste of mercurous chloride, Hg_2Cl_2 (commercially known as calomel) and mercury. This paste is in contact with KCl present in the outer jacket, through the glass frit plug fixed at the bottom of inner glass tube. The calomel electrode comes in contact with the experimental solution through a frit arranged to the outer jacket. The potential of this electrode depends on the concentration of KCl taken in the outer jacket. Some of the most popularly used concentrations of KCl and corresponding single electrode (reduction) potentials on the hydrogen scale at 25 °C are given below;

0.1M KCl | Hg_2Cl_2 (s) | Hg, Pt	0.3338V
1.0M KCl | Hg_2Cl_2 (s) | Hg, Pt	0.2800V
Saturated KCl | Hg_2Cl_2 (s) | Hg, Pt	0.2415V

The corresponding electrode reaction is;

$$\tfrac{1}{2}\,Hg_2Cl_2 + 2e^- \longrightarrow 2Hg + 2Cl^-$$

and corresponding Nernst's expression is;

$$E_{Hg_2Cl_2|Cl^-} = E^0_{Hg_2Cl_2|Cl^-} - \frac{2.303}{F}\log a^2_{Cl^-}$$

5.11 Ion Selective Electrodes – Glass electrode

Credit for the first glass sensing pH electrode is given to Cremer, who first described it in his paper published in 1906. Later several groups contributed for development of different ion selective electrodes An **Ion-selective electrode (ISE)** (also known as a specific ion electrode, or SIE) is *a sensor which converts the activity of a specific ion dissolved in a solution into an electrical potential which can be measured by some potentiometric devise like a voltmeter or pH meter.* As we know, the emf is theoretically dependent on the logarithm of the ionic activity (concentration), in accordance with the Nernst equation. Basically a concentration cell is developed with respect to the ion under observation. The sensing part of the electrode is usually made as an ion-specific membrane which is coupled with a reference electrode. So we need to have different ISEs for different ions.

The Electrochemical Circuit for an Ion Selective Electrode measurement

An ISE (with its own internal reference electrode - more details later) is immersed in an aqueous solution containing the ions to be measured, together with a separate, external reference electrode. The electrochemical circuit is completed by connecting the electrodes to a sensitive milli-volt meter using special low-noise cables and connectors. A potential difference is developed across the ISE membrane when the target ions diffuse through from the high concentration side to the lower concentration side (Fig. 5.11).

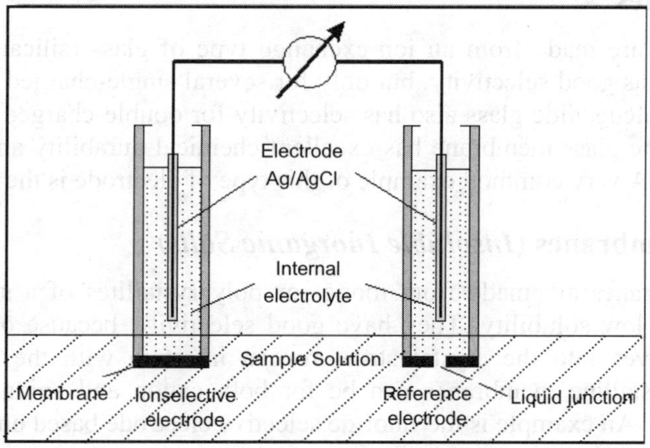

Fig. 5.11 Ion Selective Electrode (ISE) Measurement set-up.

Working of ISE

Ion Selective Electrodes (including the most common pH electrode) work on the basic principle of the galvanic cell. By measuring the electric potential generated across a membrane by "selected" ions, and comparing it to a reference electrode, a net charge is determined. The strength of this charge is directly proportional to the concentration of the selected ion. The basic formula is given for the galvanic cell:

$$E_{cell} = E_{ise} - E_{ref}$$

where the potential for the cell is equivalent to the potential of the ISE *minus* the potential of the reference electrode.

General principle of ISE analysis

At equilibrium, the membrane potential is mainly dependent on the concentration of the target ion outside the membrane and is described by the Nernst equation . Briefly, the measured voltage is proportional to the logarithm of the concentration, and the sensitivity of the electrode is expressed as the electrode Slope - in millivolts per decade of concentration. Thus the electrodes can be calibrated by measuring the voltage in solutions containing, for example, 10ppm and 100ppm of the target ion, and the Slope will be the slope of the (straight) calibration line drawn on a graph of *mV* versus Log (*concentration)*.

Unknown samples can then be determined by measuring the voltage and plotting the result on the calibration graph or by automatically converting by a devise.

Types of ion-selective membranes

There are four main types of ion-selective membranes used in ion-selective electrodes.

Glass membranes

Glass membranes are made from an ion-exchange type of glass (silicate of chalcogenide). This type of ISE has good selectivity, but only for several single-charged cations; mainly H^+, Na^+, and Ag^+. Chalcogenide glass also has selectivity for double-charged metal ions, such as Pb^{2+}, and Cd^{2+}. The glass membrane has excellent chemical durability and can work in very aggressive media. A very common example of this type of electrode is the pH glass electrode.

Crystalline membranes (*Insoluble Inorganic Salts)*

Crystalline membranes are made from mono- or polycrystallites of a single substances of inorganic salts of low solubility. They have good selectivity, because only ions which can introduce themselves into the crystal structure can interfere with the electrode response. Selectivity of crystalline membranes can be for both cation and anion of the membrane-forming substance. An example is the fluoride selective electrode based on LaF_3 crystals.

Ion-exchange resin membranes

Ion-exchange resins are based on special organic polymer membranes which contain a specific ion-exchange substance (resin). They selectively bind certain small ions from the aqueous phase and exchange them across the membrane creating the potential. Usage of specific resins allows preparation of selective electrodes for tens of different ions, both single-atom or multi-atom. They are also the most widespread electrodes with anionic selectivity. However, such electrodes have low chemical and physical durability as well as "survival time". An example is the potassium selective electrode, based on valinomycin as an ion-exchange agent.

Gas Permeable membranes

This type of membrane is used to measure dissolved gasses such as Ammonia, Carbon Dioxide, Nitrogen Oxide and Dissolved Oxygen. Gas molecules diffuse across the membrane and react with a buffer solution, changing the pH of the buffer. This pH change is measured using an internal glass electrode. This is probably the second most popular probe type after the glass membrane pH probe.

Ion-selective electrodes are used in biochemical and biophysical research, where measurements of ionic concentration in an aqueous solution are required.

Interferences

The most serious problem limiting use of ion-selective electrodes is interference from other, undesired, ions.

- No ion-selective electrodes are completely ion-specific; all are sensitive to other ions having similar physical properties, to an extent which depends on the degree of similarity.
- In some cases the electrode may actually be much more sensitive to the interfering ion than to the desired ion. This requires the interfering ion be present only in relatively very low concentrations, or entirely absent.
- The nitrate electrode is plagued by ionic interferences from chloride, fluoride, and sulfate.

Glass Electrode

Most often used pH electrodes are called glass electrodes and belong to the family of ISEs. They are sensitive only to H^+ ions. Typical glass electrode is made of glass tube ended with small glass bubble sensitive to protons. Inside of the electrode is usually filled with buffered solution of chlorides in which silver wire covered with silver chloride is immersed. pH of internal solution varies - for example it can be 1.0 (0.1M HCl) or 7.0 (Fig. 5.12).

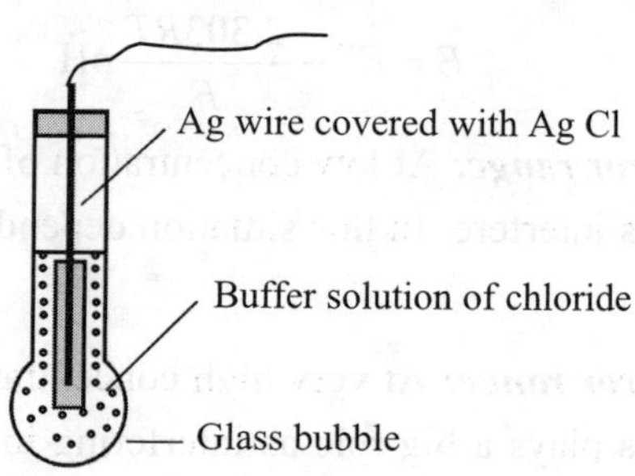

Fig. 5.12 Glass electrode.

Active part of the electrode is the glass bubble. While tube has strong and thick walls, bubble is made to be as thin as possible. Surface of the glass is protonated by both internal and external solution till equilibrium is achieved. Both sides of the glass are charged by the adsorbed protons, this charge is responsible for potential difference. This potential in turn is described by the Nernst equation and is directly proportional to the pH difference between solutions on both sides of the glass.

The majority of pH electrodes available now a days are combination electrodes that have both glass H+ ion sensitive electrode and reference electrode compartments, conveniently placed in one housing.

Range of a pH glass electrode

The pH range at constant concentration can be divided into 3 parts (Fig. 5.13):

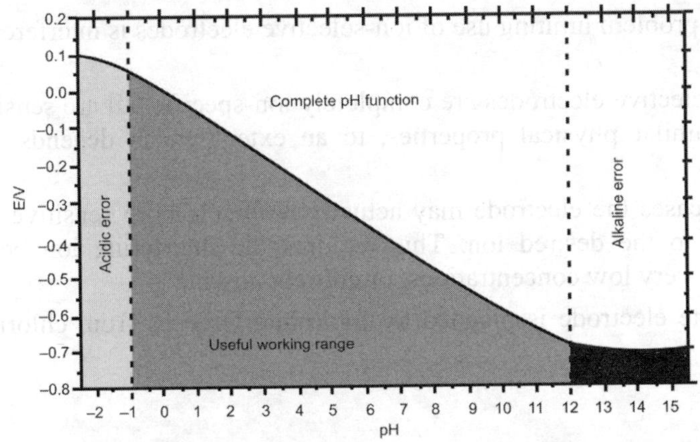

Fig. 5.13 E-pH for ion-selective electrode.

- ***Useful Working Range:*** Dependence of potential on pH has linear behavior and within which such electrode really works as ion-selective electrode for pH and obeys Nernst equation.

$$E = E^o - \frac{2.303RT}{F}\text{pH}$$

- ***Alkali error range:*** At low concentration of hydrogen-ions (high values of pH) alkali metal ions interfere. In this situation dependence of the potential on pH become non-linear.

- ***Acidic error range:*** At very high concentration of hydrogen-ions (low values of pH) the anions plays a big role as interfering ions, in addition to destruction of the glass matrix used for making glass bulb.

Almost all commonly used glass electrodes have a working pH range from pH = 1 till pH = 12. So specially designed electrodes should be used only for working in aggressive conditions.

5.12 Concentrations Cells

A **Concentration cell** is an electrochemical cell that has two equivalent half-cells of the same material *differing only in concentrations.* One can calculate the potential developed by such a cell using the Nernst Equation similar to an electrochemical cell. In concentration cell the voltage comes from its attempt to reach equilibrium, which will stop when the concentration in both half-cells are equal.

Types of concentration cells: The concentration differences could be affected in the electrode material or in the electrolyte. Further they could be subdivided into cells with transference or without transference.

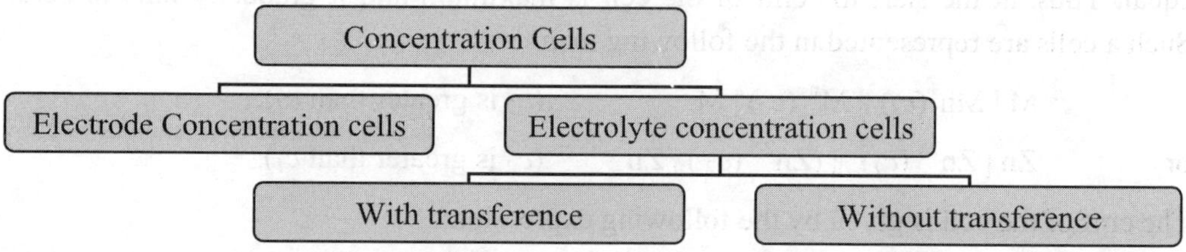

(i) ***Electrode concentration cells:***

In these cells, the potential difference is developed between two like electrodes at different concentrations dipped in the same solution of the electrolyte. For example, two hydrogen electrodes at different gas pressure in the same solution of hydrogen ions constitute a cell of this type.

$$Pt|H_2 \text{ (Pressure } p_1) \,|H^+ (a)|\, (H_2 \text{ (Pressure } p_2) \,|Pt$$

If $p_1 > p_2$ oxidation occurs at L.H.S. electrode and reduction occurs at R.H.S. electrode.

The Nernst equation can be used to derive an expression for the potential of this electrode concentration cell.

$$H_2 (p_1) = 2H^+ + 2e^-$$

$$2H^+ + 2e^- = H_2 (p_2)$$

$$H_2 (p_1) = H_2 (p_2)$$

$$E_{cell} = 0.0591/2 \log(p_1/p_2) \quad \text{at } 25^0C$$

The standard cell potential is zero because a cell cannot derive a current through a circuit with identical electrodes. Similarly we can construct different gas concentration cells, when coupled with their counter ions.

Two amalgams of the same metal at two different concentrations in the same electrolyte solution also form this class of cells.

$$\text{Na-Hg (activity } a_1) | Na^+ | \text{Na-Hg (activity } a_2)$$

$$\text{Zn-Hg (activity } a_1) | Zn^{2+} | \text{Zn-Hg (activity } a_2)$$

(ii) *Electrolyte concentration cells:*

In these cells, electrodes are identical but these are immersed in solutions of the same electrolyte *of different concentrations*. The source of electrical energy in the cell is the tendency of the electrolyte to diffuse from a solution of higher concentration to that of lower concentration. With the expiry of time, the two concentrations tend to become equal. Thus, at the start the emf of the cell is maximum and it gradually falls to zero. Such a cells are represented in the following manner:

$$M \mid Mn^+(c_1) \parallel M^{n+}(c_2) \mid M \qquad (c_2 \text{ is greater than } c_1).$$

or

$$Zn \mid Zn^{2+}(c_1) \parallel (Zn^{2+}(c_2)) \mid Zn \qquad (c_2 \text{ is greater than } c_1).$$

The emf of the cell is given by the following expression:

$$E_{cell} = 0.0591/n \log c_2/c_1 \text{ at } 25\ ^\circ C$$

The above examples are typical examples of electrolyte concentration cells with transference. Transference indicates the presence of salt bridge or liquid-liquid contact of electrolytes

Examples of electrolyte concentration cells without transference are difficult to comprehend at this stage and are not considered now.

Applications of Concentration cells:

1. The concentration cells are used to determine the solubility of sparingly soluble salts, valency of the cation of the electrolyte and transition point of the two allotropic forms of a metal used as electrodes, etc.
2. Concentration cell corrosion occurs when two or more areas of a metal surface are in contact with different concentrations of the same solution. There are three general types of concentration cell corrosion.
3. Water in contact with the metal surface will normally contain dissolved oxygen. An oxygen cell can develop at any point where the oxygen in the air is not allowed to diffuse uniformly into the solution, thereby creating a difference in oxygen concentration between two points. Corrosion will occur at the area of low-oxygen concentration which are anodic. This is the explanation for corrosion of iron piece when in contact of moisture.

5.13 Applications of Electrode Potentials – Potentiometric Titrations

(i) Determination of Ionic activities : From the knowledge of standard cell emf and standard electrode potentials, it is possible to calculate the ionic activities of ions reversible to the electrode. Suppose the standard emf of the cell Hg(s), Hg_2Cl_2(s) | KCl (satd) || Cu^{2+} (a=x) | Cu is 0.036V.

$$E_{cell} = E_{right} - E_{left}$$

$$E_{cell} = E_{Cu^{2+}|Cu} - E_{calomel}$$

$$E_{Cu^{2+}|Cu} = E_{cell} - E_{calomel}$$

$$= 0.036 + 0.242$$

$$E_{Cu^{2+}|Cu} = 0.278 \text{ V}$$

But
$$E_{Cu^{2+}|Cu} = E^0_{Cu^{2+}|Cu} - \frac{0.0592}{2} \log \frac{1}{a_{Cu^{2+}}}$$

$$0.278 = 0.337 + 0.0296 \log a_{Cu^{2+}}$$

or
$$\log a_{Cu^{2+}} = \frac{0.278 - 0.337}{0.0296}$$

$$a_{Cu^{2+}} = 0.01$$

Activity of Cu solution = 0.01

(ii) Potentiometric titrations : Potentiometric titrations is a very interesting application of electrode potentials. They involve a study of variation of *emf* with the volume of titrant added. Here we discuss the precipitation titration and acid base titration.

(a) Precipitation titrations : Consider the titration of Ag^+ ions with KCl. On addition of any chloride salt the silver ions fall out of the aqueous solution forming an insoluble precipitate of silver chloride. Hence we study the variation of silver electrode potential with the change in concentration of silver ions.

$$E_{Ag|Ag^+} = E^0_{Ag|Ag^+} - \frac{0.0592}{2} \log a_{Ag^+}$$

As the $E_{Ag|Ag^+}$ varies in a logarithmic scale of the concentration of silver ions, initially even for a hundred fold variation of silver ion concentration the *emf* variation is only 0.118V. With the approach of neutralization point, the E_{Ag^+} gets smaller and smaller, while the *emf*, a log value of Ag^+, increases rapidly.

Thus we notice a sigmoid curve with the steep portion indicating the neutralization point.

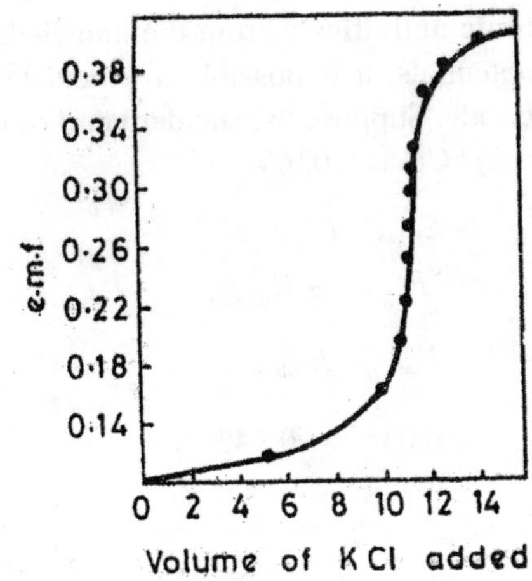

Fig. 5.14 Potentiometric titration of Ag^+ ions with KCl.

The Ag | Ag^+ electrode which is reversible to silver ion concentration is termed as the **indicator electrode.**

(b) **Acid-base titrations :** In case of acid - alkali titration a hydrogen electrode may be used as an indicator electrode. The suitable cell may be represented as

$$(Pt), Hg, Hg_2Cl_2\,(s) \mid KCl\,(satd) \parallel H^+ \mid H_2\,(1\ atm), (Pt)$$

The emf of indicator electrode is given as

$$E_{H_2|H^+} = E^0_{H_2|H^+} + 0.0595 \log\left[H^+\right]$$

Since the standard emf of hydrogen electrode is zero, we have;

$$E_{H_2|H^+} = -0.0595 \log\left[H^+\right]$$

$$E_{H_2|H^+} = -0.0595\ pH$$

From the above expression we draw the conclusion that the emf decreases with decrease in the concentration of hydrogen ions or increase in the pH of the solution. On plotting the electrode potential or cell potential against the volume of titrant added, a sigmoid curve is obtained, Fig.5.15. The volume corresponding to zero emf gives the titre value of the acid solution.

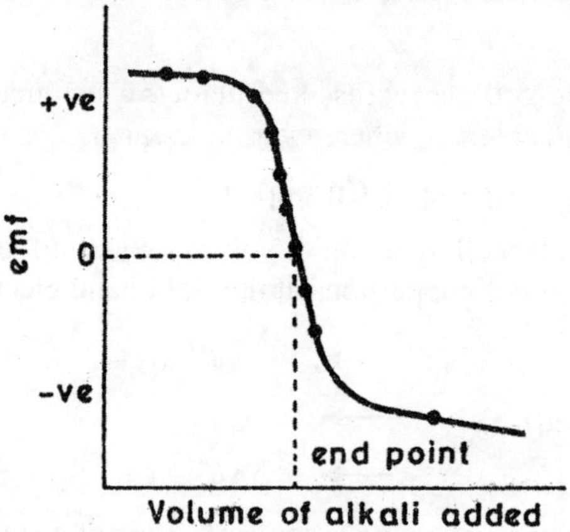

Fig. 5.15 Potentiometric acid-alkali titration curve.

In the laboratory titration generally the quinhydrone electrode is used in place of SHE.

(iii) Determination of pH : The pH of a solution could be determined by employing any electrode reversible to H^+ ions. The hydrogen electrode and quinhydrone electrode discussed above and glass electrode are popularly employed for the estimation of pH of a solution. The hydrogen indicator electrode could be coupled either to a standard hydrogen electrode (SHE) or to a standard calomel electrode (SCE). The cell diagram, when SHE is used as reference electrode, is given by

$$H_2\ (1\ \text{atm})\ |\ H^+\ (a{=}1)\ ||\ H^+\ (a{=}x)\ |\ H_2\ (1\ \text{atm})$$

The emf of the indicator electrode is given by

$$E_{H_2|H^+} = -\frac{RT}{F}\ \ln \frac{a_{H^+}}{p_{H_2}}$$

When $\quad p_{H_2} = 1$ atm and T = 298 K

Since by definition pH = –log a_{H^+}, the above expression in terms of pH is

$$E_{H_2|H^+} = 0.0592\ \text{pH}$$

Because the electrode potential of the reference SHE is zero, for the present cell diagram.

The cell diagram when SCE is used as reference electrode is given by

$$H_2\ |\ H^+\ (a{=}x)\ ||\ \text{Calomel (saturated)}$$

For this arrangement of room temperature

$$E_{H_2|H^+} = E_{cell} - E_{SCE} = 0.05916\ \text{pH}$$

Problem 17.

Verify whether the following cell, as represented below, can function as a voltaic cell (Use reduction potentials given in Table 5.5, where ever necessary).

$$Ag \mid Ag^+(aq) \parallel Cu^+(aq) \mid Cu$$

As per the convention the cell reaction involves oxidation of silver at the left hand electrode (anode) and reduction of copper ions at the right hand electrode (cathode).

Oxidation	Ag	$\longrightarrow$	$Ag^+(aq) + e^-$	$E^o_{oxid} = -0.799V$
Reduction	$Cu^+(aq) + e^-$	$\longrightarrow$	Cu	$E^o_{redn} = 0.52\ V$
Net	$Cu^+ + Ag$	$\longrightarrow$	$Ag^+ + Cu$	$E^o_{cell} = -0.279V$

Since the above cell diagram gives a negative emf, it is not a spontaneous reaction and the cell as diagrammed cannot function as a voltaic cell.

Problem 18.

Find the *emf* of the cell represented by the cell diagram $Zn \mid Zn^{2+} \mid H^+(1\ M) \mid H_2\,(1atm) \mid (Pt)$.

Given that the oxidation potential of $Zn \mid Zn^{2+}$ is 0.7618V

Emf of the cell

$$E_{cell} = \underset{(Oxid)}{E_{Zn|Zn^{2+}}} - \underset{(Oxid)}{E_{H_2|H^+}}$$

The hydrogen gas electrode is a standard hydrogen electrode, hence has a zero electrode potential.

Therefore $\quad E_{cell} = 0.7618 - 0 = 0.7618V$

Problem 19.

The standard oxidation potential of $Cd \mid Cd^{2+}$ electrode is 0.403V. Calculate the Cd^{2+} concentration if the electrode potential is 0.5V at 25 °C.

$$Cd \longrightarrow Cd^{2+} + 2e$$

$$E = E^0 - \frac{RT}{nF} \ln \frac{[Cd^2]}{[Cd]}$$

$$0.5 = 0.403 - \frac{2.303 \times 8.314 \times 298}{2 \times 96500} \log [Cd^{2+}]$$

$$0.097 = -\frac{0.05916}{2} \log\ [Cd^{2+}]$$

$$= -\,0.02958 \log [Cd^{2+}]$$

$$\log [Cd^{2+}] = -\frac{0.097}{0.02958} = -3.279$$

$$[Cd^{2+}] = 5.26 \times 10^{-4}$$

Therefore concentration of Cd^{2+} = 5.26×10^{-1} M

Problem 20.

Arrange the following species in order of their increasing oxidizing capability, MnO_4^- Li^+, Zn^{2+}, Ag^+, Co^{3+}, Ce^{4+}.

According to standard emf data.

$$MnO_4^- (aq) + 8H^+ (aq) + 3e^- \longrightarrow Mn^{2+} + 4H_2O \quad E^\circ = 1.51V$$

$$Li^+(aq) + e^- \longrightarrow Li(s) \quad E^\circ = -3.05V$$

$$Zn^{2+}(aq) + 2e^- \longrightarrow Zn(s) \quad E^\circ = -0.76\ V$$

$$Ag^+(aq) + e^- \longrightarrow Ag(s) \quad E^\circ = 0.8V$$

$$Co^{3+}(aq) + e^- \longrightarrow Co^{2+}(aq) \quad E^\circ = 1.82V$$

$$Ce^{4+}(aq) + e^- \longrightarrow Ce^{3+}(aq) \quad E^\circ = 1.61V$$

Higher the standard reduction potential higher is the oxidizing capability. Therefore the order of oxidizing strengths are;

$$Li^+ < Zn^{2+} < Ag^+ < MnO_4^- < Ce^{4+} < Co^{3+}$$

5.14 Battery Chemistry

When two or more electrochemical cells are electrically interconnected, each of which containing two electrodes and an electrolyte is called a Battery. In everyday usage, 'battery' is also used to refer to a single cell. No one battery design is perfect for every application. There are many parameters like cost, voltage, duty cycle, dimension, stability with time and temperature, shelf life, etc., on which a battery is selected for a particular operation. It is important to prioritize our list of requirements before choosing one and requires compromise. Here *duty cycle* refers to conditions the battery experiences during use like, type of discharge and current drain, e.g., continuous, intermittent, continuous with pulses, etc.

5.14.1 Classification

Batteries are classified into two categories depending on their recharging capabilities.

(i) **Primary Batteries** - *These are non-rechargeable and are meant for a single use and meant to be discarded after use.*

(ii) Secondary Batteries - *These are rechargeable and are meant for a multi cycle use.* After every use the electrochemical reaction could be reversed by external application of voltage. The cycle is reversed till the capacity fades or lost due to leakage or internal short circuit.

5.14.2 Primary Batteries (Non-Rechargeable)

These are non-rechargeable and are less expensive and are often used in ordinary gadgets like torch lights, watches and toys. Commercially many kinds of primary batteries are available but we are going to restrict our discussion to Leclanché cell, Alkaline cell and Lithium cell.

5.14.2.1 Dry Cell (Leclanché Cells or Zinc-Carbon Cell)

Anode : Zinc, **Cathode:** Manganese Dioxide (MnO_2),

Electrolyte : Ammonium chloride or zinc chloride dissolved in water

Applications : Flashlights, toys, moderate drain use equipment

The basic design of the Leclanché cell dates to 1860s, and until World War II, was the only one in wide use. It is still the most commonly used of all primary battery designs because of its low cost, availability, and applicability in various situations.

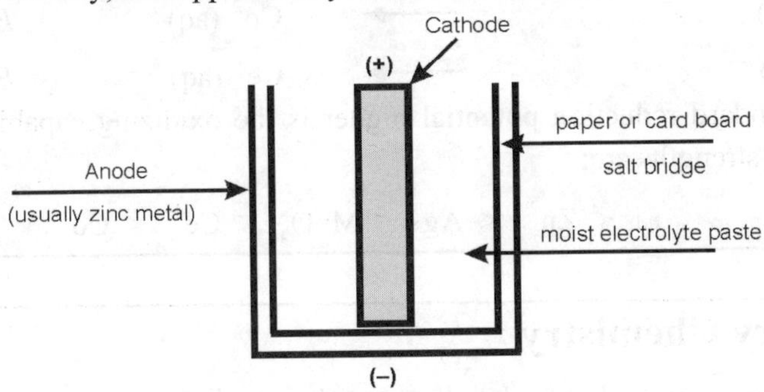

Fig. 5.16 A Cut view of Dry Cell.

In an ordinary Leclanché cell the electrolyte consists (in percent of atomic weight) of 26% NH_4Cl (ammonium chloride), 8.8% $ZnCl_2$ (zinc chloride), and 65.2% water.

At the cathode, the reduction reaction occurs within the moist paste comprised of ammomium chloride (NH_4Cl) and manganese dioxide (MnO_2) :

$$2\ MnO_2 + 2\ H_2O + 2e^- \longrightarrow 2MnOOH + 2\ OH^-$$

A thin zinc cylinder serves as the anode and it undergoes oxidation :

$$Zn\ (s) + 2NH_4Cl\ (aq) + 2OH^- \longrightarrow Zn(NH_3)_2\ Cl_2 + 2\ H_2O + 2e^-$$

The overall cell reaction can be expressed as

$$Zn + 2\ MnO_2 + 2\ NH_4Cl \longrightarrow 2\ MnOOH + Zn(NH_3)_2\ Cl_2$$

The anode Zinc is normally present as the cylindrical container (which is often surrounded by steel to prevent leakage of electrolyte). MnO_2, is not a good conductor of electricity and hence mixed with graphite to form a paste, with a central graphite rod acting as the current collector. The cell voltage increases by using synthetically produced manganese dioxide instead of that found naturally (called pyrolusite). This does drive the cost up a bit, but it is still inexpensive and environmentally friendly, making it a popular cathode. The aqueous ammonium chloride acting as electrolyte is usually thickened with polymeric material such as starch or methyl cellulose. The resultant cells are reasonably stable against discharging. The self discharging is via evolution hydrogen, a non-electrochemical reaction.

$$Zn + 2H_2O \longrightarrow Zn(OH)_2 + H_2$$

Zinc-chloride cell systems have been found to perform better than the Leelanche under heavier current drain. The electrolyte in a typical zinc chloride cell consists of 15-40% $ZnCl_2$ and 60-85% water, sometimes with a small amount of NH_4Cl for optimal performance. The overall cell reaction of the zinc chloride as the electrolyte can be expressed :

$$Zn + 2\,MnO_2 + 2\,H_2O + ZnCl_2 \longrightarrow 2\,MnOOH + 2\,Zn(OH)Cl$$

These cells are the cheapest ones in wide use, but they also have the lowest energy density and perform poorly under high-current applications.

5.14.2.2 Lithium and Other Metal Cells (Non-aqueous cells)

Applications : Pacemakers, defibrillators, watches, meters, cameras, calculators, portable, low-power use

Lithium battery chemistry comprises a number of cell designs that use **lithium as the anode**. Lithium is gaining a lot of popularity as an anode over other metals for a number of reasons as given in Table 5.6.

Table 5.6

Anode	Standard potential (V)	Density g/cm^3	Electrochemical Equivalence (Ah/g)
Li	**3.05**	**0.54**	**3.86**
Na	2.7	0.97	1.16
Mg	2.4	1.74	2.20
Al	1.7	2.7	2.98
Ca	2.87	1.54	1.34
Fe	0.44	7.85	0.96
Zn	0.76	7.1	0.82
Cd	0.40	8.65	0.48
Pb	0.13	11.3	0.26

Note that lithium, the lightest of the metals, also has the highest standard potential (>3V) of all the metals and some of the lithium cell designs have a voltage of nearly 4 V. This means that *lithium has the highest energy density*. The dry cells and alkaline cells discussed earlier have aqueous electrolyte while the lithium cells have non-aqueous medium. Aqueous systems can go upto energy densities of 700 W kg^{-1} but **Lithium non-aqueous systems** with far higher potential power densities have been reported.

Lithium cells have only recently become commercially viable because *lithium reacts violently with water, as well as nitrogen in air.* This requires sealed cells. High-rate lithium cells can build up pressure if they are short circuited and cause the temperature and pressure to rise. Thus, the cell design needs to include weak points, or safety vents, which rupture at a certain pressure to prevent explosion.

Lithium cells can be grouped into three general categories: (a) liquid cathode, (b) solid cathode, and (c) solid electrolyte.

(a) Liquid cathode lithium Cells :

Anode : Lithium, ***Cathode :*** $SOCl_2$, ***Electrolyte :*** $LiAlCl_2$

These cells tend to offer *higher discharge rates* because the reactions occur at the cathode surface. In a solid cathode, the reactions take longer because the lithium ions must enter into the cathode for discharge to occur. The direct contact between the liquid cathode and the lithium forms a film over the lithium, called the *solid electrolyte interface (SEI)*. This prevents further chemical reaction when not in use, thus preserving the cell's shelf life. One drawback, though, is that if the film is too thick, it causes an initial voltage delay. Usually, water contamination is the reason for the thicker film.

This cell consists of a high-surface area carbon cathode, a non-woven glass separator. Thionyl chloride acts as the electrolyte solvent and the active cathode material.

The materials react as follows :

Location	**Reaction**
Anode	$Li \longrightarrow Li^+ + e^-$
Cathode	$4Li^+ + 4e^- + 2SOCl_2 \longrightarrow 4LiCl + SO_2 + S$
Overall	$4Li + 2SOCl_2 \longrightarrow 4LiCl + SO_2 + S$

During discharge the anode gives off lithium ions. On the carbon surface, the thionyl chloride reduces to chloride ions, sulfur dioxide, and sulfur. The lithium and chloride ions then form lithium chloride. Once the lithium chloride has deposited at a site on the carbon surface, that site is rendered inactive. The sulfur and sulfur dioxide dissolve in the electrolyte, but at higher-rate discharges SO_2 will increase the cell pressure leading to explosion.This system has a very high energy density (about 500 Wh/kg) and an operating voltage of 3.3–3.5 V.

In high-rate discharge, the voltage delay is more pronounced and the pressure increases as mentioned before. Low-rate cells are used commercially for small electronics and memory backup. High-rate cells are used mainly for military applications.

(b) Solid Cathode Lithium Cells ($LiMnO_2$ Cell)

These cells cannot be used in high-drain applications and don't perform well under low temperature conditions. However, they don't have the same voltage delay as liquid cells and don't require pressurization. They are used generally for memory backup, watches, portable electronic devices, etc.

These account for about 80% of all primary lithium cells, one reason being their low cost. The cathode used is a heat-treated MnO_2 and the electrolyte a mixture of propylene carbonate and 1, 2-dimethoxyethane.

The half reactions are :

Anode	$Li \longrightarrow Li^+ + e$
Cathode	$Mn^{IV}O_2 + Li^+ + e \longrightarrow Mn^{III}O_2(Li^+)$
Overall	$Li + Mn^{IV}O_2 \longrightarrow Mn^{III}O_2(Li^+)$

$Li(CF)_n$ Lithium polycarbon monofluoride also falls in this category.

The cathode in this cell is carbon monofluoride. A typical electrolyte is lithium tetrafluorobate ($LiBF_4$) salt in a solution of propylene carbonate (PC) and dimethoxyethane (DME).

These cells also have a high voltage (about 3.0 V) and a high energy density (around 250 Wh/kg). All this and a 7-year shelf life makes them very suitable for low- to moderate-drain use, e.g., watches, calculators, and memory applications.

(c) Solid Electrolyte Lithium Cells

They perform best in low-current applications and have a *very long* service life. For this reason, they are used in pacemakers

Lithium iodine cells use solid LiI as their electrolyte and also produce LiI as the cell discharges. The cathode is poly-2-vinylpyridine (P2VP) with the following reactions :

Anode	$2Li \longrightarrow 2Li^+ + 2e$
Cathode	$2Li^+ + 2e + P2VP \cdot nI_2 \longrightarrow P2VP \cdot (n-1)I_2 + 2LiI$
Overall	$2Li + P2VP \cdot nI_2 \longrightarrow P2VP \cdot (n-1)I_2 + 2LiI$

LiI is formed *in situ* by direct reaction of the electrodes.

5.14.3 Secondary Batteries

These cells are rechargeable and reusable. A combination of all reversible electrochemical cells give secondary batteries. Many kinds of secondary batteries are available in the market but we are restricting our discussion to Lead acid cell, Ni/Cd cells, Ni-metal hydride cells and Lithium ion cells

5.14.3.1 Lead–acid Cells

Anode : Sponge metallic lead,

Cathode: Lead dioxide (PbO_2),

Electrolyte : Dilute mixture of aqueous sulfuric acid

Applications : Automobliles and construction equipment, standby/backup systems

Used mainly for engine batteries, these cells represent over half of all battery sales. Some advantages are their low cost, long life cycle, and ability to withstand mistreatment. They also perform well in high and low temperatures and in high-drain applications.

The half-cell reactions are :

Half-Reaction	V vs SHE
$Pb + SO_4^{2-} \longrightarrow PbSO_4 + 2e^-$	0.356
$PbO_2 + SO_4^{2-} + 4H^+ + 2e^- \longrightarrow PbSO_4 + 2H_2O$	1.685

There are a few problems with this design. If the cell voltages exceed 2.39 V, the water breaks down into hydrogen and oxygen and may lead to explosion. Another problem arising from this system is that fumes from the acid solution may have a corrosive effect on the area surrounding the battery.

These problems are mostly solved by sealed cells, made commercially available in the 1970s. This sealing in reality is only valve regulated and allowing hydrogen and oxygen recombination using catalytic recombiners. They convert these gases back into water, achieving about 85% efficiency at best. In these batteries the water lost becomes so insignificant that no refill is needed for the life of the battery. For this reason , these cells are often referred to as maintenance-free batteries. Also, this cell design prevents corrosive fumes from escaping.

These cells have a low cycle life, a quick self discharge, and low energy densities (normally between 30 and 40 Wh/kg). However, with a nominal voltage of 2 V and power densities of up to 600 W/kg, the lead-acid cell is an adequate, if not perfect, design for car batteries.

5.14.3.2 Nickel/Cadmium Cells

Anode: Cadmium,

Cathode: Nickel oxyhydroxide NiOOH,

Electrolyte : Aqueous potassium hydroxide (KOH)

Applications : Calculators, digital cameras, pagers, lap tops, tape recorders, flashlights, medical devices (e.g., defibrillators), electric vehicles, space applications

The cathode is nickel-plated, woven mesh, and the anode is a cadmium-plated net. The electrolyte, KOH, acts only as an ion conductor and does not contribute significantly to the cell's reaction. That's why not much electrolyte is needed, so this keeps the weight down (NaOH is sometimes used as an electrolyte, which doesn't conduct as well, but also doesn't tend to leak out of the seal as much). The cell reactions are as follows :

Reaction	**V vs SHE**
$Cd + 2OH^- \longrightarrow Cd(OH)_2 + 2e^-$	0.81
$NiOOH + 2H_2O + 2e^- \longrightarrow Ni(OH)_2.H_2O + OH^-$	0.45
$Cd + NiOOH + 2H_2O + OH \longrightarrow Cd(OH)_2 + Ni(OH)_2.H_2O$	1.26

Advantages include good performance in high-discharge and low-temperature applications. They also have long shelf and use life. Disadvantages are that they cost more than the lead-acid battery and have lower power densities. Possibly its most well-known limitation is a *memory effect,* where the cell retains the characteristics of the previous cycle. This term refers to a temporary loss of cell capacity, which occurs when a cell is recharged without being fully discharged. This can cause cadmium hydroxide to passivate the electrode, or the battery to wear out. In the former case, a few cycles of discharging and charging the cell will help correct the problem, but may shorten the lifetime of the battery.

5.14.3.3 Lithium Ion Cells

Anode : Carbon compound, graphite,

Cathode : Lithium oxide.

Applications : Laptops, cellular phones, electric vehicles

As discussed earlier Lithium metal batteries have safety disadvantages. These are overcome in *Lithium ion Batteries.*

Cathodes consist of a layered crystal (graphite) into which the lithium is intercalated. Experimental cells have also used lithiated metal oxide such as $LiCoO_2$, $NiNi_{0.3}Co_{0.7}O_2$, $LiNiO_2$, LiV_2O_5, LiV_6O_{13}, $LiMn_4O_9$, $LiMn_2O_4$, $LiNiO_{0.2}CoO_2$.

Electrolytes are usually $LiPF_6$, although this has a problem with aluminum corrosion, and so alternatives are being sought. One such is $LiBF_4$. The electrolyte in current production batteries is a patented liquid, and uses an organic solvent.

Membranes are necessary to separate the electrons from the ions. Currently the batteries in wide use have microporous polyethylene membranes.

Intercalation keeps the small ions (such as lithium, sodium, and the other alkali metals) into the interstitial spaces in a graphite crystal. This makes the graphite is conductive, dilutes the lithium for safety, is reasonably cheap, and does not allow dendrites or other unwanted crystal structures of Li to form.

5.15 Fuel Cells

A cell in which one or both of the reactants are not permanently contained in the cell, but are continuously supplied from a source external to the cell and the reaction products continuously removed is called a fuel cell. Unlike the metal anodes typically used in batteries, the fuels in a fuel cell are usually gas or liquid, with oxygen as the oxidant.

Around 1900 Nernst and Haber attempted for a direct carbon oxidizing fuel cells but met with a limited success. Later in 1933 Baur, a German scientist, proposed the first fuel cell that can use hydrogen as a fuel and can work at room temperature. This idea was forwarded by Bacon, a British engineer and developed a porous nickel electrode that can withstand high pressures and can give current in kilowatt range at moderate temperatures. This technology found a ready customer in Apollo program of NASA (USA). This fuel cell harnesses the chemical energy of hydrogen and oxygen to generate electricity without combustion or pollution. Hydrogen/oxygen fuel cell systems work well in space travel applications because of their *high efficiency, high power-to-weight and volume ratios, and usable reaction product (water).* In the near future, our vehicles may also be powered by fuel cells.

There are many kinds of H_2/O_2 fuel cells and are categorized on the basis of electrolyte used.

1. Proton Exchange Membrane Fuel Cells (PEMFC)
2. Alkaline Fuel Cells (AFC)
3. Molten Carbonate Feuel Cells
4. Phosphoric acid fuel cells
5. Solid oxide fuels

We are going to have a discussion of the *first two* kinds of fuel cells.

5.15.1 Proton Exchange Membrane Fuel Cells (PEMFC)

The type of fuel cell typically used in automobiles is a *Proton Exchange Membrane* (PEM)Fuel Cell , also called a *Polymer Electrolyte Membrane fuel cell* (Fig. 27.9). A fuel cell harnesses the chemical energy of hydrogen and oxygen to generate electricity without combustion or pollution.

At the anode, a platinum catalyst causes the hydrogen to split into positive hydrogen ions (protons) and negatively charged electrons.

Half-Reactions		V vs SHE
At Anode	$2H_2 = 4H^+ + 4e^-$	0
At Cathode	$4H^+ + O_2 + 4e^- = 2H_2O$	1.2

The Proton Exchange Membrane (PEM) allows only the positively charged ions to pass through it to the cathode. The negatively charged electrons must travel along an external circuit to reach the cathode, creating an electrical current.

At the cathode, the negatively charged electrons and positively charged hydrogen ions combine with oxygen to form water, which flows out of the cell.

Thus we get electricity from environmentally friendly reactants and products.

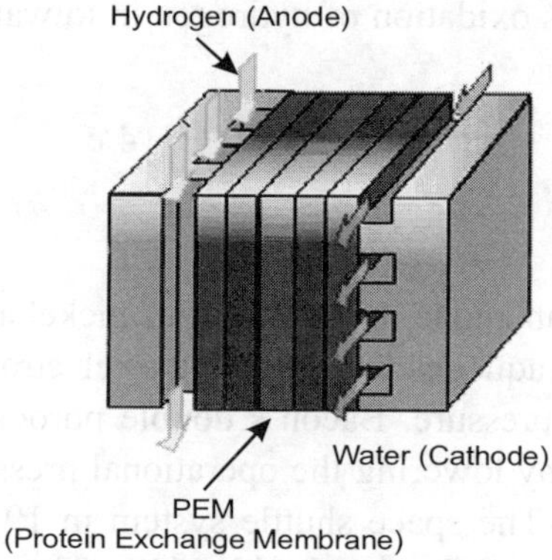

Fig. 5.17 PEM fuel cell.

5.15.2 Alkaline Fuel Cells (AFC)

In these cells, 30-45% aqueous solutions of potassium hydroxide is used as electrolyte (Fig. 27.10). So acidic impurities such as CO_2 should be avoided. Generally this is done by passing the compressed air into lime solution before allowing it to enter the fuel cell. These fuel cells yield highest voltages at comparable current densities in comparison to other classes of fuel cells.

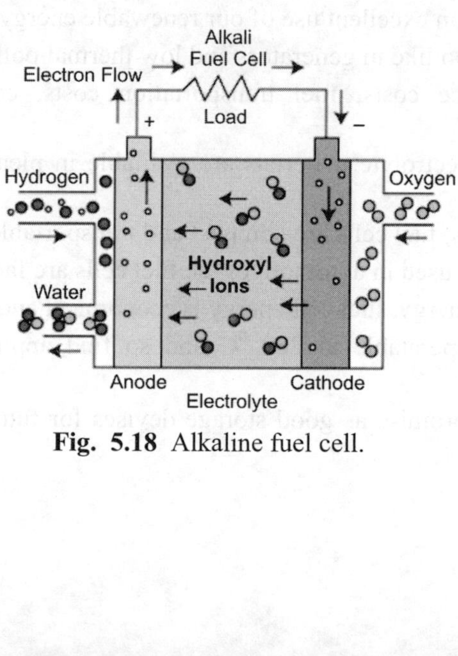

Fig. 5.18 Alkaline fuel cell.

The following half reactions take place in these cells

At Cathode : Oxygen passed is converted into hydroxide ions by consuming 4 electrons

$$O_2 + 2H_2O + 4\,e^- \longrightarrow 4\,OH^-$$

In fact, this is a two step reaction with intermediate peroxy ion formation.

$$O_2 + H_2O + 2\,e^- \longrightarrow HO_2^- + OH^-$$

$$HO_2^- \longrightarrow OH^- + \tfrac{1}{2}\,O_2$$

At Anode : Anode involves oxidation of hydrogen in to water by reacting with hydroxide anions from the electrolyte.

$$2\,H_2 + 4\,OH^- \longrightarrow 4H_2O + +4\,e^-$$

Over All Reaction :

$$2H_2 + O_2 \longrightarrow 2H_2O$$

Bacon fuel cell consists of an anode made of porous nickel and a cathode made of lithiated nickel oxide separated by 30% aqueous KOH serving as electrolyte. This cell operates at 200^0 C and 5 Mpa (about 50 atm) pressure. Bacon's double porocity fuel cell was used in 1960 NASA Apollo space program by lowering the operational pressure to 0.33 Mpa by activating the electrodes with a catalyst. The space shuttle system in 1974 replaced the heavy bipolar electrode plates with magnesium foils plated with silver. The anode now is a carbon bonded with polytetrafluoroethylene and consisting Pt-Pd catalyst and cathode consists of 90% gold and 10% platinum plated on to nickel screen. This cell operates at a pressure of 0.42 Mpa and 92^0 C and works for 2000 h.

5.15.3 Advantages of Fuel Cells

1. No emission of toxic gases and chemical wastes are in safe limits. The reactants and products are environment friendly and only we have to bother about disposal of cell material
2. High efficiency (75-85%) of energy conversion from chemical energy to electrical energy. So offer an excellent use of our renewable energy resources.
3. No noise pollution like in generators and low thermal pollution.
4. Low maintenance costs, fuel transportation costs, cell parts are modular and exchangeable.
5. The fuels and electrolyte materials are available in plenty and inexhaustible unlike fossil fuels.
6. Unlike solar cells, fuel cells are compact and transportable.
7. Unlike acid cells used in automotives the fuel cells are far less corrosive.
8. Unlike nuclear energy, fuel cell energy is economical and safe.
9. Fuel cells are operatable to 200 ^{0}C and so find application in high temperature systems.
10. Fuel cells hold promise as good storage devises for future nuclear and solar energy economies.

5.15.4 Limitations of Fuel Cells

1. The main limitation of fuel cells lie in high initial costs associated with electrode material and design costs. The later could be brought down with mass production and suitable R&D in this direction. The low maintenance and longetivity of cells could compensate some of these negative factors.
2. Large weight and volume of H_2 gas-fuel storage system.
3. Lack of infrastructure for distribution and marketing of hydrogen gas. The developing natural gas distribution net-work could pave way for this in the future.
4. High cost of H_2 gas. Technological advances should bring the costs down.
5. Most alkaline fuel cells suffer from carbon dioxide leakages and should be properly filtered and prevented from entering the cell and reacting with the electrolyte.

5.16 Electrical Conductors & Insulators (Dielectrics)

A metal rod held in hand and rubbed with fur will not seem to develop a charge. The same rod when held with a glass or rubber handle and rubbed without any human or metal contact gets charged. The explanation to this phenomena is that the metals and human body are *conductors* of electricity and glass and hard rubber are *insulators* or *dielectrics*. In a conductor, electric current can flow freely. Metals such as copper typify conductors (Fig. 5.19). Metals are classified as conductors because their outer electrons are not tightly bound. Experimental proof of *Hall effect* shows that only electrons are free to move and the positive charge is immobile in metals. For conductors like electrolyte, we find movement of both positive and negative charges.

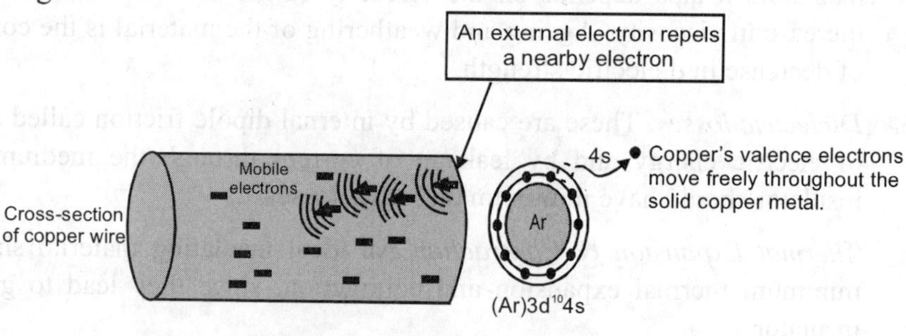

Fig. 5.19 Conduction in Copper

Insulators offer very large resistance to the flow of electric current since their outermost electrons are so tightly bound that there is essentially zero electron flow through them at ordinary voltages. Although there are no perfect insulators, the insulating ability of fused quartz is about 10^{25} times as great as copper.

Material	Resistivity (ohm m)
Glass	10^{12}
Mica	9×10^{13}
Quartz (fused)	5×10^{16}

Insulators (have high resistivity)

Material	Resistivity (Ohm M)
Copper	1.7×10^{-8}

Conductors (have low resistivity)

The materials having intermediate properties between conductors and insulators in their ability to conduct electricity are called *semi-conductors*.

5.16.1 Characteristics of Insulators (Dielectrics) :

1. *Resistivity* It is the resistance offered by a unit cube of the material. Insulators have resistivities ranging between 10^6–10^{17} ohm.m at room temperatures.
2. *Dielectric Polarisation* In the presence of applied electric field the dielectrics undergo separation of charge called dielectric polarization. Here the dielectric is internally charged and non-equipotential but prevents any flow of charge through them. If Q is the charge collected in the presence of an applied voltage V, then the capacitance of the material is C = Q/V.
3. *Dielectric Constant* The capacity of a material to store electric charge in the presence of an applied electric field is called dielectric constant. It is the ratio of the electrical capacity of the condenser filled with the material to that of the electrical capacity when evacuated. A good electrical insulator should have a low dielectrical constant.
4. *Dielectrical Strength* The ability to resist puncture or rupture of insulating material towards applied field is called dielectric strength and is expressed in volts/mm. Dielectric strength decreases with increase in thickness of material, time of application of charge, humidity and temperature. In case of liquid insulators it also depends on the viscosity of the material and increases with increase in viscosity. Ageing and weathering of the material is the consequence of decrease in dielectric strength.
5. *Dielectric losses* These are caused by internal dipole friction called absorption of electric energy and by leakage of current through the medium. A good insulator should have minimum dielectric losses.
6. *Thermal Expansion & Contraction* An ideal insulating material should have minimum thermal expansion and contraction, since they lead to gaps in the insulator.
7. *Thermal & Chemical Ageing* – Breakdown of insulator with temperature and oxygen content and contact of other chemicals is called ageing and makes it less serviceable for long time periods.
8. *Porosity* – Porosity increases moisture holding capacity of the insulator, thus leading to unwanted results.
9. *Mechanical Requirements* – Lower surface tension of the insulator permits greater wetting of the electrical component and thus giving better cooling. The surface tension of dielectrics is generally modified by the addition of wetting agents. In case of solid insulators surface roughness as well as hardness are unwanted properties of dielectrics.

5.16.2 Classification of Insulating Materials

The insulators are classified depending on their physical state into Gaseous, Liquid and Solid insulators.

(i) *Gaseous Insulators:* All the gases and air are good insulators of heat and electricity. Air is a good insulators when the voltages are not very high. For these applications generally inert gases and nitrogen are used. Power cables are filled with nitrogen at pressures ranging from 10-200 psi depending on the voltage. Presence of oxygen and halogens damage the electrical equipment through their chemical reactions and are generally avoided. Hydrogen is a light gas and has good coolant capacity. It is therefore used used in large turbo engines and synchronous condensers. Organic gases are preferred owing to their chemical reactivity and combustibility. Gases like hexafluoro sulphur , SF_6, are non-explosive, non-inflammable and non-toxic and have better cooling capacity than air and nitrogen. Its dielectric strength is 2.35 times greater than air. It has corrosive properties and is not suitable under corrosive working conditions and temperatures.

(ii) *Liquid Insulators* These insulators have better heat transfer properties than gas dielectrics but suffer from chemical breakdown. So they are often used in conjunction with solid stabilizers. We have different dielectrics of this class covering different temperature ranges (Table 5.7).

Table 5.7

Insulating liquid	Operating temperature range
Mineral oil	–50 to 110 °C
Askarels	–50 to 110 °C
Silicone liquids	–90 to 220 °C
Fluorinated liquids	–50 to 200 °C
Synthetic hydrocarbons	–50 to 110 °C
Organic esters	–50 to 110 °C
Vegetable oils	–20 to 100 °C

(a) ***Mineral oils*** Their insulating properties are influenced by the presence of hydrocarbons, oxygen, sulphur, etc. They have low dielectric constants (2.5). Used in the transformers.

(b) ***Askarel*** These are noninflammable, synthetic insulating liquids. They have excellent arc resistance and have high dielectric strength and low dielectric constant (4-6). They have excellent thermal, electrical and chemical stability.

(c) ***Silicone fluids*** They are non-corrosive to metals up to 200^{0}C. They are water like with oily consistency. On break down they yield considerable amount of gases and substances like silicon carbide, silicon dioxide. Because of this disadvantage the silicones are not suitable as switch gear oils. They are used in radio, aircraft and pulse transformers.

(d) ***Fluorinated liquids*** They are non-inflammable and chemically inert, and their vapors are non-explosive. They are highly volatile and decompose in the presence of moisture. They find use in small electric and radio devices.

(e) ***Vegetable oils*** These are cheap and oldest insulating liquids. The drying ois such as linseed oil, tung oil are used in making insulating varnishes used in the transformers. The non-drying oils such as castor oil, coconut oil, etc., are popular plasticizers in insulating resin components.

(iii) **Solid Insulators** Most of the non-metals and compounds like mica, polymers, silk, paper, asbestos, etc, fall under this category. These are capable of retaining their insulating properties up to a fixed temperature called limiting temperature.

(a) ***Cellulosic Materials*** Cellulosic materials like cotton, paper, etc., in their raw form have thermal stability with limiting temperatures of 90^0C. When they are impregnated with moderate heat resistant materials this value increases to 105 °C.

(b) ***Fibrous Materials*** Natural fibers like wool, jute, etc., and synthetic fibres like rayons, nylons, fiberglass, etc., show variable limiting temperatures depending on the impregnating and bonding materials as shown in the Table 5.8.

Table 5.8

Insulating Materials	Limiting Temperature
Cellulosic materials like cotton, silk, paper that are not impregnated.	90 °C
Silk, paper, cotton, nylons, etc., impregnated with moderate heat resistant materials.	105 °C
Fibrous materials like Asbestos, fibre glass, etc., impregnated with varnishes of less heat resistance materials.	130 °C
Mica, marble, ceramics, glass, teflon	>180 °C
Enamelled wire insulation on polymer resin bases	120 °C
Polymer resins bonded with varnishes made of high heat resistant materials	155 °C
Fibre glass, mica and asbestos impregnated with silicones and silicone rubbers.	180 °C

(c) ***Synthetic Polymers*** Polymers have low dielectric constants ranging from 2 to 5. Normal polymers like polyethylene and polystyrene are excellent insulators for low temperature operations but undergo softening at higher temperatures. They are used as thin films and find their use in chokes, capacitors, electrical bushings and telephones. Teflon is a good synthetic polymer that could withstand temperatures above 180^0C and has a dielectric constant of 2. Epoxy resins possess excellent electrical and mechanical properties and are widely used in making insulators, bushings, etc., for high voltage operations. Laminates and insulating varnishes are also prepared using epoxy resins.

(d) ***Rigid Insulators*** **Glass**, ceramics and refractories fall under this category. Glass has excellent electrical properties. Pyrex glass even has good chemical stability. Their major application is found in lighting systems. Glasses have similar coefficient of expansion as metals and are therefore used as insulating materials and metal claddings. Normal porcelains are used for low voltage applications while at high voltages glazed ceramics are used. Zircon and alumina porcelains are used in spark plug cores. High dielectric constant porcelains (15-5000) are prepared using titanates and are used for high capacity condensers with small spaces. Mylax is a mica bonded with lead glass that can be used for operations up to 600 ^{0}C and has good insulating properties with mechanical stability.

(e) ***Impregnating Coatings and Bonding Materials:*** These are synthetic high molecular weight organic and organo-metallic compounds. Bitumen, resins, drying oils etc., are known as insulating varnishes. They not only enhance the insulating properties but also help in stopping ageing of cheaply available materials like paper and cotton and reducing hygroscopicity of materials. Bonding materials are generally solutions of phenolic resins in alcohol and are used in low capacity electrical units. Other adhesives too could be used as bonding materials for low voltage and low temperature applications.

5.17 Self-Test Questions

1. The resistance of a conductor is directly proportional to
   (A) Current passed (B) Area of cross section of conductor
   (C) Length of the conductor (D) Equivalent conductance

2. Sodium chloride is a poor conductor of electricity but transforms into a good conductor in molten state because
   (A) The covalent bonds are broken in molten state
   (B) The degree of dissociation increases
   (C) It form liquid crystals
   (D) The ions get mobility

3. If "l" is the distance between the electrodes of a cross sectional area "a", then cell constant is given by the ratio
   (A) l/a (B) a/l (C) a × l (D) a – l

4. The resistance of a conductor is 5 × 10^{-2} ohms. Conductance is the equal to
   (A) 200 mhos (B) 20 mhos (C) 500 mhos (D) 50 mhos

5. The conductivity of 0.1 N KCl is 0.01120 mho cm^{-1}. If the cell constant is equal to 0.5 cm^{-1}, the conductance is equal to

(A) 2.24×10^{-2} eq^{-1} (B) 5.6×10^{-3} mhos

(C) 224 mhos (D) 560 mhos

6. The units of equivalent conductivity is

(A) ohm^{-2} cm^{-2} eq^{-1} (B) $ohm^{-1}cm^{-2}$ eq^{-1}

(C) Ohm^{-2} cm^{-2} eq^{-1} (D) Ohm^{-1} cm^{2} eq^{-1}

7. According to Kohlrausch law

(A) $\alpha = \frac{\Lambda_v}{\Lambda_\infty}$ (B) $t_+ = \frac{\Lambda_v}{\Lambda_\infty}$

(C) $\Lambda_+ + \Lambda_- = \Lambda_4$ (D) $\Lambda_v = KV$

8. Equivalent conductivity of an electrolytic solution

(A) Increases with dilution (B) Decreases with dilution

(C) Independent of dilution (D) None of the above

9. Cell constant of a conductometric cell

(A) Increases with dilution

(B) Decrease with dilution

(C) Depends on the nature of electrolyte

(D) Independent of the nature of electrolyte.

10. Equivalent conductivity of any electrolyte is algebraic sum of ________ conductivities.

11. Degree of dissociation = $\frac{\Lambda_v}{.....}$

12. Specific conductance = Cell constant x ______________

13. If the standard potential of Ca | Ca^{2+}, Ni | Ni^{2+} and Au | Au^{2+} couples are 2.87, 0.25 and -1.5V respectively, then

(A) Ni can reduce Ca^{2+} and Au^{3+} (B) Ca and Ag can reduce Ni^{2+}

(C) Ni^{2+} can oxidize Au and Cu (D) Au^{3+} can oxidize Ca and Ni

14. In a galvanic cell

(A) Chemical energy is converted into electrical energy

(B) Electrical energy is converted into chemical energy

(C) Chemical energy is converted into heat energy

(D) Electrical energy is converted into heat energy

15. The cell reaction $Zn + Cu^{2+} \longrightarrow Zn^{2+} + Cu$ is correctly represented by

(A) $Cu \mid Cu^{2+} \parallel Zn \mid Zn^{2+}$ (B) $Cu \mid Cu^{2+} \parallel Zn^{2+} \mid Z^{4}$

(C) $Zn \mid Zn^{2+} \parallel Cu^{2+} \mid Cu$ (D) $Zn \mid Zn^{2+} \parallel Cu \mid Cu^{2+}$

16. If the oxidation potential of Zn metal is 0.76V, the E_{cell} for the redox reaction

$Zn + 2H(aq, 1M) \longrightarrow Zn^{2+} + H_2(1 atm)$ is

(A) 0.76 (B) –0.76

(C) +0.38 (D) –0.38

5.17 Short and Essay Type Questions

1. State and explain Kohlrausch' law. Give its applications.
2. Write an account of acid base conductometric titrations.
3. State Kohlrausch law. Explain how it can be used to determine the Λ_4 of acetic acid.
4. What is the principle underlying conductometric titrations? Discuss titration curve obtained for titration of a mixture of HCl and CH_3COOH with NaOH. What are the advantage of conductometric titrations?
5. Define, the terms specific conductance and equivalent conductance. Explain clearly why on dilution the specific conductance decreases but equivalent conductance increases.
6. During conductometric titration of a strong acid against strong base, the conductivity decreases up to the neutralization point and then increases sharply. Explain why this happens
7. Explain the terms cell constant, equivalent conductivity. Give their units. Draw a graph showing the variation of equivalent conductance with concentration for both weak and strong electrolytes.
8. How does Kohlrausch's law of independent migration of ions help in determining the equivalent conductivity of a weak electrolyte like acetic acid at infinite dilution?
9. (a) Define the terms :
   (i) Resistivity
   (ii) Conductivity and
   (iii) Equivalent conductivity.

   (b) Describe a method for determining the equivalent conductivity of an electrolytic solution.

   (c) What is the effect of dilution on conductivity?
10. What is meant by cell constant of a conductivity cell? How is it determined? What are its units?

11. Explain the following terms
    (a) Single electrode potential
    (b) Reference electrode
    (c) Standard hydrogen electrode.
12. What is meant by pH of a solution? How is the pH of a solution determined?
13. How is the *emf* of a cell determined from a knowledge of single electrode potentials? Give a suitable example.
14. Write the cell reaction at 298K of the cell

    $Zn \mid ZnCl_2\ (a = 0.5) \parallel AgCl \mid Ag$

15. (a) Explain what do you understand by single electrode potential and redox potential?
    (b) What is hydrogen electrode? How is constructed?
16. What is the main advantage of liquid insulating materials over solid insulating materials?
17. How the thermal spalling of a refractory can be reduced?
18. Define dielectric constant.
19. On what basis solid insulating materials are classified?
20. How the strength of refractory can be measured?
21. What are advantages of SF_6 as dielectric gas?
22. What is the difference between insulator and dielectric materials?
23. (a) What are thermal insulators? What are their characteristics and engineering applications ?
24. Describe the chief applications of thermal and electrical insulators.
25. (a) What are insulators? Mention the different types of insulators.
    (b) Describe the properties and applications of
        (i) thermal insulators (ii) electrical insulators.

6 Phase Rule

6 Phase Rule

6.1 Homogeneity and Inhomogeneity of Systems

A system which exhibits identical physical and chemical properties through-out is called a ***homogenous system.*** Thus we say that copper sulphate solution is homogenous as it exhibits identical physical properties like uniform color, physical indistinguishability of copper sulphate even under ultra microscope, etc. Though milk possesses uniform white color, we classify it under heterogeneous solution, as milk is a colloidal solution and the presence of fat particles can be seen under ultramicroscope. *The particles of solute in a true solution have a diameter of few angstroms. The x-rays too have wavelengths of few angstroms, and therefore give characteristics diffraction patterns of these particles.* To solve this homogeneity and inhomogeneity of the true solutions, we have to restrict our *physical indistinguishability to visible light*, wherein the colloidal particles are large enough for detection. Keeping this boundary condition in view, we can comfortably identify all the constituents of a heterogeneous systems, which are generally termed as **phases**.

6.2 Phase Rule

Fusion, vaporization, sublimation, dissolution of a salt, liquefaction of a gas etc., all involve changes in the phases of the constituents. These phase changes are completely reversible and are said to make a **heterogeneous equilibrium**. A study of variations in the physical parameters like pressure, temperature, volume, concentration etc., of constituents on the heterogenous equilibria were discussed quantitatively by Henry's law, Clapayron-Classius equation, Raoult's law, Nernst distribution law, Le Chatelier principle, law of mass action etc. *Phase rule is a generalization of the above mentioned laws and gives an unified approach as regards to fixing the physical variables to decide the existence of the phases present in a heterogenous equilibria.* It is not meant for either to invalidate or supersede the existing laws, which mainly describe the quantitative relation among the variables.

The **phase rule** was deduced theoretically by J. Willard Gibbs (1876), but its application to physical chemistry was made by H.W.B. Roozeboom (1884). It is described by the mathematical expression;

$$F = C - P + 2 \qquad \text{.....(6.1)}$$

where F is called **number of degrees of freedom** of variables

C is the number of **components** of the system

P is the number of **phases** of the system.

The phase rule is defined as if an equilibrium is dependent only on temperature, pressure and concentration variations and is not influenced by gravity, surface forces, electrical or magnetic forces, then the **number degrees of freedom, (*F*),** exceeds the difference of number of **components, (*C*)**, and the number of **phases, (*P*)** by 2.

6.2.1 Phase

A phase is defined as any homogenous, physically distinct with well defined boundaries and mechanically separable part of a system. Boiling, freezing, sublimation and conversion of one allotropic form to other, all involve changes of phase without change of chemical composition. Ice below $0^{o}C$ and normal pressure conditions exists in mere solid phase. As we raise the temperature, it melts to give liquid water and water vapor, thus forming a system of two or three phases that are homogenous, physically distinct and mechanically separable. As we cool they once again form solid ice. Hence we refer that these three phases of water are at a heterogenous equilibrium, Fig. 6.1. Addition of few ml of alcohol to this three phase equilibria, will not generate another phase, although, the liquid phase is now referred to as a solution of alcohol in water and the vapour phase is a homogenous mixture of alcohol vapour and water vapour. But carbon tetrachloride being immiscible with water, its addition to above three phase equilibria generate a fourth liquid phase, Fig. 6.2.

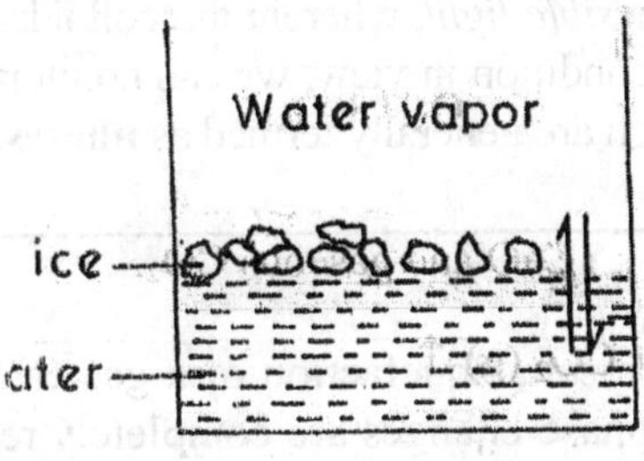

Fig. 6.1 Three phase equilibria.

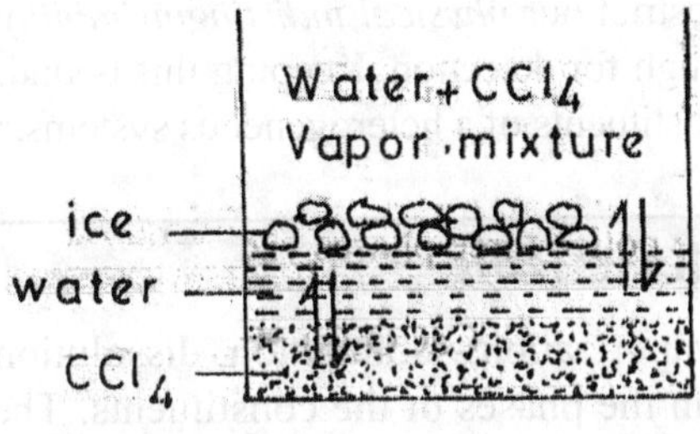

Fig. 6.2 Four phase equilibria.

Similarly addition of every solid constituent generates a separate phase as they are once again physically distinct, defined by separate boundary conditions. Thus addition of naphthalene balls to the three phase water equilibria generates a fourth phase of solid naphthalene which then involves in an exchange of naphthalene with vapor and liquid water. On the other hand addition of sand to the three phase equilibrium under consideration is not considered as a creation of fourth phase as it lacks any exchange of chemical species. For the same reason, the beaker in which the constituents are present is also not a phase of the system.

Similar to a liquid solution, a solid solution (e.g., alloy) is also considered as a single phase, no matter how many individual substances are present in it.

A gas or mixture of gases always constitute a single phase as they are physically indistinguishable and always form a true solution.

6.2.2 Components

The number of components in a system is the minimum number of independently variable chemical species necessary to describe the composition of each phase. Thus at 100 °C temperature when we take a mixture of $H_2(g)$, $O_2(g)$ and $H_2O(g)$, it forms a three component system. At arc temperatures, if this mixture is allowed to equilibrate to give water vapor,

$$2H_2 (g) + O_2 (g) \longleftrightarrow 2H_2O (g) \quad K_{eq} = \frac{[H_2O]^2}{[H_2]^2[O_2]}$$

it then becomes a two component system as the description of any two components of the system, and a knowledge equilibrium constant allows one to calculate the third component. Instead, if we start with a known amount of water vapor and allow it to dissociate to give O_2 and H_2, it becomes a one component system as the knowledge of dissociation constant of water and concentration of any one of the component fully describes all the three constituents of the vapor phase.

Similarly in the decomposition of $CaCO_3(s)$ to give CaO(s) and $CO_2(g)$, we notice three phases viz., solid $CaCO_3$, solid CaO and gaseous CO_2.

$$CaCO_3(s) \longleftrightarrow CaO(s) + CO_2 (g) \uparrow$$

At the first sight, it might appear, that this is a three component, $CaCO_3$, CaO and CO_2, system. In reality it is a single component system as one mole of decomposed $CaCO_3$ gives one mole each of CaO and CO_2 and the description of any one of the component is sufficient to describe the concentration of all the three phases present in the system. In the formation of $CaCO_3(s)$ by the combination of CaO(s) and $CO_2(g)$; we have an open

$$CO_2 + CaO(s) \longleftrightarrow CaCO_3 (s)$$

option of taking different amounts of CaO and CO_2. Hence we have to know at least, the concentration of two of the components, to fully describe the equilibrium under investigation, making it a two component system.

Another interesting example is the sodium sulphate-water system. $NaSO_4$ and H_2O combine in different proportions to give, starting from anhydrous Na_2SO_4 to aqueous solution of Na_2SO_4. Some of the equilibria are.,

$$Na_2SO_4 (s) + 0\ H_2O (l) \longleftrightarrow Na_2SO_4(s)$$

$$Na_2SO_4 (s) + 7\ H_2O (l) \longleftrightarrow Na_2\ SO_4.7H_2O (s)$$

$$Na_2SO_4 (s) + 10\ H_2O (l) \longleftrightarrow Na_2SO_4.\ 10H_2O (s)$$

$$Na_2SO_4 (s) + x\ H_2O (l) \longleftrightarrow Na_2SO_4(aq) (l)$$

$$0\ Na_2SO_4 (s) + \underset{\text{(ice)}}{H_2O (s)} \longleftrightarrow H_2O (s) \text{ or } (l) \text{ or} (g)$$

$$0\ Na_2SO_4 (s) + H_2O (s) \longleftrightarrow H_2O (s) \text{ or } (l) \text{ or} (g)$$

$$0\ Na_2SO_4 (s) + H_2O (g) \longleftrightarrow H_2O (s) \text{ or } (l) \text{ or} (g)$$

Although composition of certain phases (last three equilibria) may be stated in terms of only one of the constituents, certain others require the knowledge of the amounts of any two constituents to specify unambiguously the composition of each phase of the system. The smallest number by which the compositions of all the phases can be described being two, sodium sulphate water system is called a two component system.

Liquid water contains presumably enormous number of associated species; H_2O, $(H_2O)_2$, $(H_2O)_3 ... (H_2O)_n$

$$H_2O + H_2O \rightleftarrows (H_2O)_2$$
$$H_2O + (H_2O)_2 \rightleftarrows (H_2O)_3$$
$$\vdots \qquad \vdots$$
$$H_2O + (H_2O)_{n-1} \rightleftarrows (H_2O)_n$$

The equilibria between various polymers of water shift very rapidly, within 10^{-11} seconds at the most. Ordinary measurements require much longer time. Of the **n** species, representing (n–1) equilibria, only one species is chemically independent. Hence the system is effectively one component system.

In contrast to this behavior the dissociation of H_2O to give H_2 and O_2 in the presence of a catalyst is a three component system, while it reduces to one component system in the absence of catalyst as this dissociation equilibrium is not established for practical purpose (extremely slow).

From the above discussion it is clear that an accurate assignment of the number of components in a system pre-supposes some experimental knowledge of various equilibria involved in the system.

6.2.3 Degrees of Freedom

The number of degrees of freedom or variance (F) is the minimum number of ***intensive variables*** *(mass independent properties such as T, P, and concentration) that must be specified in order to fix the values of all remaining intensive variables*. The mass dependent properties like weight, volume, etc., are called **extensive properties** and they do not determine the equilibrium state of the phase. Physical properties like temperature, pressure, density, refractive index, viscosity, concentration (moles per unit volume), etc., are mass independent variables, and they only define the phases of a system.

Considering a system of pure water, to describe the phase and know whether water is present as solid, liquid or vapor it is just sufficient to know the pressure and temperature conditions as the other intensive variables (refractive index, viscosity etc) are automatically defined. Thus we have a freedom choice of only two variables, viz., the pressure and the temperature, of the various permitted intensive variables, and the water present in a single phase thus has only two degrees of freedom. Any change in either of these two variables could change the phase of the water. Thus water, existing in the vapor phase at 100 °C and 1 atm pressure, converts into liquid water either by increasing the pressure or decreasing the temperature.

Systems with $F = O$ are known as **invariant** or as having **no degrees of freedom**; with $F = 1$ they are called **univariant** (one degree of freedom), with $F = 2$ they are called **divariant** (two degrees of freedom); and so on. These are demonstrated in the following examples :

(i) **A pure gas :** A pure gas is completely described by just knowing the temperature and pressure. Hence it is divariant system or has two degrees of freedom. This conclusion could be arrived from the phase rule expression (6.1) as a pure gas is a one component, (C=1), one phase, (P=1) system.

Therefore;

Degrees of freedom, $F = 1 - 1 + 2 = 2$ (Divariant)

(ii) **A gaseous system consisting of two gases :** To describe the system fully, we need to define the composition of the mixture, temperature and pressure of the system. Hence it becomes a trivariant system. The same conclusion could be arrived as the gas mixture is a two component and one phase system.

Degrees of freedom, $F = 2 - 1 + 2 = 3$ (Trivariant)

(iii) **A gaseous system consisting of three gases :** Here the system could be fully described, if the composition of all three constituents, temperature and pressure are known. Now substituting the number of components as 3 and phases 1, we have from the phase rule expression.

Degrees of freedom, $F = 3 - 1 + 2 = 4$

(iv) **An undissolved solute in contact with its solution :** When CO_2 is filled in a jar containing water, the water gets saturated with CO_2 and the dissolved CO_2 is in equilibrium with free CO_2 gas. From Raoult's law, we know that the partial vapor pressure of a component at a given temperature is directly proportional to the mole fraction of that component. Just the description of temperature defines the solubility and the vapor pressure of the gas. Hence it is a univariant system. The same conclusion can be drawn from phase rule as the number of components is one and phases are two.

Degrees of freedom, $F = 1 - 2 + 2 = 1$

(v) **Ice-Water-Vapor system :** All the three phases of water, viz., ice, water and vapor, are at equilibrium at a temperature of 0.0098 ^{o}C, pressure 4.58 mm Hg. Any change in temperature or pressure disturbs the three phases equilibrium and we then get either one or two phases. Hence such a system has no degrees of freedom or invariant. As the number of components = 1, phases=3; it could be seen from the phase rule equation that;

Degrees of freedom $F = 1 - 3 + 2 = 0$

6.3 Phase Diagram

A diagram which illustrates the conditions of equilibrium between various phases of a substance is called a ***phase diagram or equilibrium diagram***. A polyphasic equilibrium which can be studied with the help of any two intensive properties, say temperature and pressure or temperature and composition or pressure and composition are represented by a two dimensional phase diagram. A trivariant system requires a three dimensional phase diagram. The importance of the phase diagrams could be understood by considering the following one component system.

6.4 One Component System

Pure water forms an example of one component system. Similarly the physical equilibria like melting, sublimation, evaporation etc., of pure substances are all example of one component heterogenous systems.

Solid $\longleftrightarrow$ Solid (Allotrophic transition)
Solid $\longleftrightarrow$ Liquid (Fusion)
Liquid $\longleftrightarrow$ Vapor (Evaporation)
Solid $\longleftrightarrow$ Vapor (Sublimation)

A one component system, should have at least one phase. Thus a one component system possess a maximum variance of two as seen by the phase rule expression.

Degrees of freedom $F = 1 - 1 + 2 = 2$

Hence a one component system is defined by at least two physical parameters viz., temperature and pressure. As we increase the number of phases of the one component system we can achieve a variance of 1 and zero.

(a) When two phase of the system are at equilibrium; degrees of freedom, $F = 1 - 2 + 2 = 1$

Hence it becomes a univariant system, allowing a variation of either temperature or pressure. This equilibrium vanishes by the variation of both the physical variables.

(b) When any three phases of the system are at equilibrium; degrees of freedom $= 1 - 3 + 2 = 0$

Hence it becomes an invariant system, allowing no variation of any of the intensive properties.

These results infer that it is not possible to have any four or any higher polyphasic equilibrium for a one component system.

6.4.1 Water System

Water system forms a typical example of one component system. Argument in support of water being a one component system has already been provided in section (6.2.2). *Water system comprises of three phases, viz., ice, liquid water and water vapor,* which show the following equilibria.

$$\text{(s)Ice} \xrightleftharpoons[\text{Sublimation}]{\text{Fusion}} \text{(l) Liquid water}$$

(s)Ice ⟷ (g) Water vapor

Evaporation

(l)Liquid water ⟷ (g) Water vapor

(s)Ice ⟷ (l) Liquid water ⟷ (g)Water vapor

Similar to other one component systems, water can have 2, 1 and 0 degrees of freedom. Because the maximum variance is 2 we can build the phase diagram of water system by studying the effects of pressure and temperature on various phase equilibria, Fig. 6.3.

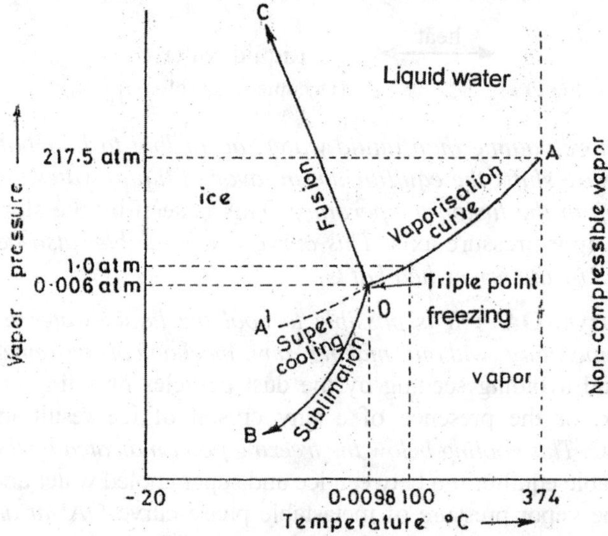

Fig. 6.3 Phase diagram of water system.

(i) Vaporization curve, OA : Liquid water at any given temperature has a precise vapor pressure. *With the increase in temperature we notice an increase in the vapor pressure,* but the rate of increase at higher temperatures is higher than at lower temperatures making *the vaporization curve, OA, non linear and slant upwards away from the temperature axis. This increasing trend continues till the system reaches a critical temperature* of 374 °C *and a critical vapor pressure* of 77.5 atm. *Any attempt to pass beyond* ***A*** *destroys the liquid vapor equilibrium leaving the water totally in incompressible vapor phase.* It is not possible to liquify water vapor beyond 374 °C critical temperature by the application of any amount of pressure.

Point O on the lower part of vaporization curve represents the freezing point of water, 0.0098 °C, under its own vapor pressure, 4.58 mm of Hg or 0.006 atm. The lower value of normal freezing point of water (0 °C) *measured under* laboratory conditions is due to :

(a) depression of freezing point by the dissolved air,

(b) the external pressure being 1atm, far higher than the critical vapor pressure of water vapor at the freezing point.

(ii) Sublimation curve, OB : *Below the freezing point the liquid phase totally converts into solid ice* and then we observe only an equilibrium between solid and vapor phases and *the curve representing this equilibrium is called* ***sublimation curve OB***. With the decrease in temperature initially *the vapor pressure drops steeply followed by slow decrease. This sublimation curve can be extended below B, but the immeasurably small values of vapor pressure bring more uncertainty than any use to the system and hence is left with an arrow head at B.*

(iii) Fusion Curve, OC : *The fusion equilibrium between liquid water and ice is represented by the curve OC*. The equilibrium may be represented

$$\underset{\text{(Occupies more volume)}}{\text{Ice}} \xleftrightarrow{\text{heat}} \underset{\text{(Occupies less volume)}}{\text{Liquid water}}$$

as ice occupies more volume than liquid water, according to Le Chatelier's principle, any raise in pressure shifts the equilibrium in favor of liquid water, accompanied by a simultaneous drop in the fusion temperature. This is seen by the sloping of the curve OC towards the vapor pressure axis. *This curve is extendable, also beyond C and this in fact is indicated by the arrow head at C.*

(iv) Super cooling curve, OA' : *It is possible to cool the liquid water at 0, by gradually lowering the temperature without making any mechanical movements like stirring, scratching etc.*, and avoiding seeding by the dust particles or a tiny crystal of ice. Any mechanical shock, or the presence of a tiny crystal of ice result in shooting up of temperature to 0 °C. *This cooling below the freezing point indicated by the broken line OA'* indicates a metastable equilibrium between ice and super cooled water and is called a super cooling curve. The vapor pressure of metastable phase curve OA' *at any temperature is greater than the vapor pressure of stable phase (curve OB). In the super cooled liquid, the solid and super cooled liquid are imagined to be separated by a thin vapor membrane and any slightest disturbance removes this insulatory vapor membrane leading to freezing of water and shooting of temperature to 0 °C.*

(v) Super heating : Similar to super cooling curve, attempts to obtain super heating of a solid beyond the freezing point O were found futile. Hence no super heating curve is shown in the phase diagram.

After obtaining the phase diagram we can make an attempt to interpret the results obtained, in terms of number of degrees of freedom or variance of water system. The phase diagram contains four curves, a point of intersection of these four curves and areas separated by these curves.

Curves : Out of the four curves OA,OB,OC and OA' present in the phase diagram of water system, the former three curves (OA,OB & OC) representing phase equilibria of any two of the stable phases. Applying the phase rule equation for this *one component, two phase equilibria represented by these curves;*

The number of degrees of freedom $F = 1 - 2 + 2 = 1$

Thus the curves represent a univariant system, ie., along the curve on *the phase diagram fixing any one of the intensive properties* (either pressure or temperature), automatically fixes the other variables. *The broken curve OA' represents a metastable equilibrium of vapor and super cooled liquid and is also an univariant system.*

Areas : Areas in a one component phase diagram indicate the existence of single phase. Applying the phase rule equation for this one component, one phase equilibria represented by the areas AOB, BOC and COA;

The number of degrees of freedom, $F = 1 - 1 + 2 = 2$

Thus the areas represent a bivariant system, suggesting that in these areas for locating any point both temperature and pressure have to be fixed.

In these areas;

(a) area covered by AOB represents only vapor phase.

(b) area covered by BOC represents only solid ice phase and

(c) area covered by COA represents only liquid water phase.

Point of inter section of curves or Triple point : All the four curves intersect at O and this point of intersection is called the **triple point**. *At the triple point all the three phases coexist and are at equilibrium.* Applying the phase rule equation for the *one component, three phase equilibrium represented by point O*;

The number of degrees of freedom, $F = 1 - 3 + 2 = 0$

Thus the triple point represents an invariant system, ie., any variation in even one of the variable (either pressure or temperature) breaks the three phase equilibrium and transforms into a two phase equilibria represented by the curves. *The temperature and pressure of the triple point are fixed* at 0.098^{o}C and 4.58 mm Hg respectively.

6.5 Two Component System

In a two component system the composition of all the individual phases available in various phase equilibria are expressed with the help of not less than two components. *Systems involving fusion of any two metals, formation of freezing mixtures and salt hydrates are some of the typical examples of two component systems.* In a two component system, minimum number of possible phases could be taken as unity and this corresponds to three degrees of freedom ($F = 2 - 1 + 2 = 3$). Hence to illustrate fully a two component system we require to define at least three physical variables, viz., temperature, pressure and composition. As explained earlier the three possible variables could only be illustrated on a three dimensional phases diagram. To simplify the situation and bring it down to a two dimensional illustration, normally one of the variables in kept constant throughout and only two of the intensive properties are varied and corresponding curves are plotted. We now illustrate the phase diagrams of silver-lead system and KI-water system and explain the finer details of these phase equilibria.

6.5.1 Silver-Lead system

The four possible phases of the silver-lead system are (a) solid silver (b) solid lead (c) solution of silver and lead in molten state and (d) vapor. Silver and lead being very high temperature boiling metals, the vapor is practically absent, therefore pressure changes have a negligible impact on the phase diagram. Negligible pressure effects on the phase equilibria led to a practice of plotting the temperature against percentage composition of one of the metals. As discussed already, this fact helped in illustrating the system by a two dimensional phase diagram, Fig. 6.4.

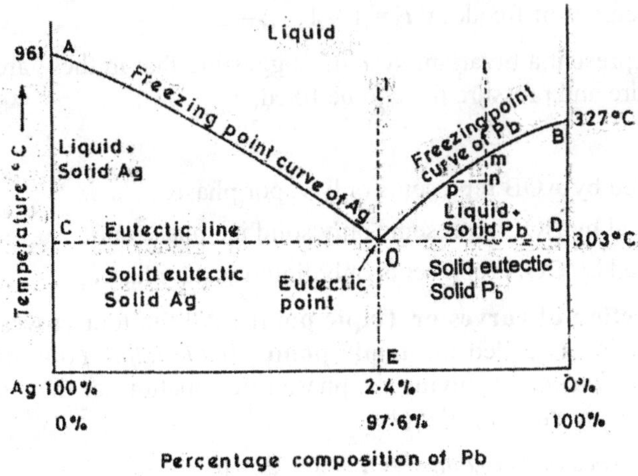

Fig. 6.4 Phase diagram of silver-Lead system.

In plotting the phase diagram, because the measurements are made *at a constant pressure of 1 atmosphere,* the variance of the system reduces by a factor of one and *corresponding phase rule equation, commonly called* ***reduced or condensed phase rule*** *equation may be written as;*

$$F' = C - P + 1$$

The freezing point curve OA : The point A represents the melting point of 100% pure silver (M.P = 961°C) where the solid silver, silver melt (liquid) and vapor coexist. The melting point of silver falls gradually by the addition of lead (depression of freezing point) till the silver melt gets saturated with lead as represented by the curve AO. At **O** the Ag solution is fully saturated with Pb and any further addition of lead is not accepted and it simply separates out as solid Pb. This saturation limit represented by O is called the ***eutectic point*** (Greek: eutectos = easy melting).

All along the curve AO, we notice three phases, solid silver, liquid and vapor in equilibrium. Applying the reduced phase rule equation for this two component, two phase equilibrium. (It is important to note that the vapor phase is already accounted for, while deducing the reduced phase rule equation);

The number of degrees of freedom F' = 2 – 2 + 1 = 1

Hence the curve AO represents a univariant system, where fixing any one of the variables automatically fixes the other physical variable and even a slight variation of one of the variables leads to the total disappearance of one of the three phases.

Freezing point curve of lead OB : In the freezing point curve OB, the point B represents the melting point (M.P = 327°C) of 100% pure lead where the solid lead melt and vapor coexist. By the gradual addition of silver, the melting point of lead falls till the lead melt gets saturated by silver, as represented by the curve OB. Any further addition of lead beyond O, simply separate out as solid lead.

Similar to the freezing point curve of silver, OB also represents a phase equilibrium between solid lead, liquid and *vapor*. Applying the reduced phase rule equation for this two component, two phase equilibrium;

The number of degrees of freedom, $F' = 2 - 2 + 1 = 1$

Hence the curve BO also represents an univariant system where defining one variable amounts to fixing of the other variables automatically.

Eutectic Point O : At this point all the four phases, viz., solid Ag, solid Pb, liquid solution and *vapor* coexist. Applying the reduced phase rule equation;

The degrees of freedom, $F = 2 - 3 + 1 = 0$

Hence point O represents an invariant system, where any minor variation in one of the variables shatters the three phase equilibrium resulting in a two phase equilibrium. The temperature and composition of the eutectic mixture corresponding to eutectic point O at 1 atmospheric pressure are 303°C and 2.4% Ag + 97.6% Pb respectively.

The salient features of eutectic point are;

(i) Eutectic point refers to lowest freezing point of a liquid mixture of any two metals at a given external pressure (conventionally fixed at 1atm).

(ii) Freezing any unsaturated liquid solution described by areas AOC and BOD leaves an eutectic mixture and pure solid metal corresponding to that area.

(iii) Similar to a molten metal, eutectic mixture freezes at a constant temperature. The solid eutectic so formed has a fixed composition and is called an alloy.

(iv) In spite of having a definite composition and fixed freezing point, eutectic mixture is never considered as a compound as the crystals of components are clearly observable under microscope.

The Areas : Drawing a horizontal line CD passing through the eutectic point, we notice three areas above the eutectic line, CD, and two areas below the eutectic line.

The Area AOB : The area AOB represents a liquid solution of Ag and Pb in equilibrium with its vapor. As the pressure is arbitrarily fixed at 1atm;

The number of degrees of freedom in the area AOB are $F' = 2 - 1 + 1 = 2$.

Hence AOB is a bivariant region. On cooling the liquid at any point *l'*, vertically above the eutectic point, eutectic solid is separated only after touching the eutectic point O. The liquid at any other point, say, ***l*** when cooled, remain unchanged until it reaches a temperature corresponding to m on the freezing curve. At *m*, solid Pb will commence to deposit keeping the liquid composition at *m*. On further cooling it enters a univariant region BOD. At any point in this region, say *n*, the composition of the liquid is given by the point *p*, corresponding to the intersection of the tie line and BO curve. Further cooling to eutectic temperature, the point *q* on the eutectic line corresponds to solid eutectic and the solid Pb.

The Area AOC : The area AOC has solid silver and liquid melt in equilibrium. The composition of the liquid melt directly corresponds to the tie line intersection point on the AO curve. This area is a univariant region ($F' = 2 - 2 + 1 = 1$).

The Area BOD : The area BOD has solid lead and liquid melt in equilibrium. The composition of the liquid melt directly corresponds to the tie line inter section point on the BO curve. Thus is also a univariant region.

The Area COE : The area COE has solid eutectic and solid Ag in equilibrium. Applying reduced phase rule equation we find that the region is univariant ($F' = 2 - 2 + 1 = 1$)

The area DOE : The area DOE has solid eutectic and solid pb in equilibrium. This is also a univariant system just like area COE.

6.5.2 Pattinson's process for the desilverization of lead

Silver is often recovered from the work up of lead and copper ores. Galena is a lead ore containing 0.1% silver in it. The enrichment of silver in this ore to 2.4% was achieved from a knowledge of Ag-Pb phase diagram. This process is called **Pattinson's process or desilverization process**. It was already discussed under the area AOB, that a molten liquid ***l***, which is rich in lead,

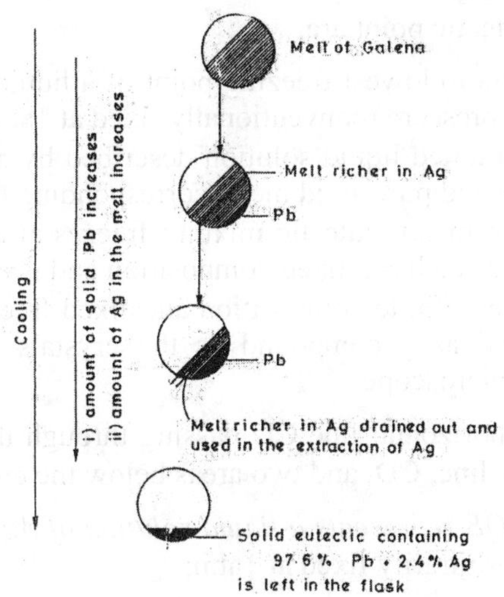

Fig. 6.5 Pattinson's process.

when cooled traces a path ***l m n q***, with a separation of lead metal. The concentration of the liquid at points ***m***, ***n***, and ***q***, correspond to the intersection point of BO curve with the tie lines passing through these points. Hence, as the liquid cools to **q**, the composition of the melt approaches the composition of the eutectic mixture (2.4% Ag and 97.6% Pb). This solid eutectic rich in Ag is profitably then used for the subsequent recovery of silver employing electrochemical techniques. The relative increase in the amount of lead in the solids and increase of silver in the melt on cooling argentiferrous lead is shown in Fig. 6.5.

6.5.3 Other applications of eutectic mixtures

(i) In the soldering of two metal pieces, often an alloy (solid eutectic) that has much lower melting point than the individual metal pieces to be joined is used and the process is called soldering.

(ii) The safety valves fitted in the pressure cookers to prevent bursting due to any malfunctioning are made up of the alloys that fuse as soon as a particular limiting temperature is reached.

(iii) Safety fuses made out of low melting alloys like Pb-Sn alloys, commonly called wood metal, (Bi 50%, Pb 25%, Sn 12½% and Cd 12½%) are used in protecting buildings against fire hazards. Wood metal melts at 65°C. It is used for plugging water sprinklers arranged in the building roof. In case of any fire in the building, the safety plug melts off, putting automatically the water sprayer in action to put off the fire.

Few eutectic systems with eutectic temperature and compositions are given in Table 6.1.

Table 6.1 Some Eutectic Systems.

Components (M.P.in brackets)		Eutectic temperature	Eutectic composition
Component A	**Component B**		**composition (% of A)**
Ag (961 °C)	Pb (327 °C)	303 °C	2.6%
Bi (273 °C)	Cd (323 °C)	140 °C	60%
Zn (419 °C)	Cd (323 °C)	270 °C	67%
KCl (773 °C)	LiCl (601 °C)	361 °C	42%
Sb (630 °C)	Pb (326 °C)	246 °C	–

6.6 Classification of Carbon Steels

Carbon steels have been classified in several ways. The most important method is on the basis of their carbon content. This method will be discussed here. The other important method is according to the process used in its making e.g., acid Bessemer steel, electric steel, and so on. Yet another classification based on uses is also not uncommon, e.g., tool steel, spring steel, etc.

Steel with only a very small amount of elements other than iron has few engineering uses, as besides the cost factor, it is soft, ductile and weak. By increasing the carbon content its hardness increases and ductility decreases. Tensile strength increases to its maximum value of about 10,000 kg/sq. cm when the carbon content is about 1%, but this steel is quite brittle. Therefore, for structural purposes lower carbon steels are used, and if it is to be welded, the carbon content should be below 0.25%. The effect of carbon content on the properties of steel is more marked in hypo-eutectoid steels (i.e., steels containing less than 0.8% carbon) than in hypereutectoid steels (i.e. containing more than 0.8% carbon). The reason is that up to 0.8% carbon the proportion of pearlite increases rapidly to 100%. On further increase in carbon content the cementite proportion, which affects the subsequent properties, does not increase as rapidly.

A common system of classification of steel on the basis of carbon content is as follows, (also see Fig. 6.6) ;

1. *Very mild and mild steels* (up to 0.25% carbon). These are used for making sheets, boiler plates, various engineering structures, rivets, bolts, soft wires, nails, etc.
2. *Low and medium carbon steels* (0.25 to 0.6% carbon). These are used for making machine parts, railway axles, rails, springs, shaftings, etc.
3. *High carbon steels* (0.6 to 2.0% carbon). These are used for making tools, files, knives, saws, wires and cables, ball bearings, etc.

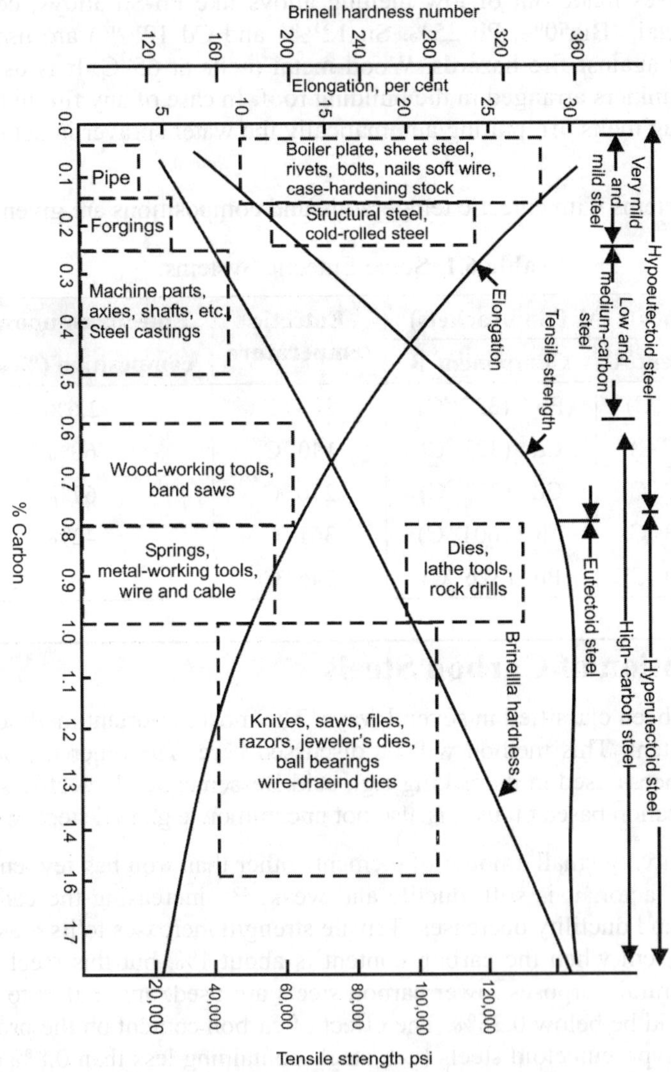

Fig. 6.6 Effect of carbon content on elongation (formability or ductility, hardness and tensile strength of steel).

(*J. Newton, an introduction to metallurgy, courtesy, john wiley & sons*).

6.7 Heat Treatment

Heat treatment may be defined as 'an operation or combination of operations of heating and cooling of a metal or an alloy in the solid state for obtaining certain desirable conditions or properties'. Heating and cooling for the sole purpose–mechanical working (e.g., hot rolling) is excluded from the meaning of this definition.

The main objective of heat treatment processes for steel are: 1. to increase strength, toughness, hardness and ductility; 2. to soften steel; 3. to relieve internal stresses and strains; 4. to refine the grain; 5. to normalize steel which has already been subjected to mechanical or heat treatment operation.

By suitable selection of a process or a combination of processes it is possible to achieve any of these objectives or a combination of them.

In general, the properties of high carbon steels are altered by heat treatment to a much greater exent than those of low carbon steels. The reason is that changes in properties are mostly brought about by changes in the form and distribution of carbon, which is obviously low in low carbon steels.

6.7.1 Principle

It was known even to primitive blacksmiths that if a red hot steel tool is plunged in water it becomes hard, but if it were left to cool slowly it would remain soft. To put it in other words, a sample of steel can be made soft, ductile and tough by one heat treatment and by another treatment it can be made hard and strong. Thus it is seen that the properties of a sample of an iron carbon alloy, e.g., steel, can be largely controlled by heating to a suitable temperature, keeping it for some time (i.e., 'soaking') at that temperature and then cooling at a certain rate. In brief these are the three steps in heat treatment by which the properties are controlled. An obvious use of this kind of treatment is to produce in an alloy the desired properties needed for different uses. Besides this, it is helpful in various operations of production engineering. For example, while making a gear the metal must be soft and machinable so that the proper shape may be given, and then it has to be made tough so that during use the shape does not change and it does not wear out. Or, when a plate is to be repeatedly pressed cold, for instance to make a shell or a tumbler, it becomes hard and brittle after a few pressings and it cannot be pressed further to give the desired shape. So it has to be softened by annealing and then further worked to get the final shape.

In Chapter XV iron-cementite equilibrium diagram has been discussed and it has been made clear that the diagram deals with *equilibrium conditions*. Heat treatment deals with non-equilibrium conditions. By heat treatment either a metastable (semistable) condition is produced in the alloy or a metastable condition which already exists (due to previous heat treatment or mechanical work) is removed. In many cases of heat treatment allotropic changes take place. There is also precipitation from the solid solution.

In the previous Chapter we have seen that at ordinary temperature carbon steel consists of Fe_3C (i.e., cementite) distributed within iron (i.e., ferrite) in different forms. As it is heated to a certain temperature, known as the *lower critical point* (or lower transformation point), α-iron *begins* to change to γ-iron. Carbon is practically insoluble in α-iron. With the appearance of γ-iron the carbon of Fe_3C begins to dissolve in it producing austenite. When heating is continued for some time at a certain temperature (which can be found from the equilibrium diagram) austenite formation is *complete*. This temperature is the upper critical point. The temperature range between the upper critical point and the lower critical point is known as the *critical range*. By heating steel past the critical range all the carbon present in different forms can be brought in a homogeneous solid solution, thus all the previous effects of the original microconstituents can be wiped off. By controlled cooling new constituents can be produced that will have the desired properties. Heating can be compared to rubbing off the old figures from the blackboard and controlled cooling is like making new figures on the blackboard.

If a sample of carbon steel, which is in the austenitic state, is *quenched* (i.e., suddenly cooled) by plunging in water or brine, the γ-iron which has face-centred cubic lattice and held carbon in solid solution, being unstable at the lower temperature, is changed to α-iron which is the stable form at that temperature. Carbon is insoluble in α-iron which has body-centred cubic lattice. Quenching lowers the temperature so rapidly that the atomic readjustments cannot fully take place and there is little time for carbon to diffuse. Thus a new constituent, known as *martensite*, results. It is very hard and brittle and is an essential constituent of the steel used for cutting tools. Obviously, if the sample is of low carbon steel the amount of martensite produced is small and the sample does not get hardened to any appreciable extent. In a sample of very high carbon steel, a very large amount of martensite is produced with the result that it becomes very brittle and may break easily. To remove brittleness the hardened sample is re-heated below the critical range. At this elevated temperature there is greater atomic mobility, consequently highly strained atoms readjust to new positions, thereby removing some hardness and brittleness. The sample is then cooled in air, water, oil, etc., to get the desired properties. This is the principle of *tempering*. It removes some of the internal stresses, yet the metal is quite hard and tough.

If steel is brought in the form of a solid solution, i.e., it is in the austenitic state and is cooled very slowly, there is enough time for equilibrium conditions to be established giving stress-free metal. γ-iron changes to α-iron and therefore the carbon which was dissolved in γ-iron diffuses and rearranges itself. This induces softness in steels and is the basis of *annealing and normalizing* operations.

The importance of proper heating cannot be over emphasised. First, the temperature to which steel is to be heated has to be decided. It depends upon the chemical composition of steel and on the structure and properties desired in the final product. If it is heated to a higher temperature or held there for a longer period larger grains will be produced. On cooling, the reverse reaction, i.e., reduction in grain size, does not take place. Other things being equal, the larger the grain size the weaker is the material. In general, this is undesirable except when greater machinability is required or for wire drawing. Hypo-eutectoid steels (i.e., $< 0.8\%$ C) or mild and medium carbon steels are generally heated to just above the *upper* critical range.

Hyper-eutectoid steels (i.e., > O.8% C) are heated just above the *lower* critical range. This enables the carbon present in the form of pearlite (but not cementite) to diffuse into the austenitic state. Higher heating temperatures cause large size grains to be produced. After heating to the required temperature the steel is held at that temperature for some time depending upon the cross section of the sample. In general, the time allowed is about half an hour per centimeter cross section. This is known as *soaking*. The aim of soaking is to allow sufficient time for the changes (or dissolution of carbon to be completed and equilibrium to be established). Heating steel to a high temperature exposes it to the danger of oxidation on the surface giving rise to oxide film and causing removal of carbon. The oxide film acts as an insulator and prevents rapid cooling. This is undesirable especially in hardening operations. Moreover, the oxide scale may not uniformly cover the entire surface, consequently soft spots may be produced at certain places. Therefore before cooling the scale needs to be removed. Generally this is done mechanically.

The mechanical properties of carbon steel depend upon its microstructure. The microstructure depends upon *carbon content* and *allotropic changes* that take place in steel during heat treatment.

6.7.2 Constant Temperature Transformation

E. C. Bain and E. S. Devenport investigated constant temperature (isothermal) transformations of austenite at temperatures below the critical temperature. The investigations resulted in time-temperature-transformation (TTT) curves. Because of their shapes like the letter 'S' these curves are also known as 'S-curves' or as 'Bain curves' after the name of the principal investigator, E. C. Bain. It has already been explained that austenite consists of carbon atoms occupying interstices of face-centred cubic lattic of γ-iron. Gamma iron is stable at higher temperatures, and on cooling body centred cubic lattice of α-iron is formed in which carbon atoms do not dissolve because of the absence of suitable interstices. For this reason the carbon atoms segregate and form cementite. This involves movement or diffusion of carbon atoms which depends on both time and temperature. That is why the study of TTT curves is important in heat treatment considerations. At higher temperatures movement of atoms is rapid and at lower temperatures it is slower. Therefore when a carbon steel is heated to form austenite and quenched and kept at a *constant* temperature the austenite transforms at a rate which is characteristic of that particular temperature.

Bain and Devenport heated a number of small specimens of carbon steels until austenite was formed. The specimens were quenched in a bath maintained at a constant temperature for different times varying from 0.5 second to 1,00,000 seconds (over a day). During this period a portion of the austenite transformed to ferrite and cementite. The specimens were withdrawn and quenched in ice-cold brine. This rapid quenching transformed the unchanged austenite into martensite. The specimens were examined under a microscope and the extent to which decomposition had taken place and the products formed were determined. The experiment was repeated with quenching done at various temperatures.

In the TTT curves, temperature is plotted as ordinate and time in seconds, on a logarithmic scale as the bascissa. Logarithmic scale is necessary for time so that both the short periods in which the transformation reactions begin and the longer periods taken for their completion can be depicted. These curves are very helpful in determining the transformation of austenite under different conditions They show what micro-structures to expect in a particular heat treatment operation. As the micro-structures decide the mechanical properties the curves help to decide the treatment to be given for obtaining the desired properties. The curves show micro-structures under non-equilibrium conditions just as the equilibrium diagrams show under equilibrium conditions.

In Fig. 6.7 is shown 'S-curve' for a 0.6 per cent carbon steel. The lines Ae_1 and Ae_3 represent the transformation temperatures under equilibrium conditions. Austenite is stable above Ae_3. It is unstable below Ae_1 and transforms in time. It is partially stable between Ae_1 and Ae_3. Ms and Mf represent temperatures at which transformation of austenite to martensite starts and finishes respectively. It will be seen from the curve that in order to obtain a completely martensitic structure, the sample has to be cooled at such a rapid rate that the time-temperature curve passes to the left of the knee of the curve (at about 550°). The constituents formed at different rates of cooling can also be known by referring to the Figure.

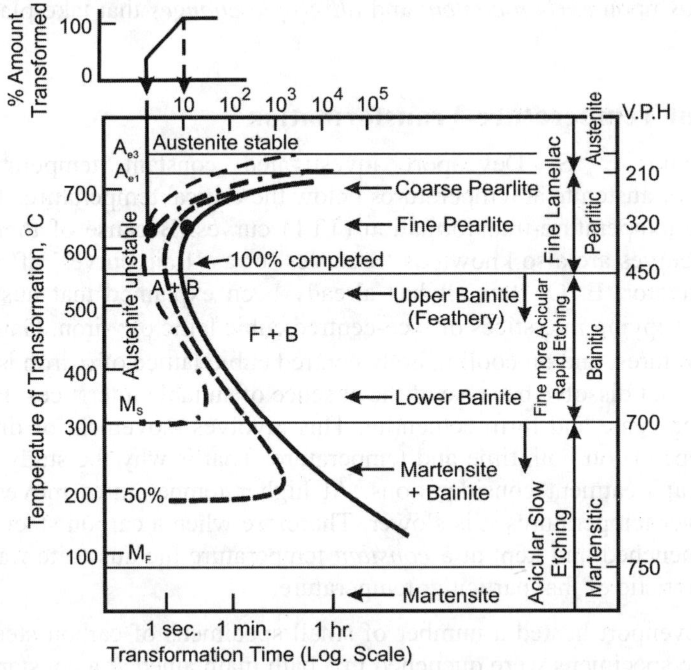

Fig. 6.7 Idealised 'S-curve' for 0.6 percent carbon steel depecting time interval required for beginning (broken curve at the left), 50 percent (dotted curve) and 100 per cent (continuous curve at the right) transformation of austenite at a constant temperature. In the right hand column the hardness (on Vickers Pyramid Number) is given.

A-Austenite F-Ferrite, p-Pearlite, B-Bainite. (E.C Rollason, *Metallurgy for Engineers*, Courtesy, Edward Arnold, London).

6.7.3 Annealing

The following description of full annealing has been given by the American Society for Testing Materials:

"A softening process in which an iron-base alloy is heated to a temperature above transformation range and after being held for a proper time as this temperature is cooled slowly to a temperature below the transformation range. The objects are ordinarily allowed to cool slowly in the furnace although they may be removed from the furnace and cooled in some medium which assures a slow rate of cooling." In simple words, annealing is a process in which steel is heated, held at that temperature for sometime to dissolve carbon in γ-iron, and then slowly cooled. The annealing process, therefore, consists of three stages, heating to the right temperature, keeping (soaking) at that temperature and slowly cooling.

The objective of annealing may be one or more of the following:

1. to soften the steel (to improve machinability);
2. to remove internal stresses (produced by mechanical working like rolling or forging or by uneven heating);
3. to change ductility, toughness, electrical and magnetic Properties ;
4. to refine the grain structure and
5. to remove gases entrapped within the mass.

A plain carbon steel with less than 0.8% carbon (hypo-eutectoid steel) is heated to just above the *upper* critical point. Steel with more than 0.8% carbon (hyper-eutectoid steel) is heated to just above the *lower* critical point–this lower heating temperature brings the pearlite in solid solution as austenite but cementite remains unchanged. Therefore, hyper-eutectoid steel is not softened as much as the hypo-eutectoid steel, but it becomes soft enough for machining. This is because heating hyper-eutectoid steel above the upper critical range will produce grains of larger size and moreover these steels are used for purposes where softness (as in mild steels) is undesirable. If the cross-section of the piece is large or the shape is complicated or carbon content is high, the rate of heating should be slow. It is then held at the annealing temperature (technically called *soaked*) for half an hour for each centimetre of cross-section.

After soaking, the steel is allowed to cool very slowly. This is generally done by leaving it in the furnace itself and shutting off the heat. Sometimes it may be buried in non-conducting material like sand, ash or lime.

The process just described is generally known as full annealing to distinguish it from *Process* or *sub-critical* annealing. When a piece is subjected to cold working (like rolling or forging steel at room temperature) some stress is produced in it and it becomes harder and more brittle. To remove this defect the piece is heated to a temperature below the lower critical point and cooled slowly. This operation is known as process annealing. This is commonly done in the sheet and wire industry and the object is to soften the metal for further cold working.

By full annealing, machinability and ductility are improved but hardness, strength and abrasion resistance become less.

6.7.4 Normalizing

Both normalizing and annealing treatments are given for making steel soft and homogeneous. Often these terms are confused with each other. There are two main points in which normalizing differs from annealing:

(i) hypo-euctectoid as well as hyper-eutectoid steels are heated above the upper critical range (in annealing hyper-eutectoid steels are heated just above the lower critical range), and

(ii) cooling is done in still air and not in a furnace. Therefore, the rate of cooling is faster. More rapid cooling becomes necessary in order to prevent the formation of coarse grains at the higher temperature.

As the name normalizing suggests, it is meant to bring back the structure and properties that are normal for the particular sample. It is in fact a corrective treatment for grain refinement in steel after it has been subjected to mechanical working like rolling, forging, etc. By normalizing finer and stronger pearlite is produced than is produced by full annealing. As compared to annealing, normalizing produces higher tensile and yield strengths and greater impact resistance though the ductility is slightly less. The hyper-eutectoid steels are rarely normalized because due to the much higher temperature employed there is a tendency to form coarser grains which reduce the strength.

6.7.5 Spheroidizing

Spheroidizing is also a softening treatment. It is applied to high carbon steels. Due to the presence of cementite high carbon steel is very hard and difficult to machine. It is heated to just below the lower critical point (e.g. 680° to 700°) when relatively larger particles of cementite are formed. These become like tiny balls or spheroids (hence the process is called spheroidizing). Thus the continuity of cementite plates is broken and the steel becomes machinable. After keeping it at the elevated temperature for some time steel is cooled slowly.

By this treatment the tool becomes less liable to crack or distort when it is subsequently hardened. On hardening, the tool will be more uniform in hardness. It makes steel more machinable, and thus saves cost of machining and increases the life of the tool.

6.7.6 Hardening

Hardening is just the opposite of annealing. The heating temperatures are similar in both cases, i.e., for hypo-eutectoid steels just above the upper critical temperatures and for hyper-eutectoid steels just above the lower critical temperatures. Hyper-eutectoid steels are not heated above the upper critical range because cementite is a very hard material and need not be completely dissolved to give austenite and then quenched to produce martensite again. The lower temperatures are sufficient to dissolve pearlite which is subsequently converted to martensite by quenching. As in annealing, the steel is soaked at the elevated temperature to allow enough time for equilibrium to be established. It is then quenched (cooled rapidly) by immersing in water, brine, oil or molten lead. Due to a quick drop in temperature there is not enough time for equilibrium to be established with the result that a forced solid solution of carbon, known as martensite, is produced. It is a very hard material. Unless it is a small and thin piece the interior does not become as hard as the outside because of slower cooling of the inner portion.

Cold water is used for rapid quenching. If the steel is quenched in water from high temperatures, there is a danger of steam bubbles forming around the metal which act as insulators and reduce the rate of cooling. Caustic soda and salt solutions have higher boiling points and better heat conductivities. Therefore they are better quenching media.

It may be noted that hardness is accompanied by brittleness. Therefore, a fully hardened piece is not so tough and may not be suitable for most commercial purposes. Naturally the hardness of low carbon steels cannot be increased greatly. Because of the lack of carbon, little cementite and martensite are produced in very low carbon steel.

When a piece of steel is heated and quenched, the temperature drops rapidly at the surface and slowly in the interior. This causes the surface to become hard but the interior is not so hard. Surface hardness obtained after quenching depends on the carbon content of steel. The depth to which the object becomes hard depends on the carbon content, the total content of the alloying elements, and the grain size. Therefore it is necessary to add alloying elements for hardness to penetrate to a certain depth. Of course, in many cases it is not desirable that hardness should penetrate deep into the interior.

6.7.7 Tempering

It has just been explained that on hardening, a forced solid solution of carbon in α-iron is produced, This is a metastable (semi-stable) state and the steel is in a stressed condition. It is brittle and likely to crack or break in use. Therefore, it is desirable that the brittleness and internal stress should be removed, though to a slight extent hardness has to be sacrificed in achieving this.

After hardening, steel is heated to a temperature *below* the lower critical range. This permits some mobility of atoms and carbide begins to be precipitated from the forced solid solution of carbon in α-iron. Steel is then slowly cooled. The temperature to which steel is to be heated and soaked and the rate of cooling depend upon the initial condition of the steel and the properties desired in the final product. Heating below 400° gives good strength and hardness. On heating from 400 to 600° good ductility and toughness are, produced but hardness is reduced. This process of removing some brittleness and relieving internal stresses, which is accompanied by a slight reduction in hardness, is known as *tempering*. If the temperature is high the carbon may diffuse, may even dissolve, and the hardness will be reduced sufficiently.

6.8 Self -Test Questions

1. Number of phases present in a mixture of nitrogen and oxygen are
   (A) 1 (B) 2
   (C) 3 (D) 4
2. Number of components present in a solution of sodium chloride is
   (A) 1 (B) 4
   (C) 3 (D) 2
3. The number of phases present in powdered tin and lead is
   (A) 1 (B) 4
   (C) 3 (D) 2
4. At melting point of ice, the system has two phase. The system is one component. The degrees of freedom of the system are
   (A) 1 (B) 2
   (C) 3 (D) 4
5. Water system is non variant at
   (A) melting point (B) boiling point
   (C) triple point (D) critical point
6. Mathematical statement of reduced phase rule is
   (A) $F' = C - P + 2$ (B) $F' = C - P + 1$
   (C) $F' = C - P + 3$ (D) $F' = C - P$
7. Maximum degrees of freedom for one component system are
   (A) 1 (B) 3
   (C) 2 (D) 4
8. Number of components in the system of Ag and Pb at eutectic temperature is
   (A) 1 (B) 2
   (C) 3 (D) 4
9. For Ag-Pb system, when it has three phases, the number of degrees of freedom is
   (A) 1 (B) 2
   (C) 0 (D) 3
10. A homogenous and physically distinct part of the system which is separated from other parts of the system is called____________.

11. The minimum number of constituents required in order to describe the composition of phases present in a system is called_______.

12. Eutectic temperature of Pb-Ag system is _______.

13. Phase diagram of Ag-Pb system is drawn between _________ and composition.

6.9 Short & Essay Type Questions

1. Draw the phase diagram for water and mention its parts.

2. Construct the phase diagram of silver - lead system and explain the importance of eutectic point

3. State phase rule and explain the terms involved. Apply the phase rule to silver-lead system.

4. What is a phase diagram ? Explain.

5. What is an eutectic ? Give one example.

6. Explain with suitable examples the following terms used in the phase rule study of heterogenous equilibrium

   (i) Triple point (ii) Eutectic point (iii) Univariant system

7. Discuss the application of phase rule to potassium iodide -water system.

8. (a) Write the phase rule equation. Discuss the application of the phase rule to a one component system.

   (b) Define the terms 'Eutectic point' and triple point.

9. Define the terms in the phase rule. Explain these terms in the following system

   (i) A gas in equilibrium with its solution in a liquid

   (ii) Ammonia gas

   (iii) $I_2(s) \longleftrightarrow I_2(g)$

10. (a) Define the terms (i) Component (ii) Phase of a system

    (b) Predict the degrees of freedom for the following

    (i) A one component system having two phases in equilibrium

    (ii) A one component system having a single phase.

    (iii) A one component system having three phases in equilibrium.

11. Discuss a method to obtain the phase diagram of Ag-Pb system and give details of the number of degrees of freedom of the curves and areas in it.

12. Find the degrees of freedom and number of components in the following systems

    (a) $CaCO_3(s) \longleftrightarrow CaO(s) + CO_2(g)$

    (b) A two phase system made by mixing arbitrary amounts of C, O_2 and CO_2 at a temperature at which the reversible equilibrium establishes

    $C_{(s)} + O_2\,(g) \longleftrightarrow CO_2(g)$

13. State phase rule and apply to water system.

14. Plot the phase diagram of a one component system and identify all the lines, areas and phases.

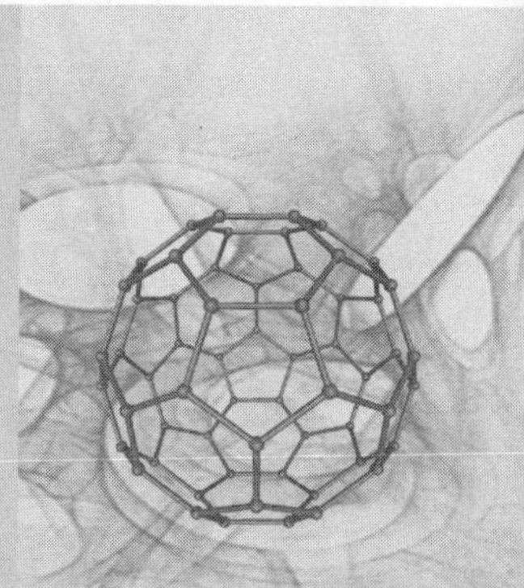

7 Fuels & Combustion

7 Fuels & Combustion

7.1 Introduction

A chemical fuel is any substance used to produce heat and power by combustion. Only *chemical* fuels will be discussed in this chapter. *Nuclear* fuels are now being used in several large plants for the generation of electrical power and propulsion of ships and will be considered in a subsequent chapter. In these cases the energy liberated is by fission reactions. By common usage the term fuel also implies that it can be stored and transported.

Fuels are primarily used for heating but they have other very important functions. For example, in an internal combustion engine petrol is not used for heating the engine. In several metallurgical operations, as in a blast furnace, coke not only supplies heat but also acts as a reducing agent.

7.1.1 A Survey of Fuel Resources

The fuel resources of the world have been estimated differently by different authorities. It gives a better picture of the problem if we consider for how many years a particular kind of fuel is likely to last rather than to give estimates of its known reserves. To give a general idea, it may the said that the world's gas supplies are likely to last fifteen years, oil eighty years, coal five hundred years, and nuclear fission materials two thousand years. Materials for fusion reaction seem to be unlimited. India is very well placed in its thorium deposits in which she holds world monopoly. By breeder reaction thorium gives an isotope of uranium which can easily undergo fission.

India has sufficiently large deposits of coal but these are small when compared with some other developed countries, and are perhaps insignificant when calculated on the per capita basis. Hence much foresight in planning and prudent use of coal is necessary. Our best grade coal, which comprises only 13% of the total reserve, contains about 17% ash. This in itself is too high an ash content when compared to coal ash of a big producer like England (which is about 5%, for comparable quality). 70% of our coal is of rather low quality containing as high as 35% ash which virtually means that one third of the coal is ash. To meet the challenge several coal washeries have been established which make several fractions of coal, the better grades being used for metallurgical purposes. Production of gas from such inferior coal needs to be given consideration.

In the early part of the 1960's the oil position of India seemed to be quite bright. Major oil fields were found between 1952 and 1964 in Naharkatiya, Cambay, Ankaleswar, Kalol, Rudrasagar and Lakwa. Some of these are drying up. Also due to the very heavy rise in oil prices the economy of the country has been upset besides the uncertainty of supply during any war or crisis. India has high technical skill in oil exploration due to which her experts are helping other countries, like Iraq, to find new oil fields. Oil exploration under the sea near the shore has received much impetus and has yielded rich dividends as evidenced by the production from the Bombay High.

When people think of petroleum, their attention naturally goes to petrol. It may be mentioned here that petrol is about one-eighth of the total petroleum products. Diesel oil, furnace oil, lubricants, etc. and numerous other materials for the manufacture of which petroleum is the basic raw material also have their importance.

The fertiliser industry is another large consumer of naphtha. It is the same fraction which gives petrol for the internal combustion engines. It is said that one part by weight of fixed nitrogen in the form of a fertiliser gives ten parts of the cereals. Hence the fertiliser industry has to be given priority. But how does naphtha come into the fertiliser scene? Ammonia is the basic material for the fertilisers for the manufacture of which atmospheric nitrogen is used and hydrogen is produced by catalytic decomposition of the hydrocarbons thus:

1. $CH_4 \rightarrow C + 2H_2$
2. $CH_4 + H_2O \rightarrow CO + 3H_2$
3. $CO + H_2O \rightarrow CO_2 + H_2$

(CO_2 being removed by washing with water under pressure).

In the coal-based fertiliser plants the hydrogen and nitrogen will be produced by passing steam and air through hot coal. This has the advantage of utilizing poor grade coal and the saving of naphtha.

A very large number of useful products are made from petroleum fractions. Some of the major consumers, to name a few out of a large number, are plastics, rubber, organics, synthetic fibres, pharmaceuticals. Our earlier Five Year Plans were more oil-oriented than coal-oriented, as is clear from the Dieselisation of the railways, oil based fertiliser plants, encouragement given to small consumers to have Diesel engines, and oil-fired furnaces. This approach has undergone radical rethinking. As petroleum has become a material of strategic significance, several countries are planning to have stocks equivalent to '90 days' use. Our country should take immediate steps to do this. Serious attention is being given to the production of synthetic petrol also to basic raw materials for petro-chemicals utilising low grade coal.

A sizeable amount of our electric power comes from hydroelectric power plants which depend upon rainfall. Paradoxically, when there is less rainfall, with concomitant less electricity available more water pumps need to be operated for irrigation.

It is absolutely necessary to economise and efficiently use the fossil fuels, but that is not enough. To burn away coal and petroleum products, which are essential raw materials for many chemicals is dangerous. These must be preserved for certain vital chemical industries. All this forces one's attention to nuclear power stations. One major apprehension against nuclear fuels is the safety factor. The present safety record of the nuclear power plants is good. An expert has suggested that the plants may be built underground at suitable impermeable geological formations so that in case of accidental wreckage, the radio-active wastes may not spread. According to him, the underground construction will add only a small fraction to the total cost. The safety and security of handling and transporting a fuel like plutonium will remain. According to the famous Nobel Laureate, Glenn T. Seaborg, even the large scale use of nuclear energy for power production will increase the radioactivity of the environment by less than one percent of the natural background radioactivity. He further says that this is less dangerous to man than billions of tonnes of carbon dioxide, millions of tonnes of sulphur dioxide and large quantities of other harmful matter emitted by plants using coal and petroleum products. When nuclear fusion plants are established, radioactive wastes will be much less. Processes have been suggested for the disposal of the radioactive wastes mainly by converting them to solids and burying them safely. It is easier to exercise control over a few large nuclear power plants producing electricity for the entire country than to control the activities of a large number of thermal power plants scattered all over the country.

7.2 Classification

Fuels are classified according to whether they occur in nature (primary fuels) or are prepared (secondary fuels) and according to whether they are in the solid, liquid or gaseous state. The classification is summarized in Table 7.1. They are classified according to their occurrence so that it is known whether a material is available in nature and can be used directly or it has to be specially prepared. Each physical state has its advantages and disadvantages over the other states, hence it is convenient to classify according to the physical state.

Table 7.1

State	Natural (or primary)	Prepared (or secondary)
Solid	Wood, Peat Lignite, Coal	Charcoal Coke
Liquid	Petroleum	Petroleum distilates, Alcohols, Tar
Gaseous	Natural gas	Coal gas, Coke oven gas, Producer gas, Water gas, Blast furnace gas

7.3 Solid Fuels – Wood, Chemical, Coal

In India wood is used to a considerable extent as a domestic fuel. Its use as an engineering fuel is restricted. On a moisture-free basis, wood contains about 50% C, 6% H_2, 43% O_2 and less than 1% ash. Freshly cut wood contains 25 to 40% moisture and air-dried wood contains about 15% moisture. The calorific value of common air-dried wood is about 3,500 cal/g, but in some cases it may be as high as 5,000 cal/g.

A reference to bagasse, as a fuel, may be made here. Bagasse is the residue left after extracting the juice from sugar cane. It is an important fuel in sugar mills. Its calorific value on 15% moisture basis is 3,800 cal/g. Attempts are being made in India to make paper on a large scale out of bagasse instead of using it as a fuel in the sugar mills.

A recent study shows that wood has surpassed nuclear power as an energy source. Brazil, which is the eleventh largest steel producing country of the world, meets 40% of its energy needs of the steel industry by burning wood. Of course, Brazil is a vast country with thick forests and low population density. Sweden, a leading country in paper and pulp industry, meets 60% of the energy needs of this industry by burning wood wastes. Several countries are experimenting with mixing methanol (produced from wood) which is pollution free, with petrol.

A warning must be given against indiscriminate use of wood. Cutting down of forests has serious environmental effects. Although attention has been directed to wood due to shortage of petroleum and also the increasing price, methods of increasing wealth by planting fast growing trees are being explored. As a matter of fact there is already a need in many countries, like India, to have more forests.

7.3.1 Wood Charcoal

Wood charcoal is obtained by the destructive distillation (heating out of contact with air) of wood. Water and various kinds of volatile matter are expelled. If the process is carried out carefully the original form and structure of wood is left. Charcoal consists of about 80% C, 15% O_2 and N_2, 2% H_2 and 3% ash. Its calorific value is high being 8,000 cal/g. Usually the weight of charcoal obtained does not exceed 20% of the weight of wood. Its volume is 50-75% of the original volume. The physical properties largely depend upon the temperature at which the destructive distillation was carried out and on the rate of heating.

Because of its purity and low ash content, charcoal was the most important metallurgical fuel. Its chief defect is its poor mechanical strength which results in its being crushed to powder in the operation. The powdered charcoal can easily be swept away in a current of gases and also it may prevent the proper flow of gases in the furnace. Consequently it has been largely replaced by coke. Because of its very large surface area, its main industrial use is in the adsorption of gases, colouring matter, etc., in the various purification processes of the chemical industry.

7.4 Coal

7.4.1 Introduction

Coal is a complex substance which originated several million years ago from the various plants that existed profusely on the surface of the earth. The plants were caught up in swamps, buried underground and slowly changed due to heat, pressure and complex biochemical and chemical processes. Now it is found as beds or layers in between clay or sediments which were deposited in course of time. Coal is an important fuel. The fact that on proper processing it yields gas, oil, fertilizers and chemicals increases its usefulness. As coal is easy to mine and handle, its use will continue even against severe competition from mineral oil, hydroelectric power and nuclear fuels.

7.4.2 Origin

Coal has been defined as 'a stratified rock consisting of the more or less altered remains of plants, together with associated mineral matter', The process which changed plant into coal has taken millions of years, and has been accomplished by bacteria, heat and pressure.

Geologists have put forward two theories of coal formation. Millions of years ago when the temperature was moderate and rain-fall was heavy, vegetation was thick and grew to far greater size than it does at present. This was so, especially in low-lying wet land. The first theory, known an *'in situ'* theory, contends that the coal formation took place where the vegetation grew and accumulated. The second theory, known as *'drift'* or *'transportation'* theory, holds that the trees, etc., were uprooted and the material drifted or was transported by rivers to lakes and estuaries and deposited there. Thus huge piles of wood collected and got buried underground. There are arguments for and against each of these theories but it is now accepted that in some cases the first theory and in other cases the second theory is more applicable. For example, the great purity of many coal seams can be better explained by the *'in situ'* theory and great thickness of coal seams appears more reasonable on the basis of the *'drift'* theory (About 20 m depth of vegetation is needed to give 1m depth of coal seam). The Gondwana coals of India show more evidence in favour of the drift theory.

As the plants accumulated under water, some silt also settled, along with them. Thus, at a later stage some mineral matter was mixed with the coal as it formed. The initial transformation of wood is believed to be due to bacterial action. In the earlier stages aerobic bacteria caused some oxidation but mostly anaerobic bacteria brought about the degradation of organic matter and the removal of oxygen. The bacterial action produced acidity, which if accumulated, prevented further action and coal formation could not proceed. In some situations the soil was alkaline which removed the acidity and the bacterial action continued to take place bringing about further degradation of matter. This explains why coal is found associated with certain kinds of rocks. There is still some difference of opinion among the authorities regarding the exact mechanism of the bacterial action. But many now accept that the bacterial action took place only in the early stages and is responsible for the breaking up of organic matter into simpler molecules. The degraded matter, as time elapsed, gave rise to peat. Beyond this stage heat, pressure and time became the chief factors responsible for further conversion. Coal formation cannot be fully explained on the basis of heat and pressure theory, even though they are high at the depths at which coal is found. The process is extremely slow and hence coal formation must have been completed only in the course of millions of years. It is interesting to note that in most cases, although there are some exceptions, with the increase in vertical depth, the carton content increases and volatile matter decreases. Anthracite and graphite are formed under extreme conditions.

This process of coal formation may be complete or may be stopped at any stage, thus giving rise to material of varying maturity. This is expressed by the term 'rank' of coal.

The stage in the formation of coal from vegetation can best be summarised thus :

Plant debris $\rightarrow$ peat $\rightarrow$ lignite $\rightarrow$ brown coal $\rightarrow$ sub-bituminous coal $\rightarrow$ bituminous coal $\rightarrow$ semi-anthracite coal $\rightarrow$ anthracite coal $\rightarrow$ graphite.

This series is known as *'peat to anthracite'* series. Each member is more *'mature'* or is of higher *'rank'* than the one which precedes it in the series. Thus anthracite coal is of higher 'rank' than semianthracite coal. With the progress of coal forming reactions, moisture and oxygen were removed and the percentage of carbon increased. Heating value increased from 4,500 cal/g for lignite to 8,600 cal/g for bituminous coal. There is a slight fall in the heating value of anthracite coal because of the removal of hydrogen.

7.4.3 Coal Classification

Coals have been classified in several ways. The most common method of classification is on the basis of rank. From the origin of coal it is clear that wood, after a long interval of time under certain conditions, was converted into coal. Rank represents the extent to which this process of coal formation has proceeded or, in simple words, maturity or age of coal. The important ranks are described in the following pages.

7.4.4 Peat

In the process of the formation of coal from wood, the first stage is the formation of peat. Strictly, it is not regarded as coal. Peat is mostly dug by manual labour, not much machinery is employed and it is air-dried. It is uneconomical to obtain peat lying buried deep under the earth's surface. One principal difficulty in its use is that peat deposits are associated with large amounts of water. There may be as much as 90% water and only 10% solid. To make it an economical fuel, its moisture content should be reduced to 30% or less. For this purpose it is dried in the sun for about one to two months. Its calorific value is only slightly higher than that of wood. Large quantities of peat are used as domestic fuel in Europe. In some parts of Russia, due to the shortage of coal, a considerable amount of peat is also used for power generation. It is not one of the important fuels used in India. Small amounts of peat deposits are found in Nilgiri Hills (Tamil Nadu) and Sunderban (W. Bengal).

7.4.5 Lignites and Brown Coal

In lignites the transformation of wood to coal has taken place to a greater extent than in peat. Their properties are intermediate between peat and coal. Lignites are found near the earth's surface and usually in thick seams, sometimes as much as 30 m in thickness. These are classed into two groups.

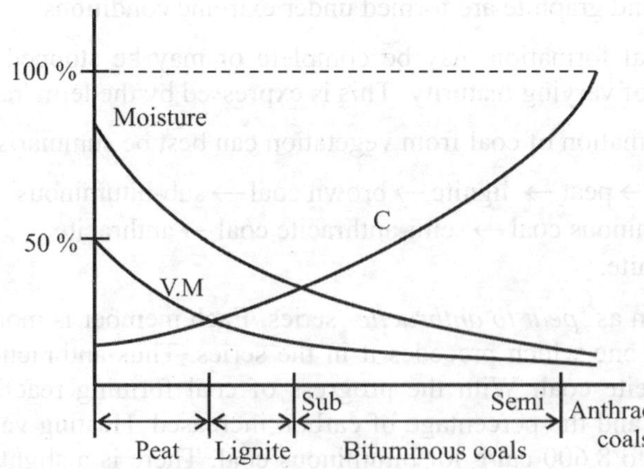

The true lignites are those which contain high proportions of relatively unchanged woody matter, have a high percentage of moisture, say, up to 60% and have low calorific value. The other form is commonly known as brown coal. Brown coals contain much less moisture, around 20% and have a higher calorific value.

Lignites are usually amorphous and break easily, they are therefore difficult to transport. They burn with long smoky flames.

Large amounts of lignites are consumed in Germany and Russia. In German plants large quantities, about 2,000 tonnes per day, are continuously dried by steam or hot gases obtained by the combustion of a portion of the lignite. The moisture content is thereby reduced to about 15% and then the product is made into briquettes by pressing. On the basis of a moisture content of 15%, the calorific value is about 4,500 cal/g.

Lignite is found scattered in several parts of India but the most important deposits are at Neyveli, about 200 km south of Madras. These are spread over an area of about 250 sq. km. The deposits in this area are estimated to be about 200 million tonnes and 5.9 million tonnes of lignite was mined in 1981. These deposits are at a depth of 55 m below the ground level and the average thickness is about 16 m. One of the main difficulties faced in the working of the project has been that water exists in the mines under high pressure which can completely flood the mines. This difficulty has been overcome by drilling a number of artesian wells and constantly pumping out the water.

The Neyveli Lignite Project is a complex project and plays an important role in the economy of the area which lacks fuel deposits. A 250 MW thermal power station is operational at Neyveli. The project manufactures 1.5 million tonnes of urea, to be used as fertilizer and 3.8 million tonnes of carbonized briquettes for use in the homes and the industries.

Out of the total estimated lignite deposits of 2,100 million tonnes in India, 1,900 mt are in Tamil Nadu alone. Smaller amounts of lignite are also found in Kutch, Rajasthan and Kashmir. The Kashmir lignite is of low calorific value (about 200 cal/g).

7.4.6 Sub-bituminous Coals

These are black in colour and are more homogeneous and smooth in appearance. The moisture and volatile matter contents are high, and on exposure to air, these slowly crumble into smaller pieces. These are non-caking coals. The calorific value of sub-bituminous coals on a mineral free basis, is 7,000 cal/g or a little higher. Carbon content varies from 75 to 83% and oxygen content from 13 to 20%.

7.4.7 Bituminous Coals

The term bituminous coal covers coals with a wide range of properties. Therefore the term is used after prefixing caking, non-caking, low volatile, medium volatile, high volatile, etc. (e.g., caking bituminous coal). The carbon content varies from 78 to 90% and volatile matter from 20 to 45%. Coals of this class are used in the largest quantities in industries for making metallurgical coke, coal gas and for domestic heating. The calorific value, on a mineral free basis goes up to about 9,000 cal/g.

Bituminous coals are black, comparatively hard and brittle and burn easily with a yellow smoky flame. The moisture content is below 6 percent. The bituminous coals are of two broad types. The 'caking coals' are rich in hydrocarbons. On heating, these swell and fuse together forming larger lumps. More air therefore has to be supplied for their combustion. The 'non-caking coals' are usually called 'steam coals'. These do not fuse and are largely used for industrial purposes and are especially suitable for steam raising.

7.4.8 Anthracite Coals

These are of the highest 'rank' and in them the coal formation process has proceeded up to the maximum extent. The quantity of these coals available is very limited and hence they are expensive. Small amounts of anthracite are found in Darjeeling and Jammu. The carbon content is 90 to 98% and volatile matter is less than 8%. Therefore they ignite with difficulty. Anthracite coals are hard and brittle and produce very little dust. These are black but can be handled without soiling the fingers. They are non-caking and burn without smoke and with a short non-luminous flame giving intense local heating.

7.4.9 Indian Coals

In the late 18th and early 19th century, exposures of coal were noticed in different parts of India. In 1774 the first colliery of India was opened in Raniganj coalfield and in 1775 the first consignment of 100 tonnes of coal was received in Calcutta. Systematic exploration of coalfields was started 1845.

The coal deposits in seams more than 1.2 m thick and occurring down to depth of 600 m (which is the present working limit) are estimated to be about 1,20,000 million tonnes. Out of this only about 4,000 million tonnes are Tertiary coal and the remainder is Gondwana coals. The coking coals are in short supply, being only about 2,000 million tonnes. Deposits of inferior variety coals are very vast. The quantity of coking coal is insufficient for the requirements of Indian metallurgical industries, therefore the inferior coals are being washed to remove ash and coking coal is being suitably blended with weakly coking coal.

Most of the Indian coal is what is known as Gondwana coal. 'Gondwana' is a geological term which refers to a certain rock system which is about 200 million years old. Gondwana coal is a laminated bituminous coal with alternate dull and bright layers. In India, anthracite coal is found to a negligible extent. Coal deposits of the Tertiary era (60 million years old) are found in Assam, Rajasthan and Jammu. There are nearly 850 collieries are working.

Coalfields of Bihar and Bengal produce about 80% whereas the coalfields of Madhya Pradesh 10%, Orissa 5% and the rest of the country produce about 5% of the coal. This shows the uneven geographical distribution of coal and the consequent distribution problems. The important coal-fields of Bihar and Bengal are Barjora, Bokaro, Giridih, Karanpura, Jharia and Raniganj. The last two named are the biggest and best coalfields of the country.

The coalfields have seams or layers of coal beds from which coal is mined. Most of the coalfields have less than half a dozen seams but Raniganj, Jharia and Karanpura fields have more than twenty seams. Most of the coal seams are less than 10 meters thick. Some of the thickest coal seams of India are the Upper Seam No.1 (45 m) of the Talchir coalfield, the

Argada Seam (47.5 m) of the south Karanpura and the Jhingurda seam (152 m) of Singrauli which is the thickest seam of the world.

Most of the Indian coals have a high ash content. This is a serious disadvantage. Coal suitable for metallurgical industry is not at all sufficient for the needs of the country and is mostly available from some of the seams of Jharia, Raniganj and Bokaro coalfields.

The Gondwana coal, which contributes about 98% of our coal, has been classed into the following two broad groups

Raniganj coal has greater moisture content (about 5%) and more volatile matter (30-38%) and therefore burns easily with a long flame. It has lower calorific value, the ash does not form clinkers, and it is unsuitable for making coke but is good for making gas.

Jharia or Barakar coal has less than 2% moisture and the volatile matter content is about 28%; consequently it burns with a shorter flame, and is more suitable for steam raising in locomotive boilers (because less combustion space is needed). Its calorific value is higher and ash clinkers at higher temperature. It is suitable for making coke.

Singrauli coalfield situated in M. P. on the border of U. P. is becoming the largest coalfield of the country. An estimate puts its reserves at 9,000 million tonnes. It is mined mostly by opencast mining, which is a great advantage. In this coalfield Jhingurda and Gorbi projects are already producing large quantities of coal. Bina and Jayant mines have also been developed. Bina has a reserve of 130 m. t. It has placed U. P. on the coal map of India.

Coal from Singrauli coalfield has high ash content and most of it cannot be transported economically to long distances. Therefore, several super thermal power plants have been built in this area so that coal has not to be transported to a long distance and electricity is distributed over large parts of the country.

The Tertiary coal is much younger than Gondwana coal and is found in comparatively small amounts in widely separated regions like Assam, Jammu and Kashmir, Rajasthan and Tamil Nadu. The seams are quite thin, being 0.5 to 3 m thick. The coal being younger in age is mostly of lower rank like the lignite. Due to the geological upheavals, some coalfields of Jammu and Kashmir and Assam are of higher ranks. These have low moisture (< 3%) and ash (2 to 10%) contents. The Tertiary coal has higher sulphur content–Assam coals have up to 7% sulphur–making them unsuitable for metallurgical purposes.

7.4.10 Grading of Coal

The basis of grading is different from classification of coal. Classification is, generally, based on the origin of coal or the rank, which indicates how far coal formation has proceeded. The objective of grading is fixing its price for marketing. In India up to June 1975 low moisture coal was graded on the basis of ash content and high moisture coal on the basis of ash and moisture content. Coal containing more than 2% moisture is regarded as high moisture coal. This system did not take into account the calorific value of coal which was assumed to be practically the same in a particular grade. Actually this has been found to vary by 10% or so even in a particular grade. As coal is primarily used for heating purposes, a new grading system has been introduced from July 1, 1975. In this system *useful heat value* of coal has been incorporated in the basic system which was in force till that time. This new system is given in Table 7.2.

Table 7.2

Grading Criteria	Grade	Useful heat K cal/kg
NON-COKING COALS		
(a) *High moisture:*		
Ash + Moisture percentage		
(i) Below 17.5	Selected A	6135
(ii) 17.5 to below 19	Selected B	6135-5930
(iii) 19 to below 24	Grade I	5930-5240
(iv) 24 to below 28	Grade II	5240-4600
(b) *Low moisture:*		
Ash percentage		
(v) Below 15	Selected A	6620
(vi) 15 to below 17	Selected B	6620-6350
(vii) 17 to below 20	Grade I	6350-5940
(viii) 20 to below 24	Grade II	5940-5340
(ix) 24 to below 28	Grade III A	5340-4710
(x) 28 to below 35	Grade III B	4710-3800
COKING COAL (Bengal & Bihar)		
Ash percentage		
Below 13	Grade A	7110
13 to below 14	Grade B	7110-7010
14 to below 15	Grade C	7010-6880
15 to below 16	Grade D	6880-6760
16 to below 17	Grade E	6760-6605
17 to below 18	Grade F	6605-6480
18 to below 19	Grade G	6480-6350
19 to below 20	Grade H	6350-6190
20 to below 24	Grade HH	6190-5560
OUTLYING FIELDS (M, P., Orissa, Maharashtra; Gujarat)		
Ash + Moisture percentage		
Below 19	Selected Grade	5960
19 to below 24	Grade I	5960-5170
21 to below 28	Grade II	5170-4570
28 to below 35	Grade III	4570-3770

(Coals of Assam and Andhra are not graded but Andhra coal may be grouped under "Outlying fields". For Assam coals also useful heat value is calculated as in other cases)

Out of the gross calorific value of coal, as determined in a bomb calorimeter a certain amount, which depends on the ash content of coal, is not available for useful purposes. As the percentage of ash increases the amount of heat not available increases. Accordingly certain deductions, as summarized in Table 7.3, are to be made from the gross calorific value for different percentages of ash to obtain the calculated useful calorific value.

Table 7.3 Deductions to be made from gross calorific value to obtain the useful calorific value of coal.

Ash %	Deduction K cal/kg	Ash %	Deduction K cal/kg
2	20	30	1030
5	65	35	1200
10	145	40	1370
15	260	45	1540
20	480	50	1710
25	790		

In order to preserve and make proper use of coal resources of the country, the Coal Controller allots coal of certain grades to the consumers. The consumer is not allowed to use coal of a grade higher than that fixed for him by the Coal Controller. The consumers are classified according to the industry to which they belong and are further subdivided on the basis of the use to which the coal is finally put. For example, the railways are allotted Selected A non-coking steam coal for mail and express trains, Selected B non-coking steam coal for passenger and through goods trains, and so on. The boilers using pulverized coal are allotted Grade III A and III B. Such restrictions are imposed so that superior coal is not used for purposes for which inferior coal can be used. Thus superior coal, which is in short supply, is reserved for specialized use.

With a view not to exhaust the limited reserves of coking coal a limited amount of coke is being imported.

7.5 Calorific Value

As the most important use of a fuel is in heating, the calorific value determination is of importance. The *gross* or *higher* calorific value of a fuel is the amount of heat, expressed in calories per gram, which is liberated when the fuel is completely burnt in an enclosed vessel at constant volume in the presence or excess of oxygen, the products of combustion having been cooled to the room temperature. The heat evolved due to the formation of nitric and sulphuric acids during combustion are subtracted from the heat evolved. In such a case, water vapour produced by the combustion of hydrogen and evaporation of moisture will condense and will also give heat. This is what happens in calorific value determination in the laboratory but in actual use water vapour is not condensed; it escapes along with hot combustion gases, and a lesser amount of heat is available. The *net* or *lower* calorific value is the gross calorific value minus latent heat of the water vapour. In the net calorific value it is assumed that the products have cooled to room temperature (15°) but the water vapour has not condensed.

Gross of Higher Calorific Value (HCV) is defined as the total amount of heat produced when a unit mass / volume of the fuel has been burnt completely and the products of combustion have been cooled to room temperature (15 °C or 60 °F) .

Net or Low calorific value (LCV) is the net or actual heat produced when a unit mass / volume of the fuel is completely burnt and the products are permitted to escape i.e. HCV includes the latent heat of water vapour for med (587 Kcal/kg)

$$LCV = HCV - \text{mass of Hydrogen} \times 9 \times \text{latent heat of water vapour formed}$$

$$LCV = HCV - 0.09\ H \times 587 \text{ cals/g or Kcals/kg}$$

Dulong proposed a formula for the calculation of the calorific value from the chemical composition. Several modifications to it have been proposed, one of the formulae is:

$$\text{Gross C. V.} = \frac{1}{100}\ (8{,}080\ C + 34{,}500(H - O/8) + 2240\ S)\ \text{cal/g.}$$

(C, H, O and S are percentages of these elements in coal.)

In the formula oxygen is assumed to be in combination with hydrogen and only surplus hydrogen is available for giving heat. Calorific values can easily be converted from one unit to another as follows :

1 cal/g = 1 K cal/kg = 1 G.H.U/lb = 1.8 B.Th.U/lb.

7.5.1 The Bomb Calorimeter

Several manufacturers have put on the market bomb calorimeters differing slightly in design. The simple sketch, given in the Fig. 7.1, shows only the bomb of a calorimeter used for determining the calorific values of solid and liquid fuels.

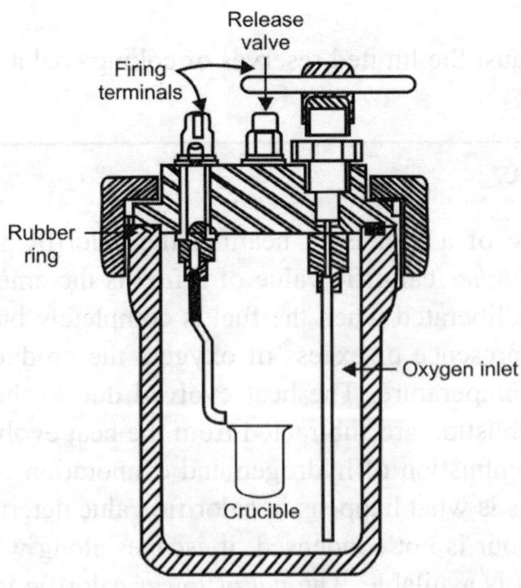

Fig. 7.1 Bomb for determining calorific values of solid and liquid fuels.

The calorimeter has a carefully made, strong cylindrical bomb in which the combustion takes place. It is made of stabilized austenitic steel which is resistant to corrosion and can withstand a pressure of at least 50 atmospheres. The bomb has a cap which is screwed to the body making a perfect gas-tight seal. Two stainless steel electrodes are fitted to the cap. One of these is a tube through which oxygen is introduced. A small ring is attached to this electrode on which the crucible containing the fuel is supported. A valve for releasing the pressure is also provided. The capacity of the bomb is about 300 ml. The bomb is placed in a calorimeter which is surrounded by an outer jacket to prevent heat losses. A stirrer is provided for stirring water in the calorimeter.

Procedure. About 1 g of the fuel is weighed into the crucible and a weighed piece of firing wire is stretched between the electrodes in such a manner that it is in close contact with the fuel so that it may be ignited. Often the powdered coal is made into a small pellet with the fuse wire inserted in it, weighed and placed in the crucible. 2 ml of water is poured in the bomb so that the products of combustion of sulphur and nitrogen may be absorbed in it. The cap is screwed down on the bomb and oxygen is filled in up to a pressure of about 25 atmospheres. The bomb is then placed in a weighed amount of water taken in the calorimeter. Electrical connections are made, stirring is started and temperature reading is taken with a thermometer reading to 0.01°. When the thermometer shows a steady temperature, the electrical contact is made to fire the fuel and temperature readings are continued for 5 minutes after the maximum temperature is attained. The water is stirred during the experiment. The bomb is then removed and allowed to stand so that the acid mist may settle down. The pressure is slowly released and the contents of the bomb are carefully washed, the washings are collected and analysed for sulphuric and nitric acids formed. As sulphur and nitrogen do not give sulphuric acid and nitric acid in actual practice, correction needs to be made for the heat given out in their formation. Heat from the fuse wire has also to be taken into account and cooling corrections have to be made. The calorific value is then calculated from the following formula:-

$$\text{Calorific value} = \frac{(T+t) \times (W+w) - \text{corrections}}{\text{Wt. of fuel in g}} \text{ cal/g}$$

Where T = observed rise in temp; t = radiation correction;

W = weight of water; w = water equivalent of calorimeter.

Correction includes heats of formation of nitric and sulphuric acids and heat given out by the fuse wire while igniting the fuel.

7.6 Liquid Fules

7.6.1 Sources

The main source of liquid fuels is petroleum which on distillation gives several fractions. The lower boiling fractions find use in internal combustion engines and the higher boiling fractions in Diesel engines and in oil-fired furnaces. Alcohols, tar and colloidal fuels are used to a lesser extent.

7.6.2 Advantages

The use of liquid fuels has increased in recent times because of the following advantages :

1. Transportation, storage and handling are easier. Liquid fuels can be transported by pipes to distant places and in a plant the storage may be at one place and it may be piped to points of consumption. For this reason liquid fuels can be stored in a place not otherwise useable. This is of special advantages in ships where storage tanks of even irregular shapes can be made and otherwise useless space utilised. Labour charges are also less.
2. For the same storage space the heating value of a liquid fuel is 1.5 to 2 times more than that of a solid fuel. Difference in storage space needed by liquid and gaseous fuels is great. Liquid fuels take much less storage space than is needed for the storage of gaseous fuels. This is a distinct advantage of liquid fuel over gaseous fuels.
3. Combustion can be started immediately. The rate of combustion can be changed readily and stopped as soon as heating is not needed. This results not only in better control over the process but also in the economy of fuel.
4. Liquid fuels have several other advantages, for instance, cleanliness, absence of ash, no danger from spontaneous combustion, less excess air needed for combustion, facility in refueling, etc. They also require less furnace space for combustion.

The chief disadvantages of many of the liquid fuels are that they are highly inflammable and are volatile. Of course some of the liquid fuels, like fuel oils, are not so inflammable and volatile.

7.7 Petroleum

7.7.1 Origin

The theories of origin of petroleum have been more controversial than that of coal. It is now accepted that it originated from organic matter like the bodies of fish or other aquatic and plant life. They fell to the bottom of a very calm, shallow sea where they got covered by a layer of silt. Under such conditions, the organic matter which was settling continuously decayed under almost anaerobic (i. e., in absence of air) conditions. The steady accumulation of silt above the organic matter led to an increase in temperature and rise in pressure. As with coal, bacterial action played an important role. The oil produced moved to porous sandstone and limestone rocks. For petroleum deposits to form it is necessary that there be an overlaying rock which is impervious so that the petroleum cannot migrate upwards, and also that escape in other directions be blocked by various rock formations. There are several types of rock formations for oil trap. It is easy to see that in the same deposit one could strike oil, or gas or even salt water. Often there is an accumulation of gas above the oil and salt water underneath. Any natural gas or oil moved to the surface would be evaporated and lost, leaving the high boiling fraction as a natural deposit of pitch.

Petroleum is brought to the surface by drilling wells. There is a possibility that the well may strike the water layer, Fig. 7.2, and no oil be brought to the surface by that particular well, or that it may strike oil or gas. Initially the gas pressure may be enough to force some of the oil to the surface but later it has to be pumped. The oil so obtained is known as crude oil.

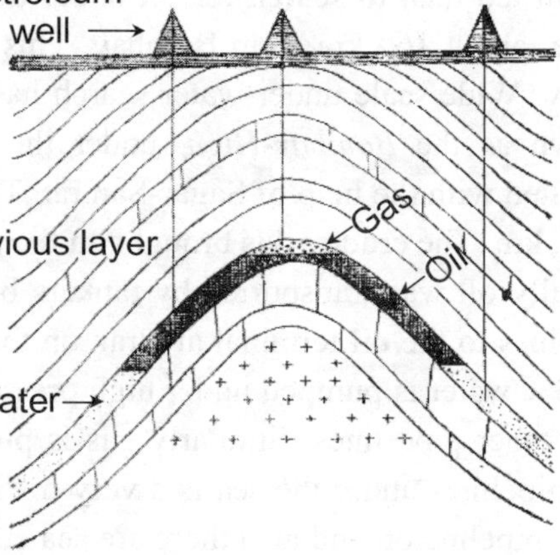

Fig. 7.2 Oil, gas and water in a petroleum reservoir.

7.7.2 Indian Resources of Petroleum and Petrochemicals

The history of petroleum in India begins with its accidental discovery in Digboi, Assam, in 1889-90. The Digboi oilfield has so far produced over 9 million tonnes of the crude oil for which about a thousand wells have had to be drilled. A few years after getting independence, extensive oil exploration began which changed the picture of the petroleum industry in India faster than could have been imagined in pre-independence days. Due to our heavy dependence on imported crude oil, the country's economy received a serious jolt in the latter part of 1973 when supplies from the Middle East were interrupted and prices were increased several times. Upto August 1977 under the auspices of Oil and Natural Gas Commission 1,468 wells were drilled. Of these 781 were found oil-bearing and 95 gas producing.

In 1981 the exploration cost, which includes cost of the seismic and other surveys and test drilling was about Rs. 8 per barrel of petroleum produced whereas the international price is more than, Rs. 200 per barrel (as compared to Rs. 15 in 1973).

After independence, the Government of India revised the petroleum policy completely. Instead of importing the more expensive refined products, it started importing crude petroleum for distillation and refineries set. With the discovery of newer oilfields, mostly in Gujarat and Assam and the construction of long pipelines for carrying crude oil and distillates to different parts of the country, the number of refineries set up in the private and public sectors and their capacities have expanded. The chief oil-producing areas are Ankleshwar,

Nawagam and Kalol in Gujarat and Rudrasagar and Lakwa in Assam. Crude oil production was 0.45 million tonnes in 1960, which increased to 3 m. t. in 1965, 7.2 m. t. in 1973 and 12.8 m. t. in 1979 with import of about 14 m. t. In 1980-81 India imported 16.7 million tonnes of crude oil and about 8 m. t. of petroleum products at an estimated cost of Rs. 5,600 crores.

Shortage of petroleum led man to search for oil under the sea. In 1974 oil and gas were found in the Arabian sea, about 160 km from Bombay. This opened a new chapter and high hopes in Indian economy. Wide scale under-water search has been made but so far the main source is what is known as the *Bombay-High,* under the high and rough sea. The first exploratory well was drilled with the help of Sagar-Samrat. The Bombay-High-oilfield covers an area of about 1500 sq. km. The crude oil is brought on the process platform and most of the water is separated. Initially oil was transported by tankers but since June 1979 both oil and gas are brought by pipelines to the oil terminal at Uran on the main land of our country (Fig. 7.3). Oil with less than 5% water is pumped under high pressures through the trunk pipe lines to which are connected other pipe lines. Similarly gas is pumped under high pressure after dehydration. Laying of pipelines under the sea is a very difficult work. There is no hard and steady ground to rest the pipeline on and also there are sea currents which tend to wash it off. Moreover, sea water is highly corrosive. The under-sea technology, a very challenging technology, is still in the developing stage. From 1982, when production is reached the maximum production the Bombay High oil field is expected to produce 1,60,000 to 2,00,000 barrels per day of oil for the next 15 years. (20,000 barrels per day means 1 million tonnes per year). The off-shore oil production of 5 million tonnes in 1980 is to be increased to 12 million tonnes. The oil is associated with gas, which is called associated gas, as it is a co-product with oil. Its production is expected to reach 3 to 4 million m^3 per day. It is to be noted that production of associated gas is not under independent control as it is produced along with the oil. The Bassein oil field produces gas by itself and is known as non-associated gas and its production can be increased or decreased independently, according to requirement. Typical composition of these gases is given in Table 7.4.

Table 7.4 Typical Composition of gas from Bombay-High.

Composition Hydrocarbons	Associated gas % by vol.	Non-Associated gas % by vol.
C_1	69.1	78.0
C_2	13.1	6.7
C_3	9.3	4.0
C_4 & higher	5.3	2.0
Others (O_2, N_2, CO_2 etc.)	3.2	9.3

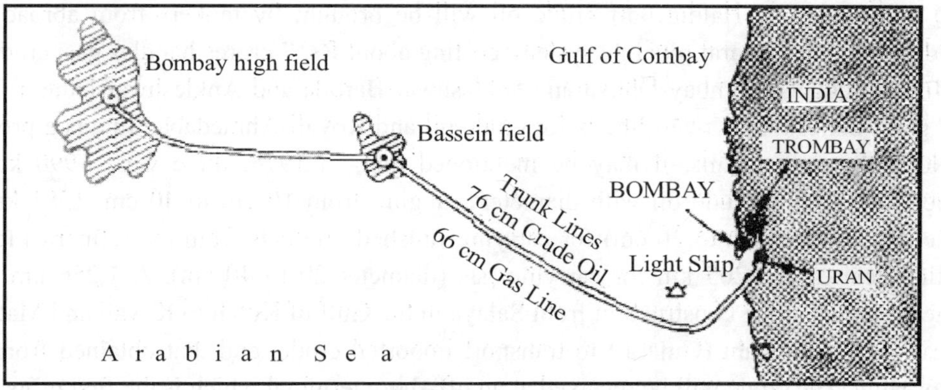

Fig. 7.3 Location of offshore oil fields.

As petroleum obtained from the Bombay-High is of low sulphur content, not much hydro-treatment of sulphur removal is needed. It gives oil with cyclic naphtha and high aromatic content. It is being processed at Koyali (Gujarat) and Mathura (U. P.) besides at the refineries around Bombay. The large quantities of the resultant benzene, toluene and xylene will help in the development of organic chemical industries.

As a very large quantity of gas became available unexpectedly there were no facilities to utilize it in any way other than burning in the thermal power plants as an interim measure. It is a wasteful process but there was no other way. The gas production from the Offshore sources is expected to be 20 million cubic meters per day for the next 30 years. Ten fertilizer plants are being built which will use methane to produce about 13,500 tonnes of ammonia and 22,500 tonnes of urea per day. Thus fertilizer production of the country will be doubled.

C_2 and C_3 fractions of the gas on cracking yield ethylene and propylene which are used for the manufacture of various synthetic organic chemicals like low and high density polyethylene, polyvinyl chlorides (PVC), ethylene glycol (to be used by the aromatic plants for making polyester fibers), polystyrene, acetone, etc. C_4 and the higher fractions are liquified and transported to the various bottling centres for supply in cylinders as cooking gas for domestic use.

The cheapest mode of transporting oil and gas is by taking them through pipelines. Several large networks of pipelines have been built in India. The Gauhati product pipeline, constructed at a cost of about Rs. 7 crores, is designed to transport petroleum products from Gauhati Refinery. Rudrasager-Gauhati-Barauni pipeline system transports crude oil from Rudrasagar oil fields to Gauhati and Barauni refineries. It brings oil from Assam through 1200 km of difficult terrain crossing many rivers. The Barauni refinery pipeline system is very long and has cost Rs. 26 crores. It is in two sections, Barauni-Kanpur and Barauni-Haldia. Barauni to Kanpur is a product line, i.e. finished products are pumped from Barauni

(Bihar) to Kanpur. At Haldia port crude oil will be brought by tankers from abroad and pumped to Barauni. Gujarat pipeline system, costing about Rs. 7 crores handles gas, crude oil and refined products. Cambay-Dhuvaran, Ankleshwar-Baroda and Ankleshwar-Uttaran lines are for gas; Ankleshwar-Koyali line is for crude oil and Koyali-Ahmedabad line is a product line. Not going into details, it may be mentioned that, in 1974, there were 1990 km of pipelines for carrying crude oil with diameters ranging from 10 cm to 40 cm, 1581 km of pipeline (diameter 10 cm to 30 cm) for carrying finished products from the refineries to the marketing centres and 205 km for carrying gas (diameter 20 to 40 cm). A 1,256 km long pipeline system is under construction from Salaya in the Gulf of Kutch to Koyali and Mathura Refineries via Virumgam (Gujarat) to transport imported crude, and that obtained from the Bombay-High. The crude will be received at an off-shore terminal which is the first of its kind to be set up in the country.

In the public sector there are nine refineries. The oldest refinery is at Noonmati near Gauhati, Assam, and has an annual capacity for handling 0.88 million tonnes of crude oil. The second refinery at Barauni in Bihar has a capacity of 3 million tonnes. The third refinery constructed is at Koyali, near Baroda (Vadodara) in Gujarat with an annual capacity of 7.3 million tonnes of crude oil. The capacity of Cochin Refinery is 3.3 million tonnes. The refinery at Madras (3 mt) is specializing in the production of lubricants. Other refineries in the public sector are at Bombay, Trombay, Vishakhapatnam and Haldia. The Assam Oil Company, in the private sector operates a refinery at Digboi. One of the biggest refineries (6 mt) of the country has started production near Mathura. This site has been chosen because it is in the interior of the country and is safe from defence point of view. Also it is well connected by rail transport with different parts of the country so that the products can be cheaply marketed. Crude oil will be brought to it from Gulf of Kutch by 1200 km long pipeline.

The Petrochemical Complex of the Indian Petrochemical Corporation Limited, near Vadodara, Gujarat, was inaugurated on the 30th March 1979. It has 14 individual production units constructed at a cost of Rs. 400 crores. It is helping the production of petrochemicals, thermoplastics, elastomers, synthetic fibres and other organic intermediates for drugs, dyestuffs, and paints. The IPCL was established in 1969. The Aromatics Complex for the production of xylenes, and dimethyl terephthalate (DMT) was commissioned in 1973. Orthoxylene is used for the manufacture of phthalic anhydride which is the chief raw material for plasticizers used with polyvinyl chloride (PVC) resin. DMT is the raw material for polyester fibre and films used in textile and industrial fabrics, electrical industry and manufacture of X-ray films. The production of xylenes was about 60,000 tonnes and of DMT, 24,000 tonnes in 1978.

The construction of the Olefins Complex was started in 1974. In 1978 it produced about 2,00,000 t of ethylene, 70,000 t of propylene and 31,000 t of butadiene. Its downstream units produce a number of chemicals used in the plastics, rubber, synthetic fibre and allied industries.

It is estimated that aromatic and olefin units will ultimately yield about 3,00,000 tonnes of products which, after conversion to the final products will be sold for about Rs. 3,000 crores per annum. Some of the widely used products are polypropylene (30,000 t), low density polyethylene (80,000 t) and polybutadiene rubber (20,000 t). Details of products and their preparations are beyond the scope of this book but it is sufficient to give a glimpse into its importance.

7.7.3 Classification of Crude Petroleum

Petroleum consists of a mixture of gaseous, liquid and solid hydrocarbons together with small amounts of nitrogen and sulphur compounds. The hydrocarbons may be of the paraffinic, naphthenic (cyclo-paraffinic), olefinic or aromatic series. According to the nature of the main constituents present, the crudes are classified into three groups:

1. *Paraffin-base* petroleum mainly consists of hydrocarbons of the paraffin series. On distillation solid paraffin wax is left behind.
2. *Asphalt base* petroleum is rich in non-paraffinic hydrocarbons, like aromatic and naphthenic hydrocarbons, and on distillation asphalt or bitumen is left behind.
3. *Mixed-base has* a composition in between the above two types and on distillation some wax appears in the distillates and the residue mostly consists of asphalt.

From the standpoint of lubrication engineering, it is noteworthy that a lubricant from paraffin-based crude is of lower specific gravity, and change in its viscosity with change in temperature is less than that of a lubricant from the naphthenic series. At lower temperatures more of waxy solid separates from the paraffin-base lubricant. The petroleum base also affects the performance in the internal combustion engines, as will be presently explained.

7.7.4 Petroleum Refining

In general, crude petroleum is not marketed. Crude petroleum is distilled and refined to give several fractions which are more valuable. A landmark in the distillation process was the use of the bubble-cap distillation column in 1920. Catalytic conversions have been increasingly dominating the petroleum industry since 1940.

The crude oil is obtained in the form of a water-in-oil emulsion and the emulsion is broken up by adding de-emulsifying agents or by electrical treatment. Care is taken to remove sodium and magnesium chlorides as these corrode the refining equipment and deposit scales in the heating pipes.

The emulsion-free crude is heated in pipes in a furnace heated by burning gas or oil. The vapours are passed through a tall cylindrical fractionating column which is provided with a number of carefully designed fractionating plates, see Fig. 7.4. As the vapours move upwards, the higher boiling fractions condense and fall back and only lower boiling vapours move to

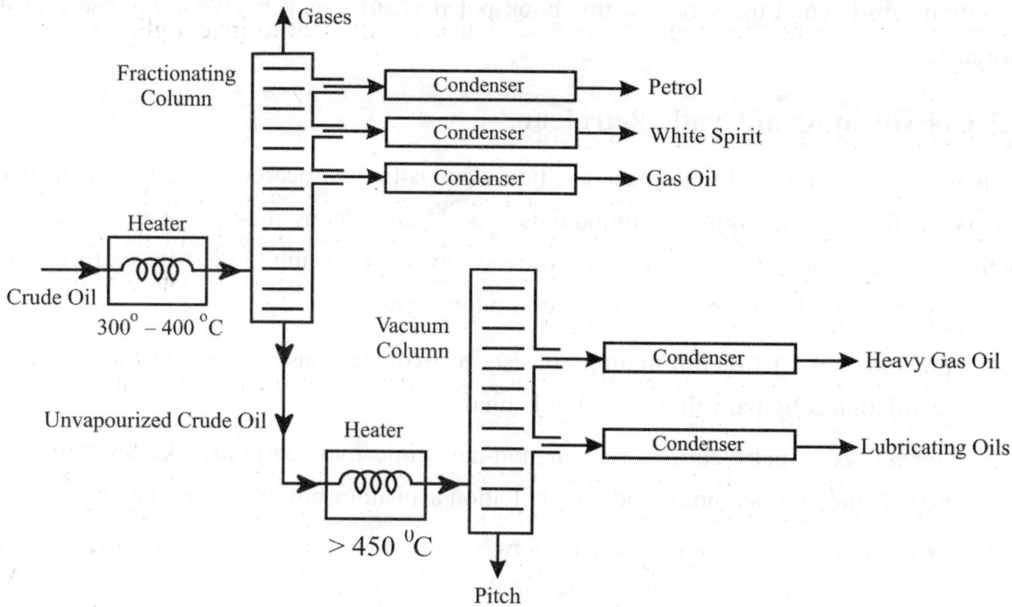

Fig. 7.4 Petroleum distillation.

higher plates where further condensation takes place and some more liquid falls back. As the freshly condensed liquid comes to the lower plates, it gets heated and vapourized and again condenses on higher plates. This process of condensation and vapourization takes place many times causing separation of hydrocarbons according to their boiling points. Thus the higher boiling fractions condense towards the lower part of the column. The lower the boiling point the higher is the position of the plate on which a fraction condenses. Pipes are led off from different points taking out mixture of vapours and liquids of the desired boiling points which are condensed separately and different Fractions are obtained. This gives *straight-run* petrol, kerosene (or white oil) and gas oil fractions. Operations from drilling to the final use are shown in Fig. 7.5.

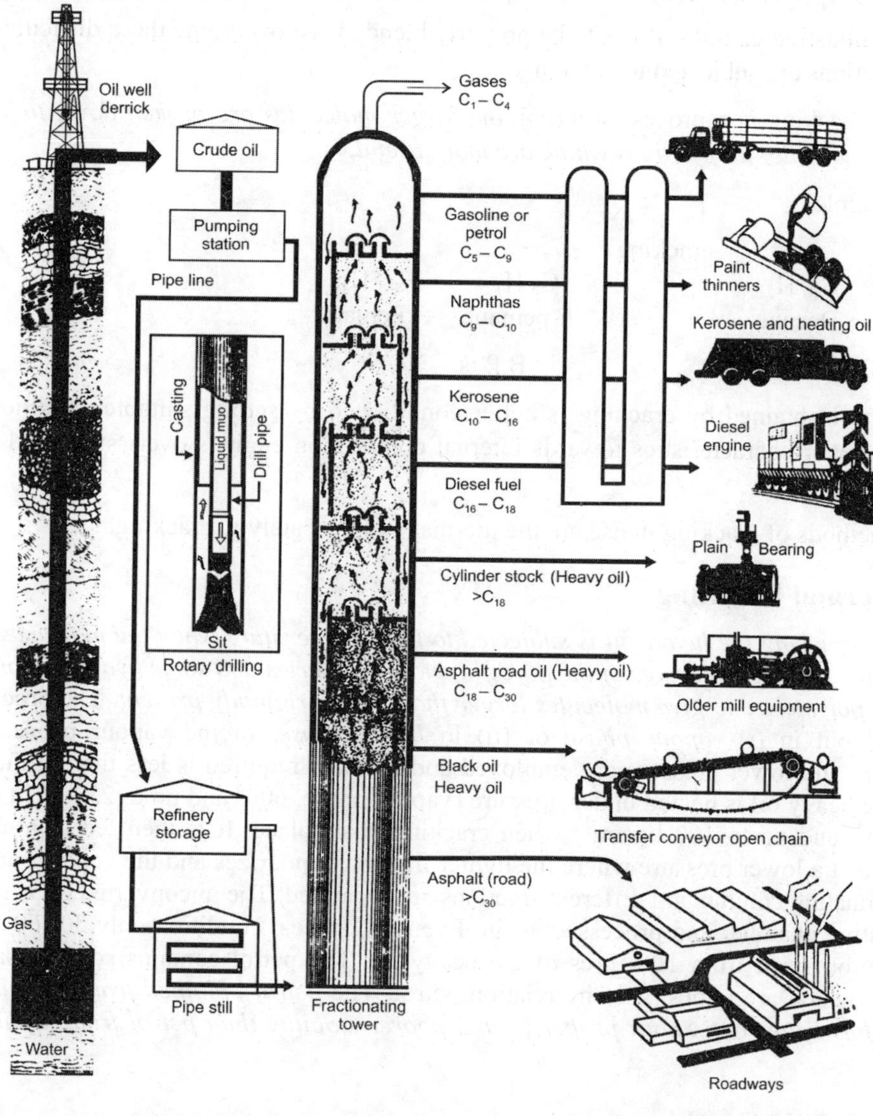

Fig. 7.5 Flow-sheet showing pumping and refining of petroleum and applications of different fractions, (Courtesy, U.S. Steel Corporation)

If the entire crude is distilled at atmospheric pressure, as described above, considerable thermal decomposition of the higher boiling hydrocarbons would take place. About half of the unvapourized crude oil is removed from the bottom and distilled under reduced pressure in a similar fractionating column and heavy gas, oil and lubricating oil fractions are separated. The residue gives pitch.

Of all the fractions collected petrol has the largest demand, but the yield, by the above-described method, is only 20% of the crude. The result is that there is surplus of heavier fractions. Moreover, the straight-run petrol is not of sufficiently high quality for direct use in internal combustion engines. It has to be properly blended. To overcome these difficulties, the heavier fractions are subjected to cracking.

Cracking: Cracking is a process *whereby the larger molecules are broken down to smaller molecules giving lighter fractions which are more useful.*

For example

$$\underset{\text{Decane}}{C_{10}H_{22}} \xrightarrow{\text{cracking}} \underset{\text{n-pentane}}{C_5H_{12}} + \underset{\text{Pentane}}{C_5H_{10}}$$

B.P = 174 + °C B.P is ≃ 36 °C

The petrol obtained by cracking is better conditioned by setting suitable conditions and hence has better characteristics towards internal combustion engines over 'struit and petrol' from refinery.

Two methods of cracking in use are the thermal and the catalytic cracking.

7.7.5 Thermal Cracking

In thermal cracking the heavy oil is subjected to high temperature when the molecules break down to give smaller molecules of the paraffin and olefin series and some hydrogen, but some molecules polymerize to give molecules larger than those originally present. The process can be carried out in (i) *vapour phase* or (ii) in *liquid phase.* In the vapour phase, higher temperature and lower pressure are employed and the time required is less than in the liquid phase. The heavy oil is heated under pressure (vapour phase, 600° and up to 23 kg/cm^2; liquid phase, 450° and up to 100 kg/cm^2) when cracking takes place. It is then led to a chamber maintained at a lower pressure where the lighter fraction vapourizes and the vapours are taken to a fractionating column and different fractions are separated. The unconverted oil is returned to the heating chamber and processed again. Even under ideal condition only 50- 60 litres of □etrol are obtained from 100 litres of the heavy oil. The petrol contains some unsaturated molecules which are undesirable in aviation petrol. The *petrol obtained from vapour phase cracking has better anti-knock properties but poorer stability than petrol from liquid phase cracking.*

7.7.6 Catalytic Cracking

In catalytic cracking reactions similar to those in thermal cracking are brought about in the presence of catalyst. The result is that *lower temperature* and *much lower pressure* is necessary and the process is under better control. It produces some aromatic and branch-chained hydrocarbons. *The amounts of unsaturated hydrocarbons produced and the sulphur content are lower than in thermal cracking.* The petrol produced has better anti-knock characteristics than the petrol from the thermal cracking.

The process is almost similar to thermal cracking. The heated oil is brought in contact with the catalyst. There are several methods of making the catalyst bed. The bed may be a fixed or moving. The powdered catalyst is fed along with the oil in the reactor wherein the catalyst continuously moves alongwith vapours of oil and effective cracking can be achieved in case of moving bed methods. The vapours escaping the distillation tower for condensation are filtered through cyclones which drop the catalytic powder and fall at the bottom. This is again heated and regenerated and sent into the reactor under pressure. Like this the catalyst powder will be continuously moving along with oil between reactor tower and regenerator tower as shown in the flow diagram. Essentially the catalyst consists of certain almunium silicates. These are either naturally occurring clays or are synthetically made. After use for sometime deposits of tarry matter are formed which poison the catalyst. It is regenerated by burning off the deposits. The oil is then led to a flash chamber where the product vapourizes due to the lower pressure (Fig 7.6). Vapours are led to a fractionating column and different fractions are separated in the usual manner.

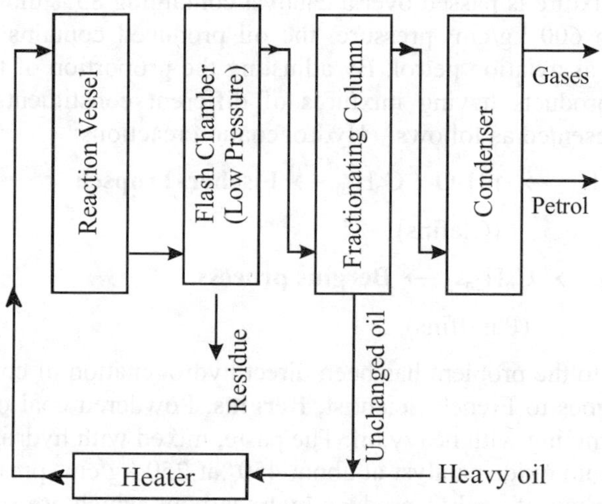

Fig. 7.6 Cracking of heavy oil.

7.7.7 Synthesis of Petrol

A large quantity of gaseous hydrocarbons, saturated as well as unsaturated are produced by primary distillation and cracking processes. These can be reformed by various processes to give higher molecular weight portions suitable for being used as fuel for internal combustion engines. There are several processes which are quite complex in nature and are beyond the scope of the present volume, and will only be described briefly.

An important variant of catalytic cracking is known as *hydrocracking*. Here the cracking is done in the presence of hydrogen and a suitable catalyst under very high pressure (200 kg/cm^2) and at high temperature (550°). Hydrogen combines with the unsaturated products and ultimately gives better petrol.

Another technique adopted is *dehydrogenation* i.e. removal of hydrogen, thus creating more unsaturation, followed by polymerization to give the desired kind of hydrocarbons. The refinery gases consist of saturated as well as unsaturated hydrocarbons. The saturated ones are less reactive and have to be converted to unsaturated ones by dehydrogenation and then all the hydrocarbons are polymerized. If necessary, after polymerization the product can be hydrogenated to remove residual unsaturation. In *aromatization* the straight-chain hydrocarbons are converted into aromatic hydrocarbons which are more useful for blending. Isomerization is used to improve anti-knock properties.

Synthetic petrol from coal. Much work has been done in Germany for obtaining petrol and other useful substances from coal by passing steam through a hot bed of coal. For converting coal to synthesis gas (syngas) or liquid hydrocarbons five major items to be controlled are pressure, temperature, steam, hydrogen and catalyst. The most important of the processes is the **Fischer-Tropsch process** in which first water gas ($CO + H_2$) is made from low grade coal. Water gas is mixed with more hydrogen and passed over a suitable catalyst whose main constituents are cobalt, nickel and iron. By regulating the composition of the gas mixture, the temperature and the catalyst, a variety of products can be produced at atmospheric pressure. When the same gas mixture is passed over a catalyst containing 85% thoria and 15% alumina at 400° and at 300 to 600 kg/cm^2 pressure, the oil produced contains high proportions of isobutane and is used as aviation petrol. By adjusting the proportion of the reactants and the working conditions products having mixtures of different constituents are obtained. The reactions may be represented as follows – Hydrogenation reactions:

(i) $nCO + 2nH_2 \rightarrow nH_2O + C_nH_{2n} \rightarrow$ **Fischer-Tropsch**

(Water gas) (Olefins)

(ii) $C_nH_{2n} + H_2 \rightarrow C_nH_{2n+2} \rightarrow$ **Bergius process**

(Coal) (Paraffins)

Another approach to the problem has been direct hydrogenation of coal, a French method, the credit for which goes to French Scientist, **Bergius**. Powdered coal of low ash content is made into a paste by mixing with heavy oil. The paste, mixed with hydrogen, is passed over a tungsten or molybdenum oxide catalyst at about 450° at 250 kg/cm^2 pressure when hydrogen combines with the carbon of coal to produce hydrocarbons which are separated into various fractions. The heavy oil produced is used to make paste with a fresh batch of coal powder.

The Fischer-Trópsch process can use a low grade coal, which is abundant in India, but requires a much higher pressure for conversion. The advantage of the Bergius process is the lower pressure employed for carrying out the operation, but it requires high grade coal, which is in short supply. In the world market, synthesis of petrol from coal has not been economically competitive. One large coal-based synthesis gas (syngas) and liquid hydrocarbon plant has been in operation from the midfifties in South Africa near Johannesburg. Another plant, three time the size of the first plant, has gone into production. Coal-based projects require much more capital than petroleum based. For example, it is estimated that a coal-based ethylene plant required three times the investment as compared to the naptha based of comparable size. In the Fischer-Tropsch process about two-thirds of the coal is wasted in one form or the other. Due to the repeated increase in the petroleum prices after 1973, the interest in such processes has been revived and various countries including India, are considering setting up synthetic petrol plants. Due to abnormal increase in the petroleum prices and by improving the technology, it is hoped that in the future the process may become economical. At any rate, it has become necessary to pay attention in this direction with a view to reducing dependence on foreign countries.

There is no commercial synthetic petrol plant in India. At the Central Fuel Research Institute, Dhanbad, one 450 litres/day pilot plant has been working for the vapour phase hydrogenation of suitable fraction of coal tar as well as coal hydrogenation products. A 500 kg/day slurry phase coal hydrogenation pilot plant has been sanctioned for installation at CFRI, Dhanbad.

7.7.8 Refining of Petrol

The petrol obtained by straight distillation or by synthesis contains some unsaturated hydrocarbons which get oxidized and polymerized causing gum and sludge formation on storing. Also, there are some sulphur compounds which adversely affect tetra ethyl lead which needs to be added for obtaining better anti-knock properties.

The petrol is treated with concentrated sulphuric acid which removes these constituents. It is then treated with sodium hydroxide to remove the acid and finally washed with water. Some compounds, known as inhibitors, are also added which retard the oxidation reaction and thus improve storing quality.

After refining, the products are blended to give petrol that meets the various specifications like octane number, volatility, heating value, etc.

7.8.9 Petrol or Gasoline

Straight run petrol consists of a mixture of paraffinic, naphthenic and aromatic hydrocarbons, with 5 to 8 carbon atoms, in varying proportions. Cracked and reformed petrol is a more complex mixture. Its specific gravity is about 0.73 and the boiling range is 40° to 200°. The most important use for petrol is in internal combustion engines. The following factors should be considered in evaluating the quality of petrol for the purpose :

1. In spark-ignition engines petrol has to be partly vapourized on mixing with air. If it does not vapourize, there is difficulty in starting the engine. On the other hand too high a volatility will cause vapour formation in the fuel line thereby reducing fuel supply and stopping the engine. This is known as vapour-locking.
2. In the engine it should not form gum or cause corrosion. It should be stable in storage.
3. It should have proper 'anti-knock' properties which are explained in detail in the following paragraphs.

As calorific value is almost directly related to specific gravity a fuel of higher specific gravity is preferred where space is limited as in aeroplanes.

7.8 Knocking

Although theoretically the power output of an internal combustion engine increases with increase in the compression ratio, in actual practice it falls rapidly after a certain compression ratio. If the fuel-air mixture is compressed beyond this point two difficulties arise. Firstly, because of the high temperature due to the higher compression ratio, the fuel ignites even before the regular spark occurrence. This is one of the causes of pre-ignition. The second difficulty arises if after the start of the combustion and the flame proceeding at a particular rate, the last portion of the fuel burns instantaneously. This instantaneous combustion causes a shock wave to be set up. This is known as 'knocking', 'pinking', 'detonation'. It depends on the characteristics of the fuel, besides design factors.

7.8.1 Octane Number

The most common way of expressing the knocking characteristics of a fuel is by octane number. It has been found that n-heptane knocks very badly and its anti-knock value has been arbitrarily taken as 0 and *iso-octane* gives very little knocking and its anti-knock value has been taken as 100. The octane number is numerically the percentage by volume of *iso-octone* in a mixture, of *iso-octane* and n-heptane which just matches the knocking characteristics of the fuel. Thus, if a sample of petrol gives as much knocking as a mixture of 80 parts of *iso-octane* and 20 parts of n-heptane its octane number is 80. Now-a-days fuels with octane numbers above 100 are available and the value is determined by extrapolation.

$$CH_3-CH_2-CH_2-CH_2-CH_2-CH_2-CH_3$$

n-heptane (C_7H_{16})

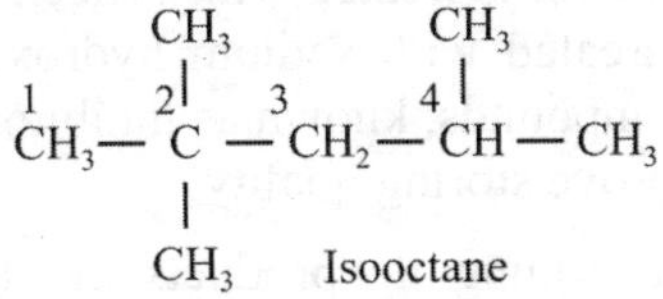

Isooctane

(2, 2', 4-trimethylpentane)

The maximum permissible power increases with increase in the octane number. Based on experimental results graphs have been drawn to show this relationship. Another unit, known as performance number, is now being used to enable direct deduction of the power output that can be expected of a fuel.

7.8.2 Chemical Structure and Knocking

A detailed relationship between the knocking tendency of a fuel and the chemical structure has been worked out. This is important because the maximum power that can be developed is related to the anti-knock property of the fuel. In general, the anti-knocking property increases with increase in the compactness of the molecule, cyclic structure and double bonds. Thus in the paraffin series butane has an octane number of about 90 whereas heptane has 0. Short side chains increase antiknock properties. Cycloparaffins have better anti-knock properties but aromatic hydrocarbons have still better properties. Olefins of the same chain length as the paraffins have better properties.

It is because of these reasons that cracked petrol gives better engine performance than straight run petrol. The olefins, and the diolefins in particular, do not have very good keeping qualities. They are more easily oxidized and tend to form objectionable gums. This tendency, however, can be checked by the addition of inhibitors. The aromatic compounds are more susceptible to pre-ignition.

7.8.3 Anti-knock Agents

From the foregoing discussion the necessity of improving anti-knock properties is obvious. At the same time it is clear that it is not always easy to alter the chemical structure so that its performance is improved. To improve anti-knock properties some organo-metallic compounds are added of which tetra ethyl lead, (TEL), $Pb(C_2H_5)_4$, is most widely used. TEL is a colourless oily liquid of sp. gr. 1.62 and B. P. 200°. It has a sweet odour but is extremely poisonous and a fuel containing it should be handled with great care lest it enter the body through a cut or otherwise. In motor spirit about half ml and in aviation fuels one ml or a little more of TEL are used per litre of petrol. Because of its poisonous nature petrol containing TEL is always coloured.

It has already been explained that knocking is due to the spontaneous ignition of the compressed mixture of petrol vapour and air. In the cylinder tetra ethyl lead decomposes giving a fog of lead oxide particles. This fog influences combustion reactions of petrol and air mixture and prevents knocking. On burning, TEL deposits lead oxide and metallic lead on the spark plug and on the cylinder walls and piston head (necessitating more frequent cleaning). This results in ignition at these points. To remove this difficulty ethylene dibromide is also added so that lead is removed as volatile bromide along with the exhaust gases.

Tetra ethyl lead is more effective on alkanes than on aromatic compounds. Sulphur compounds reduce its effectiveness.

7.9 Gaseous Fuels

7.9.1 Advantages

The use of gaseous fuels is on the increase because of the following advantages :

1. Combustion is under better control. Gases burn instantaneously and the rate of burning can be changed very easily. Gases are miscible with air, therefore, the excess air needed is much less than that needed for liquid or solid fuels. Oxidizing or reducing atmosphere can be easily maintained. All these contribute to fuel economy and efficiency of furnace operation.
2. Heat from hot flue gases can be recovered and gaseous fuel can be pre-heated in regenerative furnaces, as is done in steel making by the open hearth process in glass manufacture. This gives economy of heat and at the same time a higher temperature is produced.
3. Gaseous fuels may be prepared at a central place and distributed over a wide area through pipelines. In some cases solid fuels unsuitable for direct use can be converted into gaseous fuels and valuable by-products recovered.
4. Neither smoke nor ash is produced. Thus gaseous fuels are very clean.
5. If the supply is obtained from a public supply there is no problem of storage.

The principal disadvantage of gaseous fuels is that they occupy large volumes, therefore, storage is difficult.

7.9.2 Coal Gas and Coke-oven Gas

Coal gas and coke-oven gas may be considered together because both are obtained by the destructive distillation of coal and their composition and calorific values are similar. Coal gas is made in horizontal or vertical retorts by heating-coal rich in volatile mater and gas is the main product. Coke-oven gas is made in coke-ovens by heating coking coal. Metallurgical coke is the main product and gas is the by-product. In India coke-oven gas is made to a greater extent than coal gas and is used in steel factories. In European countries coal gas is widely used as 'city gas' for lighting and heating purposes. The average calorific value is around 5,000 kcal/m^3. Coal gas has a slightly higher calorific value than coke-oven gas. It has about 50% hydrogen, 30% hydrocarbons (mainly methane), 5% carbon monoxide and smaller amounts of other gases.

The changes taking place on heating coal have been described in detail in this chapter under 'Coke'. It only needs to be added that the temperature of carbonization is governed by the end products needed. A very low temperature (300-400°) gives a higher yield of volatile products, like benzene. As the temperature is increased the yield of gaseous products increases and the yield of liquid products decreases.

The volatile products obtained from the retorts of coke-ovens are drawn off through ascension pipes connected to a collecting main which runs horizontally along the entire row of ovens or retorts. Here the volatile matter gets slightly cooled and some of the tar and water is condensed. The tar is led off to a tar well. The hot gas passes up the condensers where further condensation takes place and the liquid separates into tar and ammonia liquor layers. The gas is then washed by a spray or water or sulphuric acid to remove the remaining ammonia. The cooled gas is finally washed with oil, and benzol is recovered. The various portions so obtained are given elaborate treatment for by product recovery.

The gas has hydrogen cyanide, hydrogen sulphide and carbon disulphide as impurities. It is passed through purifiers which consist of a series of chambers having several shelves loosely packed with ferric hydroxide which retains hydrogen sulphide.

$$Fe_2O_3.\ xH_2O + 3H_2S = Fe_2S_3 + (x + 3)\ H_2O$$

Hydrogen cyanide also gets removed as ferric cyanide. Iron sulphide is exposed to air when sulphur separates and is recovered. The presence of carbon disulphide is undesirable because on oxidation, it forms corrosive oxides of sulphur. This is removed by passing the gas over a heated nickel catalyst when some of the hydrogen of the gas reduces carbon disulphide to hydrogen sulphide and carbon. The gas is stored in huge gas holders floating on water.

7.9.3 Producer Gas

Producer gas is essentially a mixture of carbon monoxide and nitrogen but often steam is blown along with air, thereby giving some hydrogen also. A typical producer giving the reactions is illustrated in the Fig. 7.7. The producer consists of an air-tight cylindrical furnace made of steel plates lined with fireclay bricks. There is water-seal at the base which helps the removal of ash. Air, or a mixture of air and steam, is blown through a pipe at the base. At the top there is an inlet for coal and also an outlet for the gas generated.

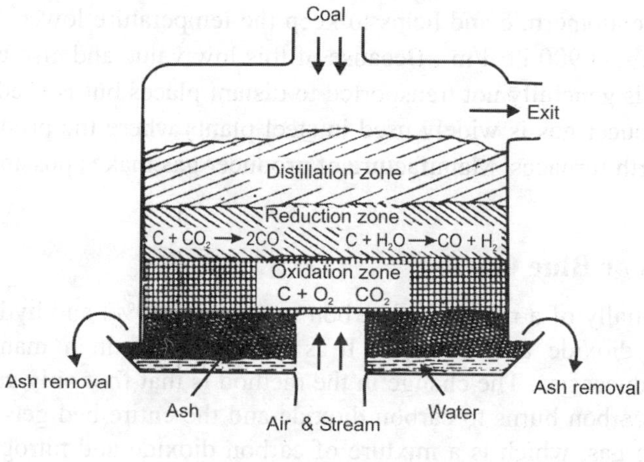

Fig. 7.7 Producer gas plant.

Coke or even inferior coal of high ash content is generally used for making producer gas. Bituminous coal gives gas of higher calorific value. The producer is filled with the fuel and the fuel is ignited. Air is forced in through the hot fuel and in the lower zone, known as tie oxidation zone, carbon burns to give carbon dioxide. The reaction is strongly exothermic.

$$2C + O_2 = 2CO + 52.8 \text{ kcal}$$

$$2CO + O_2 = 2CO_2 + 135.3 \text{ kcal}$$

(The positive sign indicates exothermic reaction and the negative sign indicates endothermic reaction).

Carbon dioxide passes upwards through the bed of coal and is reduced to carbon monoxide. The reaction is endothermic.

$$CO_2 + C = 2CO - 41.2 \text{ kcal}$$

In the uppermost part of the fuel bed there is fresh coal which is at a much lower temperature and no reaction takes place but the coal gets pre-heated and volatile matter distils. Gas goes out through the outlet and ash is removed through the water-seal. In efficient plants the producer gas so obtained contains about 33% carbon monoxide, 64% nitrogen and the rest is carbon dioxide and hydrogen.

In working the producer in the manner described above, some serious difficulties are faced. The combustion of carbon in air heats the lower part with the result that the ash forms clinkers which are difficult to remove. The firebars, on which the coal rests, and the refractory lining of the plant get heated to a high temperature with the result that their lives are reduced. In addition, the gas coming out is at a high temperature and if it is transported to distant parts of the factory much of its sensible heat is lost.

To reduce these difficulties some steam, about 0.35 kg per kg of coal, is blown in along with the air. In the reduction zone, steam reacts with carbon, producing carbon monoxide and hydrogen;

$$C + H_2O = CO + H_2 - 41.9 \text{ kcal}$$

This reaction is endothermic and helps to keep the temperature lower. The calorific value of producer gas is about 900 kcal/m^3. Because of this low value and also as it comes out at a high temperature it is generally not transported to distant places but is used on the spot where it is generated. Producer gas is widely used in steel plants where the producers are built just below the open hearth furnaces. Manufacture of producer gas makes possible the utilization of lower grade coal.

7.9.4 Water Gas or Blue Gas

This consists essentially of a mixture of carbon monoxide (40%) and hydrogen (50%) along with some carbon dioxide and nitrogen. It is manufactured in a manner similar to the manufacture of producer gas. The change in the method is that first only air is passed through the coal bed when carbon burns to carbon dioxide and the entire bed gets heated to 1000° or more. The outgoing gas, which is a mixture of carbon dioxide and nitrogen is not used as a fuel but is used to preheat the incoming air and steam. Steam is then blown through red hot coal when carbon monoxide and hydrogen are produced and are collected. This reaction is endothermic, therefore the bed cools down. Steam supply is stopped and air is again blown in to raise the temperature. Water gas is also called 'blue gas' because both carbon monoxide and hydrogen burn with a blue flame. It is of little value for lighting purposes. Its calorific value is about 2,700 kcal/m^3.

The calorific value of water gas being low, it is also not suitable for supplying to distant places. The calorific value can be increased by mixing oil gas. This is done by first heating a checker work of firebricks by burning a portion of the gas. Then oil is sprayed on this heated checker work when it gets vapourized and partly cracked and the water gas is blown through it. Thus *carburetted water gas* is obtained which contains about 35% each of carbon monoxide and hydrogen, 20% of various lower hydrocarbon gases and 10% of carbon dioxide and nitrogen. Its calorific value is about 4,500 kcal/m^3.

7.9.5 Blast Furnace Gas

The reactions taking place in a blast furnace are such that gas similar in composition to producer gas is obtained. It contains about 25% carbon monoxide and has a calorific value of 800 kcal/m^3. The quantity of gas produced is very large but its calorific value is so low that it cannot be used for many purposes nor can be supplied to distant places.

The gas contains a considerable quantity of dust and is passed through dust catchers where the heavier particles drop down. It is then washed with a spray of water and passed through electrical precipitators to remove the remaining particles. Gas purification is important because the dust is deposited in the pipelines, burners and furnaces and thus reduces efficiency.

In the past a quantity of the gas was used for gas-driven blower engines for the blast furnace but this use is on the decline. The present tendency in the iron and steel industry is to have integrated steel plants. 25% of the gas is used up in heating the blast air, and the balance finds important and large scale use for steam raising in the power plant and heating the coke ovens, open-hearth furnaces and reheating furnaces. The use of blast furnace gas for heating coke ovens makes available a larger quantity of coke-oven gas, which has a higher calorific value, for other purposes. This has resulted in savings on fuel cost in steel plants.

7.9.6 Natural Gas

Almost every petroleum well gives out some gaseous product consisting mostly of methane along with some ethane, propane and other hydrocarbons. In addition, there are large deposits of the gas in gas wells without any liquid petroleum being present. The gas is given the name "natural" gas. In the U.S.A. 4,00,000 million m^3 of the gas is produced and distributed through about 0.8 million km of pipelines.

Natural gas is 'wet' as it contains hydrocarbons which could condense in the pipeline. These are removed by compression and cooling or by absorption in liquid oil. The 'dry' gas so obtained contains over 80% methane, the balance being mostly ethane. Its calorific value is about 10,000 kcal/m^3.

Wet gas is treated to remove propene, propane, butane, and butane which are used as LPG. Before use the natural gas in purified by removing objectionable ingradients such as dust, grit, water H_2S, CO_2, N_2 and heavier liquefiable hydrocarbons (LPG). The composition of natural gas is largely 70-90% methan,e 5-10% ethane, 3% of H_2 and traces of CO and CO_2 with C.V of 12,000-14,000 Kcals/m^3. H_2S is removed by heating or scrubbing with monoethanalamine.

As the facility of pipelines is lacking in many place, easily condensable hydrocarbons like propane, isobutane and butane are liquified at room temperature by compression. The liquified petroleum gases (known as L. P. G.) are sold in cylinders for use as domestic fuel.

The use of natural gas supplied through pipelines and in gas cylinders is increasing rapidly in India.

7.10 Determination of Calorific Value

7.10.1 Junker's calorimeter

Figure 7.8 illustrates a gas calorimeter originally designed by Junker. It consists essentially of an inner combustion chamber in which the gas burns. The chamber is surrounded by a water jacket through which a number of tubes, called flues, pass. The gas on combustion heats the water jacket and then passes down the flues in the water jacket, thus complete transfer of the heat of combustion to water is ensured.

Water enters the calorimeter through a constant head and its temperature is noted by a thermometer. The gas whose calorific value is to be determined passes through a gas meter, which records its volume, and burns in a burner, as shown in the figure. The hot products of combustion move upward in the chamber and then downward through the flues provided in the water jacket and escape through the exit. A thermometer at the exit indicates its temperature which should be as near the room temperature as possible so that the transfer of heat to water is complete. Water, moving in a direction opposite to that of the gases in the flues, goes out from the upper portion. Its temperature and volume are recorded. Water formed by the condensation of steam is also collected.

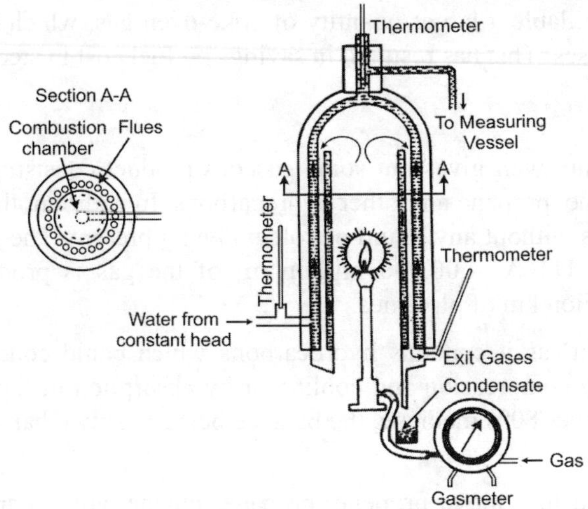

Fig. 7.8 Junker's gas calorimeter

The volume of gas burnt during the experiment is converted to volume at 15° and 760 mm pressure. Knowing the weight of the water flowing and the rise in its temperature the heat given out by the gas is known. The gross calorific value is then obtained by dividing the heat given out by the corrected volume of the gas. For obtaining the net calorific value, the condensed water is weighed and the weight multiplied by the latent heat of vaporization of water. This is subtracted from the total heat and the result is then divided by the corrected volume of the gas.

7.11 Combustion

Combustion is defined as a chemical reaction which is accompanied by heat and light. In a narrow sense it is the union of an element or a compound with oxygen. For practical purposes, it is used to designate only those cases in which the development of heat is rapid and is accompanied by a considerable rise in temperature. In the common fuels it involves the burning of carbon and hydrogen in air and also to a much smaller extent of sulphur. For proper combustion, the substance must be brought to its *kindling* or *ignition temperature*, which is defined as the temperature at which it ignites and continues to burn without further addition of heat from outside. The ignition temperature depends upon several factors; the presence of a catalyst, for instance, lowers the temperature.

7.11.1 Spontaneous Combustion

Some materials like coal, especially powdered coal, oily rags or wastes, are known to burn by themselves. If such materials are stored in such a manner that they are almost insulated and the loss of heat from them is slow, then heat produced by their slow oxidation gradually raises the temperature. As the temperature begins to raise the rate of oxidation increases with a further rise in temperature. The result is that the ignition temperature is soon reached and the material begins to burn. This is an example of spontaneous combustion and must be carefully guarded against in storage of such materials, as it is highly risky.

7.11.2 Some Practical Aspects of Combustion

When a fuel like bituminous coal is fired into a furnace, its volatile constituents are vapourized. Moreover, if the furnace is already not at a sufficiently high temperature, the volatile gases (which mostly consist of hydrocarbons) will not be heated to their ignition temperatures and thereby escape unburned, causing wastage. Besides the proper temperature, there should be a sufficient supply of air for combustion of the gases to take place. The amount of air that will pass through the grates also depends upon the size of the coal fired. Smaller pieces, like coal dust, restrict the flow.

Theoretically, the presence of *moisture* in coal is undesirable as it causes waste of heat, but in practice a small quantity of it is advantageous. Moisture may be present in the coal itself or may be added by moistening the coal before use. Water coming in contact with red hot carbon vapourizes and then reacts with carbon giving carbon monoxide and hydrogen. This reaction is endothermic and lowers the temperature. Thus, moistening of coal helps to keep the temperature of the fire bars low and also prevents the formation of clinkers. If sufficient air and combustion space are available the gases produced will burn to give heat and form water again, which generally escapes as steam. There is some loss of heat in the form of the latent heat of steam, but the other advantages outweigh this disadvantage. It is estimated that if 10% moisture is present, loss of heat due to it will be about 1% of the total heat of the fuel but if moisture was not present about 10% of heat might be wasted, due to loss in the ash-pit and incomplete combustion. The process of the addition of moisture to coal just before it is fed into the furnace is known as *coal tempering*. Optimum free moisture content is 7 to 9% when coal has minimum density. For this reason air flow increases, thereby combustion is facilitated. Free moisture also reduces the escape of the fines (i.e., fine particles of coal). Moisture may be added by sprinkling water on coal or, better still, by admitting steam through the bottom of the coal hopper.

In practice, the efficiency of combustion is judged by the *percentage* of CO_2 in the flue gas. As air contains 21% by volume of O_2 and 1 volume of O_2 gives rise to 1 volume of CO_2, there can be up to 21 % of CO_2 in the flue gases by the combustion of carbon. Naturally, attempts are made to get as high a percentage of CO_2 in the flue gas as possible. If there is excess of air, there will be loss of heat as some heat is carried away as sensible heat. Hence excess air is usually reduced. This practice, carried on without proper consideration, may lead to lower efficiency. By cutting down excess air, the combustion may be incomplete, resulting in the formation of CO. In this reaction only 2,416 calories of heat per gram of carbon is produced as against 8,100 calories when a gram of carbon burns to give CO_2. Hence the percentage of CO is a more important consideration, as greater saving can be made by eliminating CO in flue gases than trying to prevent some loss of sensible heat in excess air. In actual practice, some CO is present even when there is a small amount of excess air.

The optimum excess air varies from furnace to furnace as it depends upon its design and the fuel burnt. The optimum excess is that which has the minimum of heat, both potential (e.g., in the form of CO which could burn to give more heat) and sensible heat going in the flue gases. In coal fired boilers usually 50 to 100% excess air is supplied.

7.11.3 Calculations

Simple combustion calculations can be studied in three categories :

(A) Burning of solids or liquids.

(B) Burning of gases.

(C) Calculation of air supplied.

When the composition of the fuel is known, the theoretical amount of air needed for its combustion and the products of combustion can be easily calculated with the help of a simple chemical equation. For this purpose the composition of air is taken to be:

	By weight	By volume
Oxygen	23%	21%
Nitrogen	77%	79%

This means that when 1 kg of O_2 is npeeded for a reaction

$$\frac{1\times 100}{23} = 4.35 \text{ kg of air is to be supplied.}$$

In the same manner for

$$1 \text{ m}^3 \text{ of } O_2, \frac{1\times 100}{21} = 4.76 \text{ m}^3 \text{ of air is to be supplied.}$$

This gives the amount of N_2 that will go along with the products of combustion.

Hydrogen burns to give water which is supposed to be in a liquid state and the volume, being comparatively very small, is neglected. The volumes of solids are also neglected.

If the analysis of fuel shows O_2, it is supposed to be in chemical combination; therefore, its amount is to be subtracted from the calculated total requirement.

Some useful conversions are :

(i) Gram molecule of gases occupy 22.4 litres at N. T. P.

(ii) Pound molecule of gases occupy 359 cu. ft. at N. T. P.

(iii) 1 lb. of air occupies 12.39 cu. ft. at N. T. P.

Since air contains 23% by weight of oxygen, the weight of air needed is the weight of oxygen divided by 0.23. Let C, H, O and S represent the weights of the elements in kg per kg of fuel, the weight of air theoretically needed is :

$$W = 11.6\,C + 34.8\,(H - O/8) + 4.35\,S$$

(A) Burning of Solid or Liquid Fuels : In such cases the analysis of solid or liquid fuel is given and amount of O_2 needed for combustion is to be calculated. From the amount of O_2 other requirements can easily be found. The combustion of each element is calculated separately.

(a) *Combustion of Carbon.*

(i) $2C+O_2 = 2CO$

In terms of weights $2 \times 12 + 2 \times 16 = 2 \times 28$

or 24 parts of C need 32 parts of O_2 for combustion to give 56 parts of CO,

or

1 g of C needs $\frac{32}{24} = 1.333$ g of O_2 giving 2.333 g of CO.

(ii) Complete combustion of C to CO_2, can be calculated as follows

	Carbon		Air	Flue gases
	C	+	$O_2 + (N_2)$	$= CO_2 + (N)$
Wt. in g or Ib.	12	+	32 + (107)	= 44 + (107)
	1	+	2.667 + (8.93)	= 3.667 + (8.93)
	1	+	11.6	= 12.6
Volume in liters at 0° and 760 mm.	12g	+	22.4 + (84.3)	= 22.4 + (84.3)
	1g	+	1.87 + (7.0)	= 1.87 + (7.0)
Volume in cu.ft. at 0° and 760 mm.	12 lbs	+	359.0 + (1350)	= 359.0 + (1350)
	1 lb	+	30.0 + (112.5)	= 30.0 + (112.5)
			142.5 cu. ft.	142.5 cu. ft.

(b) *Combustion of Hydrogen.*

$2H_2 + O_2 = 2H_2O$; 4 + 32=36

or 1 g of H_2 needs 8 g of O_2 to produce 9 g of water.

making use of Avogadro's hypothesis;

2 vols. of H_2 need 1 vol. of O_2 to produce water (which occupies negligible volume).

(c) *Combustion of Sulphur.*

$8 + O_2 = SO_2$; 32 + 32=64

or 1 g of S needs 1 g of O_2 to produce 2 g of SO_2.

Example 1.

Calculate the weight and volume of air needed for the combustion of 1 kg of carbon. Give the composition of the combustion products.

Solution :

$$C + O_2 = CO_2;\ 12 + 32 = 44$$

$\therefore$ Wt. of O_2 needed for 1 kg of C $= 1 \times \frac{32}{12} = 2.667$ kg.

As 23 parts by wt. of O_2 is present in 100 parts of air, the wt. of air needed

$$2.667 \times \frac{100}{23} = 11.6 \text{ kg}$$

Calculation of volume of air needed:

As 1 gram mole of a gas occupies 22.4 litres at N. T. P.

32 g of O_2 occupies 22.4 litres at N. T. P.

$\therefore$ 32 of O_2 is Present in $22.4 \times \frac{100}{21}$ litres of air at N. T. P.

$\therefore$ Volume of air at N. T. P. needed for the combustion of 1000 g of

$$C = \frac{2667}{32} \times \frac{22.4 \times 100}{21} = 8890 \text{ litres.}$$

As 1 m^3 = 1000 litres, the volume (at N, T. P.) of air needed for the combustion of 1 kg of carbon = 8.89 m^3

Wt. of N_2 present in 11.6 kg of air = 11.6 – 2.667 = 8.933 kg

Wt. of CO_2 produced in the reaction $= \frac{1 \times 44}{12} = 3.667$ kg,

$\therefore$ 1 kg of C needs 11.6 kg of air for combustion and the products contain 8.933 kg of N_2 and 3.667 kg of CO_2 or 70.9% N_2 and 29.1% CO_2 by weight are present in the flue gases.

Volumetric composition :

1 g mole of a gas occupies 22.4 litres at N, T. P.

$\therefore$ Volume of $N_2 = \frac{22.4 \times 8933}{28} = 7145$ litres.

and volume of $CO_2 = \frac{22.4 \times 3667}{44} = 1866$ litres.

Total volume = 9011 litres.

$\therefore$ Volume of $N_2 = \dfrac{7145 \times 100}{9011} = 79.3\%$

and volume of $CO_2 = \dfrac{1866 \times 100}{9011} = 20.7\%$

This principle can be used to solve actual problems as illustrated.

Example 2.

The percentage composition of a sample of bituminous coal, by weight was found to be C = 76%; H_2 = 5.2%; O_2 = 12.8%; N_2 = 2.7%; S = 1.2%; Ash = 2.1%. Calculate the minimum (a) weight and (b) volume at N. T, P. of air necessary for complete combustion of I kg of coal and percentage composition by weight of dry products,

Solution : The calculations maybe made in a tabular form:

Wt. of element per kg of fuel (in kg)	**Wt. of O_2** needed (in kg)	**Wt. of dry products** of combustion (in kg)
C = 0.76	$0.76 \times \frac{32}{12} = 2.026$	$0.76 \times \frac{44}{12} = 2.786\ (CO_2)$
H_2 = 0.052	$0.052 \times \frac{16}{2} = 0.416$	
O_2 = 0.128	-	-
N_2 = 0.027	-	0.027 (N_2)
S = 0.012	$0.012 \times \frac{32}{32} = 0.012$	$0.012 \times \frac{64}{32} = 0.024(SO_2)$
Ash = 0.021	-	-
Total = 1.000	2.454	

Wt. of oxygen to be supplied = Wt. of oxygen needed-Wt. of oxygen present =2.454 - 0.128 = 2.326 kg,

$\therefore$ Wt. of air needed = $2.326 \times \dfrac{100}{23} = 10.113$ kg.

$\therefore$ Wt. of nitrogen supplied with air = wt. of air - wt. of oxygen

= 10,113 – 2,326

= 7.787 kg.

The term "products of combustion" refers to the gaseous products of combustion (i.e., not the ash, which has not taken part in combustion).

∴ products in the combustion gases, excluding water, are

CO_2 = 2.786 kg; SO_2 = 0.024 kg; N_2 = 0.027 kg (from coal) + 7.787 kg (from air supplied) =7.814 kg. Total = 10.624 kg.

∴ Percentage composition by wt.:

$$CO_2 = \frac{2.786 \times 100}{10.624} = 26.22\% \qquad N_2 = \frac{2.786 \times 100}{10.624} = 73.55\%$$

$$SO_2 = \frac{0.024 \times 100}{10.624} = 0.23\%$$

Example 3.

The percentage composition of a sample of anthracite coal is C, 90; H_2, 3.5; O_2, 3.0; N_2, 1.0; S, 0.5; the remainder being ash. Estimate the minimum weight of air required for the combustion of 1 kg of this fuel, and the composition of the dry products of combustion, by volume, if 50% excess air is supplied.

Solution :

Wt. of element kg	Wt. of O_2 needed kg	Wt. of dry product kg
C= 0.9	$0.9 \times \frac{32}{12} = 2.40$	$0.9 \times \frac{44}{12} = 3.30\ (CO_2)$
H_2 = 0.035	0.035 × 8 = 0.28	...
O_2 = 0.03	...	...
N_2 = 0.01	...	0.01 (N2)
S= 0.005	$0.005 \times \frac{32}{32} = 0.005$	$0.005 \times \frac{64}{32} = 0.01\ (SO_2)$
Total	2.685	

Wt. of O_2 to be supplied = Wt. of O_2 needed - Wt. of O_2 present

= 2.685 - 0.03 = 2.655 kg,

$$\text{Wt, of air needed for combustion} = 2.655 \times \frac{100}{23} = 11.54 \text{ kg.}$$

As 50% excess air is supplied, the wt. of excess air supplied

$$= 11.54 \times \frac{50}{100} = 5.77 \text{ kg.}$$

$\therefore$ Wt. of air supplied = 11.54 + 5.77 = 17.31 kg.

Wt. of present in air $\dfrac{17.31 \times 77}{100}$ = 13.33 kg.

Wt. of N_2 from the fuel = 0.01 kg,

$\therefore$ total wt. of N_2 = 13.34 kg.

Wt. of free O_2 (due to excess air) $= \dfrac{5.77 \times 23}{100} = 1.33$ kg.

$\therefore$ The Wt. of dry products in flue gases are: CO_2, 3.30 kg;

N_2, 13.34 kg; O_2, 1.33 kg and SO_2, 0.01 kg.

Molecular wt. of these gases are CO_2 – 44; N_2 – 28; O_2 – 32 and SO_2 – 64 respectively.

Let 1 kg mole of a gas occupy V m^3

$\therefore$ The volumes of the gases are:

$$CO_2 = \frac{3.30 \times V}{44} = 0.0750 \text{ V m}^3; \quad O_2 = \frac{1.33 \times V}{32} = 0.0416 \text{ V m}^3$$

$$N_2 = \frac{1.334 \times V}{28} = 0.4764 \text{ V m}^3; \quad SO_2 = \frac{0.01 \times V}{64} = 0.00015 \text{ V m}^3$$

$\therefore$ total = 0.5932 V m^3

$\therefore$ Percentage by volume of the gases is:

$$CO_2 = \frac{0.0750V \times 100}{0.5932V} = 12.64\% \qquad O_2 = \frac{0.0416V \times 100}{0.5932V} = 7.01\%$$

$$N_2 = \frac{0.4764V \times 100}{0.5932V} = 80.32\% \qquad SO_2 = \frac{0.00015 \times 100}{0.5932V} = 0.03\%$$

Example 4.

A sample of petrol was found to contain 15.2% H_2 and 84.8% C by weight. Calculate (i) the weight of air needed for the combustion of 1 kg of the fuel; (ii) the volumetric composition of the products of combustion if 15% excess air is supplied,

Solution :

I kg of petrol contains 0.152 kg of H_2 and 0.848 kg of C

Wt. of O_2 needed for 0.152 kg of H_2 = 0.152 × 8 = 1.216 kg

Wt. of O_2 needed for 0.848 kg of C $= 0.848 \times \frac{32}{12} = 2.261$ kg

Total wt. of O_2 needed for combustion $= 3.477$ kg

$\therefore$ Wt. of air needed $= 3.477 \times \frac{100}{23} = 15.12$ kg

Excess air supplied $= 15.12 \times \frac{15}{100} = 2.268$ kg,

Wt. of O_2 in excess air $= 2.268 \times \frac{23}{100} = 0.522$ kg,

Total air supplied $= 15.12 + 2.268 = 17.388$ kg.

$\therefore$ Wt. of N_2 in flue gases $= 17.388 \times \frac{77}{100} = 13.39$ kg.

Wt. of CO_2 produced per kg of fuel $= 0.848 \times \frac{44}{12} = 3.109$ kg.

$\therefore$ The flue gases from the combustion of 1 kg of petrol contain 3109 g of CO_2; 522 g of O_2 and 13390 g of N_2

Let 1g mole of a gas occupy V litres.

$\therefore$ The relative volumes of the gases are:

$CO_2 = \frac{3109\text{ V}}{44} = 70.7$ V litres $\quad O_2 = \frac{522\text{ V}}{32} = 16.3$ V litres

$N_2 = \frac{13390\text{ V}}{28} = 565.2$ V litres

Total vol. $= 565.2$ V litres

$\therefore$ Percentage of CO_2 by vol. $= \frac{70.7\text{V} \times 100}{565.2\text{ V}} = 12.5\%$

Percentage of O_2 by vol. $= \frac{16.3\text{V} \times 100}{565.2\text{ V}} = 2.9\%$

Percentage of N_2 by vol. $= \frac{478.2\text{V} \times 100}{565.2\text{ V}} = 84.6\%$

(A) Burning of gases : These calculations are comparatively easy and their principles are well worked out in junior classes. The following example illustrates how to solve the problems:

Example 5.

The volume analysis of a producer gas is: H2 – 14%, CH4 – 2%, CO – 22%, CO2 – 5%, O2 – 2%, N2 – 55%. Find the air required for the perfect combustion of 1 m3 of the gas. If 40% excess air is supplied, find the volume analysis of the dry products.

Solution : The Combustion reactions are :

1. $2H_2 + O_2 = 2H_2O$

   2Vol. 1 Vol. (nil)

2. $CH_4 + 2O_2 = CO_2 + 2H_2O$

   1 Vol. 2 Vol. 1 Vol. (nil)

3. $2CO + O_2 = 2CO_2$

   2Vol. lVol. 2Vol.

Applying these reactions to the combustion of 100 m^3 of the gas:-

Gas	Vol. in m^3	Vol. of O_2 needed	Vol. of dry CO_2	products N.
H_2	14	7	-	-
CH_4	2	4	2	-
CO	22	11	22	-
CO_2	5	--	5	-
O_2	2	[-2]	-	-
N_2	55	-	-	55
Total	100	(22-2 = 20)	29	55

∴For the perfect combustion of 100 m^3 of the gas, 20 m^3 of O_2 is needed and the dry products contain 29 m^3 of CO_2 and 55 m^3 of N_2.

$$\therefore \text{ Vol. of air needed per } 100 \text{ m}^3 = 20 \times \frac{100}{21} = 95.24 \text{ m}^3$$

$$\therefore \text{ Vol. of air needed per m}^3 = 0.95 \text{ m}^3$$

Vol. of N_2 in the air supplied = 95.24 - 20 = 75.24-m^3

$$\text{Vol. of excess air supplied} = 95.24 \times \frac{40}{100} = 38.1 \text{ m}^3$$

$$\text{Vol. of } O_2 \text{ in excess air} = 38.1 \frac{21}{100} = 8.0 \text{ m}^3$$

$$\therefore \text{ Vol. of } N_2 \text{ in excess air} = 38.1 - 8.0 = 30.1 \text{ m}^3.$$

Dry combustion products of 100 m^3 of gases contain 29 m^3 of CO_2 + 55 m^3 of N_2 (as calculated in the table) + 75.24 m^3 of N_2 (from air supplied for perfect combustion) + 30.1 m^3 of N_2 (from excess air supplied) + 8.0 m^3 of O_2 (from excess air)

$$= 29.0 \text{ m}^3 \ CO_2 + 160.34 \text{ m}^3 \ N_2 + 8.0 \text{ m}^3 \ O_2 = 197.34 \text{ m}^3.$$

$$\therefore \quad CO_2 = \frac{29.0 \times 100}{197.34} = 14.7\% \qquad O_2 = \frac{8 \times 100}{197.34} = 4.1\%$$

$$N_2 = \frac{160.34 \times 100}{197.34} = 81.2\%$$

(B) **Calculation of Air Supplied :** It has already been stated that if theoretical minimum air is supplied in a furnace, the combustion reactions will not be complete and there will be considerable wastage. In coal furnaces 50 to 100% excess air is generally supplied. In practice it is difficult to measure the air supplied. For proper control and economy it is also necessary to know the actual state of combustion, i.e., whether or not complete combustion is taking place, how much of CO is being formed, how much of free oxygen goes in the flue gas, etc. All this information can be had from the flue gas and fuel analyses, as will be shown in the examples that follow.

7.12 Flue Gas Analysis by Orsat Apparatus

The principle of analysis is that the gas to be analysed is taken in a burette which is connected to several pipettes containing suitable solutions. The gas is forced through these pipettes one after another. Thus certain constituents get absorbed and the volumes of the unabsorbed gases are found, from which the volume of the particular constituent is known.

The most widely used apparatus for the purpose is the Orsat apparatus, Fig. 7.9. It consists of a capillary glass tube manifold connected at one end to a gas burette *B* and to a three way stop cock *d* at the other end. The manifold has three side tubes connected to three stop cocks *a*, *b* and *c*, through which are connected three absorption pipettes, p_1, p_2 and p_3, each of which has two bulbs as shown in the figure. Just below the cocks there are marks to show the levels of the solutions. The gas burette is surrounded by a water jacket so that the temperature of the gas remains unchanged during the experiment. The lower end of the burette is connected by a rubber tube to an aspirator bottle *L*, filled with water. Three-way stop cock d is connected to the flue gas source through a U-tube containing glass wood for retaining smoke particles. The working of the three-way cock should be well studied before use because one position connects it to the flue gas source, the second to the atmosphere and the third shuts off the main apparatus.

The pipettes are usually filled with glass tubes so that the surface area of contact between the gas and the solution is increased. The three pipettes p_1, p_2 and p_3 have a solution for the absorption of CO_2, O_2 and CO respectively. p_1 has potassium hydroxide solution (200 g in 500 ml of water). It absorbs only CO_2. p_2 has a solution of pyrogallic acid (25 g) and potassium hydroxide (200 g) in water (500 ml). It can absorb CO_2 and O_2. p_3 contains cuprous chloride (100 g) solution in liquor ammonia (125 ml) and water (75 ml). It can absorb CO_2, O_2 and CO. Hence it is necessary that the flue gas is passed first through pipette *Pi* when CO_2 is removed, then through p_2 when only O_2 will be removed (as CO_2 has already been removed in p_1) and finally through p_3 when only CO will be removed. The whole apparatus is readily available in compact wooden cabinets.

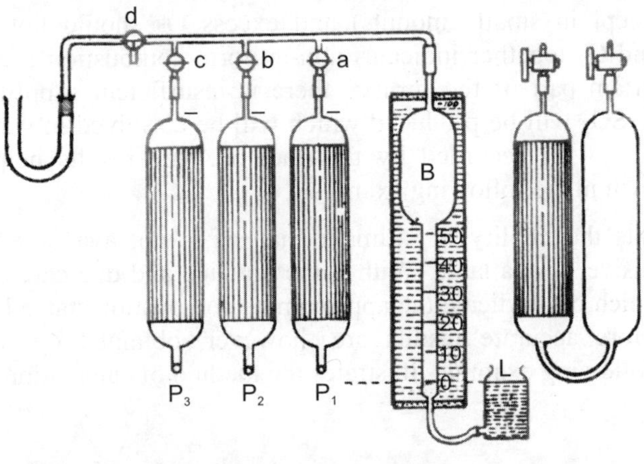

Fig. 7.9 Orsat apparatus

The stoppers are greased and the solutions taken as described. Cocks *a*, *b* and *c* are closed. Cock *d* is opened to the atmosphere and aspirator *L* is raised till the burette is filled with water but water does not enter the capillary tube. Cock d is closed, cock a is opened and the leveling bottle *L* is lowered. Water from the burette runs into *L*, thereby sucking out gas from the apparatus which is now completely closed except for an opening through pipette p_1, the back limb of which is open to the atmosphere. The result is that air enters the back limb and the solution is pushed to the front limb of the pipette. Thus the level of the solution in the pipette is raised up to the fixed mark and cock a is closed. In a similar manner the levels of the solutions are brought to fixed marks in pipettes p_2 and p_3.

The apparatus is now ready for taking in the gas for analysis. Cock *d* is opened to the atmosphere and *L* is raised till the burette is filled with water. Cock d is now connected to the gas supply and *L* is lowered drawing in the gas to be analysed in the burette. This sample is mixed with the air which was present in the apparatus. Cock *d* is opened to the atmosphere and the gas expelled by raising aspirator bottle L. This process of taking in and expelling the gas is repeated twice. Finally the burette is completely filled with the gas, cock *d* is opened to the atmosphere and bottle *L* is very carefully raised till the level of water in it is the same as in the burette which stands at zero mark. Now the burette has 100 ml gas at atmospheric pressure. Cock d is closed and a is opened and L raised to drive the gas into p_1. Bottle *L* is lowered to take the gas back in the burette. The gas is again sent into pipette p_1. This process is repeated several times to ensure full absorption of CO_2 in KOH solution. Bottle *L* is lowered till the solution in the pipette stands at the fixed mark and cock *a* is closed. The levels of water in the bottle *L* and burette *B* are equalized and the volume noted. To be certain that all CO_2 has been absorbed the process is repeated with the same sample till the two volume readings are the same. The decrease in volume gives the volume of CO_2 in 100 ml gas sample. The volumes of O_2 and CO are determined in a similar manner by passing the remaining gas through pipettes p_2 and p_3 respectively.

The Orsat apparatus is simple and easy to work. It gives valuable information for regulating combustion in the furnace. Presence of free CO shows incomplete combustion and air supply has to be increased. If there is a large amount of free O_2 it shows a large excess of air. Both CO (except in small amounts) and excess O_2 should not be present together. Presence of CO and O_2 together indicates non-uniform combustion i. e., excess air is being sent in, and in certain part of the furnace there is insufficient supply of air. If sulphur is present in the fuel, SO_2 will be produced which will be dissolved in water over which gas is collected and will not be recorded by this method. The results help to make important calculations as shown in the following examples.

In smaller plants the facility of ultimate analysis is not available but Orsat apparatus, which is not expensive, is available. With some practice and due care one should be able to get good results which can indicate the approximate combustion state. This is satisfactory for most purposes. More accurate results are, however, obtained by detailed analysis and calculations. The following example illustrates the method of calculation.

Example 6.

A sample of flue gas has the following analysis CO_2, 12.5%; O_2, 7.1%; N_2, 80.4%. Calculate the excess air supplied.

Solution. On the assumption that there is no nitrogen in the fuel, oxygen and nitrogen must have come from air.

100 m^3 of flue gas has 80.4 m^3 of N_2

$\therefore$ Vol. of air supplied per 100 m^3 of flue gas

$$= 80.4 \times \frac{100}{79} = 101.8 \text{ m}^3$$

$$\therefore \text{ Vol. of } O_2 \text{ in air actually supplied} = 101.8 \times \frac{21}{100} = 21.4 \text{ ml}$$

Vol. of excess O_2 present in 100 m^3 of flue gas 7.1 m^3

$\therefore$ Vol. of O_2 used in combustion = 21.4 - 7.1 = 14.3 m^3

$$\therefore \text{ Percentage of excess air} = \frac{7.1}{14.3} \times 100 = 49.7\%.$$

Example 7.

The analysis of the coal in a boiler trial was:

C – 88%, H_2 – 3.6%, O_2 – 4.8%, other matters 3.6% and the flue gas analysis by volume was CO_2 – 10.9%, CO – 1%, O_2 – 7.1%, N_2 – 81%. Find the proportion of carbon burned to CO, and the air required per kg of fuel for the combustion as it actually occurred and also the weight of flue gas per kg of fuel burned.

Solution: For converting the volumetric composition of flue gas into composition by weight, the percentages are to be multiplied by the molecular weights of the respective gases. This gives their relative weights, which are:

CO_2, 10.9 × 44; CO, 1 × 28; O_2, 7.1 × 32; N_2, 81 × 28.

Wt. of C in CO_2 present = $10.9 \times 44 \times \dfrac{12}{44} = 10.9 \times 12$ parts

Wt. of C in CO present = $1 \times 28 \times \dfrac{12}{28} = 1 \times 12$ parts

∴ Total wt. of C = (1O.9 +1) × 12 parts

∴ 11.9 × 12 parts by weight of C is associated with 81 × 28 parts of N_2 by weight.

Wt. of C in 1 kg of the coal = 0.88 kg.

∴ In the flue gases 0.88 kg. of C is associated with

$$0.08 \times \frac{81 \times 28}{11.9 \times 12} \text{ kg of N2}$$

∴ Wt. of air supplied per kg of coal

$$\frac{0.88 \times 28}{11.9 \times 12} \times \frac{100}{77} = 18.15 \text{ kg.}$$

$$\text{Percentage of C burning to CO} = \frac{\text{Wt. of C in CO}}{\text{Wt. of C in CO}_2 \text{ and Co}} \times 100$$

$$= \frac{1 \times 12}{10.9 \times 12 + 1 \times 12} \times 100 = 8.4\%$$

Weight of flue gas per kg of fuel burnt can be calculated thus-

11.9 × 12 parts by Wt. of C is present in flue gas whose wt. is 10.9 × 44 (CO_2) + 1 × 28 (CO) + 7.1 × 32 (O_2) + 81 × 28 (N_2) parts. or 142.8 parts by wt. of C is present in 3003.8 parts by wt. of flue gas.

∴ 1 kg of fuel contains 0.88 kg of C

∴ the wt, of dry flue gas from it

$$= \frac{0.88 \times 3003.8}{142.8} = 18.51 \text{ kg.}$$

As the analysis of the flue gas does not include moisture, this is the weight of dry gases. Generally, in problems, the weight of dry flue gas is asked, but in this particular problem 'dry' flue gas has not been mentioned. Therefore the weight of moisture needs to be added. The moisture produced per kg of fuel is obtained from the weight of hydrogen.

Wt. of H_2 per kg of fuel = 0.036 kg

∴ Wt. of water = 0.036 × 9 = 0.324 kg.

∴ Wt. of flue gas per kg of fuel, including moisture

$$18.51 + 0.32 = 18.83 \text{ kg.}$$

Example 8.

In a boiler trial the dry coal as burned contained 84% of carbon and 3% of free hydrogen. The flue gas analysis gave 11.5% of CO_2, 8.4% of O_2 and 80.1% of N_2. Calculate per kg of dry fuel, the weight of necessary air and the weight of excess air.

Solution: Multiplying the volumetric results of analysis by the respective molecular weights of the gases the following relative weights are obtained:

$$CO_2 = 11.5 \times 44;\ O_2 = 8.4 \times 32;\ N_2 = 80.1 \times 28$$

Relative wt. of C in 11.5 × 44 parts of CO_2 is

$11.5 \times 44 \times \frac{12}{44} = 11.5 \times 12$ parts which is associated with 80.1 × 28 parts of N_2.

1 kg of fuel contains 0.84 kg of C, and the wt. of N_2 associated with it

$$= 0.84 \times \frac{80.1 \times 28}{11.5 \times 12} \text{ kg.}$$

$$\therefore \text{Wt. of air} = \frac{0.84 \times 80.1 \times 28}{11.5 \times 12} \times \frac{100}{77} = 17.73 \text{ kg per kg of fuel.}$$

Wt. of necessary air can be found by calculating the wt. of O_2 needed for C and H_2 in 1 kg of fuel.

Wt, of C in 1 kg of fuel = 0.84 kg.

$\therefore$ Wt. of O_2 needed for it $= 0.84 \times \frac{32}{12} = 2.24$ kg.

Wt. of H_2 in 1 kg fuel = 0.03 kg.

Wt. of O_2 needed for it $= 0.03 \times \frac{16}{2} = 0.24$ kg.

$\therefore$ Total wt. of O_2 needed per kg. of fuel = 2.48 kg,

$\therefore$ Wt. of air needed $= 2.48 \times \frac{100}{23} = 10.78$ kg.

Wt. of excess air supplied.

= Wt. of air supplied – wt. of air needed

= 17.73 – 10.78 = 6.95 kg.

Alternative method for calculating weight of excess air:

N_2 associated with 8.4% excess $O_2 = 8.4 \times \frac{79}{21}$ parts by volume

Ratio of excess N_2 to total $N_2 = \frac{8.4 \times 79}{21} \times \frac{1}{80.1}$

$\therefore$ Excess air $= 17.73 \times \frac{8.4 \times 79}{21} \times \frac{1}{80.1} = 7.00$ kg.

$\therefore$ Wt. of air used = wt. of air supplied - excess air

$= 17.73 - 7.00 = 10.73$ kg.

The following is much simpler method of calculating the percentage of excess air but if nitrogen and sulphur are present in the fuel the results are slightly incorrect, In the above example, 100 m^3 of flue gas has 80.1 m^3 of N^2

$\therefore$ Vol of air actually supplied per 100 m^3 of flue gas

$= 80.1 \times \frac{100}{79} m^3$

Vol. of O_2 in the air supplied $= 80.1 \times \frac{100}{79} \times \frac{21}{100} = 21.3 m^3$

Vol. of excess O_2 present in 100 m^3 of flue gas = 8.4 m^3

$\therefore$ Vol. of O_2 used in combustion = 21.3 - 8.4 = 12.9 m^3

$\therefore$ Percentage of excess air $= \frac{8.4}{12.9} \times 100 = 65.1\%$

(Excess air on the basis of previous method $= \frac{7}{10.73} \times 100 = 65.2\%$)

Note : Whatever method of calculation is applied the results must be same provided the question has also been framed correctly. Often the setter of the question does not do detailed and full calculations and simply alters the figures here and there e.g., one may lazily make the percentages of CO_2 as 12.0% and O_2 9% and N_2, 79% in the above question. This is wrong because by altering percentage of CO_2 the other percentages get altered according to the laws of combustion and not arbitrarily, and answers by different correct methods would be different. If this happens the obvious conclusion is that the question is wrong. The teachers framing the question have to be careful.

7.13 Examples

1. A sample of coal contains 87% carbon, 2% hydrogen, 1% oxygen, 1% sulphur and 9% ash. Calculate the theoretical weight and volume (at N. T. P.) of air required for complete combustion of I kg of the sample of coal.

   (Answer : 10.78 kg; 8.27 m^3).

2. A boiler is fired with coal having the following percentage composition by weight: C, 74.0; H_2, 6.8; O_2, 13.2; N_2, 2.1; S, 1.0 and ash, 2.9.

   Calculate the minimum quantity of air required for the complete combustion of 1kg coal and the percentage composition by weight of the dry products of combustion if 20% excess air is supplied.

   (Answer : air, 10.4 kg; CO_2, 21.11; O_2, 3.74; N_2, 74.99; SO_2, 0.16)

3. A sample of coal has 80% C; 5% H_2; 0.5% Sand 14.5% ash. Calculate the theoretical quantity of air necessary for the combustion of 1 kg of coal. Find the analysis of the flue gas by weight if 20 kg of air are supplied per kg of coal.

   (Answer : air, 11.04 kg; N_2, 73.83%; CO_2, 14.07%; O_2, 9.89%; H_2O, 2.16% and SO_2, 0.05%).

   Note: This question does not say, "dry flue gas."

4. A sample of dry coal has the following composition: C, 84%; H_2, 4%; O_2, 6%; S, 1% and the remainder ash. If 50% excess air is supplied estimate the percentage composition by volume of the dry products of combustion.

   (Answer : CO_2, 12.51; O_2, 7.04; N_2, 80.40; SO_2, 0.05).

5. A sample of coal was found to have the following percentage composition:

   C, 81%; H_2, 5%; S, 0.5%; N_2, 3% and ash, 10.5%. Calculate the percentage composition by volume of the flue gases, when air, 30% in excess by weight is supplied for combustion.

   (Answer : CO_2. 13.74%; N_2, 81.33%; O_2, 4.90% and SO_2, 0.03%).

6. The fuel oil used in a Diesel engine has 85.8% C; 12.5%. H_2; 1.5% Sand and 0.2% O_2. Analysis of the products of combustion by Orsat apparatus shows 8% CO_2 and 10% O_2 by volume. Calculate the weight of air supplied per kg of the oil.

(Answer : 26.7 kg.)

7. During a boiler trial, the average analysis of the flue gas gave the following results; CO_2, 10.75%; CO, 1.16%; O_2, 6.8%; N_2, 81.29%. The chemical analysis of the coal fired gave: C, 88.1%; H_2, 3.5%; O_2, 4.8%; ash, 3.6%. Determine the number of kg of air supplied to the boiler furnace per kg of coal. Calculate also the weight of flue gas per kg of coal.

(Answer : air, 18.23 kg; dry flue gas, 18.49 kg. (with moisture 18.805 kg).

8. The flue gas analysis by volume in a boiler trial was CO_2 10.5%; CO, 1%; O_2; O_2, 8%; N_2 80.5%: and the coal analysis, as burned, was C, 82%; H_2, 4.2%; O_2, 48%; other matter, 9%. Calculate the minimum weight of air needed for complete combustion and the weight of excess air supplied per kg of coal.

(Answer : air, 10.76 kg; excess air 6.64 kg).

9. The percentage composition of a sample of a producer gas, by volume is as follows: H_2, 16; CH_4, 1; CO, 22; CO_2, 6; O_2, 1; N_2, 54. Find the minimum volume of air necessary for the complete combustion of 100 m^3 of this gas and the composition of the dry products of combustion by volume.

(Answer : air, 95.23 m^3, CO_2, 18.3%; N_2, 81.7%)

10. A producer gas has the following composition by volume; CH_4, 3.5%; CO, 25%; H_2, 10.4%; CO_2, 10.8%; N_2, 50.3%. Calculate the theoretical minimum quantity of air required per m^3 of the gas. If 20% excess air is supplied, find the percentage composition of the dry products of combustion.

(Answer : air, 1.175 m^3 ; CO_2, 19.07%; N_2, 78.53%; O_2, 2.40%)

11. A gas used in an internal combustion engine had the following composition by volume; H_2, 45%; CH_4, 36%, CO, 15%; N_2, 4%. Find the volume of air required for the combustion of 1 m^3 of the gas.

(Answer : 4.86 m^3.)

12. In a boiler trial the following analysis of the flue gas was obtained, by volume: CO_2, 13%; O_2, 6%. Find the kg of air per kg of coal if the carbon content of coal was 85%.

(Answer : 16 kg.)

13. A sample of coal was found to contain the following: C, 81%; H_2, 4%; O_2, 2%; N_2, 1%; the remaining being ash. Estimate the quantity of minimum air required for complete combustion of 1 kg of the sample. Find the composition of dry flue gas by volume if 40% excess air is supplied.

(Answer: 10.7 kg; CO_2, 13.23%; N_2, 80.77%; O_2, 6%)

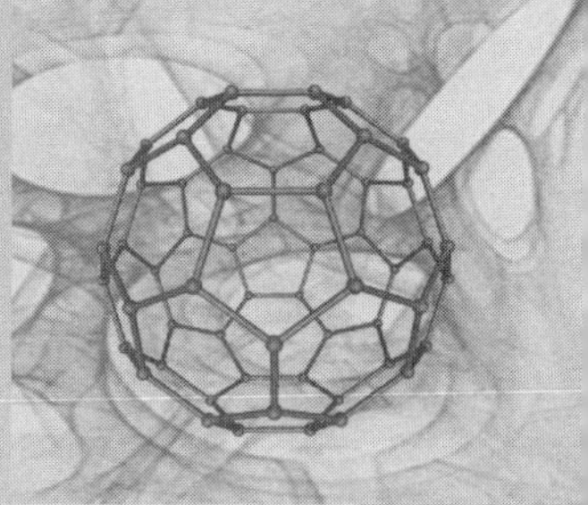

8 Building Materials

(Cement and Refractories)

8 Building Materials (Cements and Refractories)

8.1 Introduction to Cements

The term 'cement' and 'concrete' are often loosely used. One frequently speaks of a cement road. It is really speaking about a concrete road, not a cement road. Portland cement is the most important ingredient of concrete. Concrete is used as a material of construction more than any other material, e.g., weight of concrete used in construction is 5 to 10 times the weight of steel used. About a tonne of concrete is used per human being per year. 33 million tonne of Portland cement was manufactured in India in 1985 as compared to 2.7 million tonnes in 1951. To meet the increased demand of the cement, attempts are being made to increase the installed capacity to 60 m.t.

Various limes and mortars have been used for construction since ancient days. The earlier materials did not set under water. It is said that the Romans were the first to prepare cement, a hydraulic lime, by mixing lime paste and volcanic ash, that would set under water. In some pyramids lime plaster used by the Egyptians 4,500 years ago is still in good condition. In 1824 Joseph Aspdin, an English bricklayer of Leeds, prepared a cement in his kitchen by strongly heating a mixture of limestone and clay. The product on powdering and mixing with water would set under water. Aspdin patented it and gave it the name Portland cement because its colour resembled the colour of stones from Portland, England.

Portland cement is a product obtained by calcining a suitable mixture of lime bearing (calcareous) and clay bearing (argillaceous) materials and adding small amounts of other approved materials (like gypsum) and finally powdering the mixture. This powdered mixture hardens into a solid mass on reacting with water and gives a water resistant product. Thus the basic raw materials for the cement manufacture are lime, silica and alumina. The use of Portland cement has increased so greatly because of its excellent properties, cheapness and abundance of the raw materials used in its manufacture.

The uses of concrete are too well known to be narrated here. Large structures for which no other material is available have been possible on account of the unusual and varied properties of concrete. To give just one example, the tallest chimney constructed in India, at Singrauli, is 220 m high. Concrete is such a versatile construction material that it is used in quantities larger than any other manmade construction material. An estimate puts its use in India in 1986 as 70 million cubic meters.

The biggest cement factory in India, with an annual capacity of 1.7 million tonnes of cement is at Chunar, Dist. Mirzapur, U. P. Clinkers are produced at Dhala, which is surrounded by limestone deposits and is very close to the Singrauli coalfields. The clinkers are transported to Chunar and ground with blast furnace slag (8 lakh tonnes per year) brought from Bokaro steel plant and gypsum from Rajasthan. There are already two other cement factories, in the public sector, producing 8 lakh tonnes of cement annually.

8.1.1 Manufacture of Portland Cement

Portland cement: In this brief review of the changes taking place in setting and hardening, the heat-changes which take place have been given in later. The hydration reactions are exothermic. Tricalcium aluminate releases the maximum heat. In a massive structure, like a dam, heat release is undesirable as the temperature of the interior may rise, say, 50° above the outside temperature and set up high thermal stresses. For such purposes low-heat cement is needed. It has a smaller amount of tricalcium aluminate but large amount (more than 50%) of dicalcium silicate which has a low heat of hydration. For making concrete blocks it is desirable that the concrete sets quickly and gains strength and 'high early strength' cement is used which has about 60% of tricalcium silicate. Thus for specialized uses five general types of cements are made.

Without going into details of the types, the behaviour of the constituents can be summarised thus : Tricalcium aluminate gives early set and high heat of hydration (207 cal/g). It has poor resistance to sulphates, therefore cement for use in huge structures like dams or for use in sea waters should have a minimum of tricalcium aluminate. High alumina cement is also made which is resistant to sea water. It attains high strength in a short time and has high heat evolution. This does not contain tricalcium aluminate but other aluminates are present.

Tricalcium silicate gives early strength and makes the clinkers easy to grind. Its heat of hydration (120 cal/g) is high but is less than that of tricalcium aluminate. Dicalcium silicate has minimum heat of hydration (62 cal/g) and is therefore, an essential constituent of low heat cement. Its early strength is low and full strength is attained in a year or more. It is hard, therefore it makes clinkers difficult to grind.

'Rapid setting' cement contains a much smaller amount of gypsum. It reacts quickly with water and is used for emergency work like sealing water-tanks, repairing busy roads, etc.

Raw materials for the manufacture of Portland cement fall into three groups :

1. *Calcareous*, i.e., those which contain lime. Lime-stone is mostly used for this purpose. The limestone or chalk should not contain more than 5% of magnesium carbonate as it gives free MgO in the final product. Magnesium oxide very slowly reacts with water and expands causing cracking of the concrete. Presence of free silica in lime-stone is undesirable as it does not react easily but silica combined with alumina is helpful. Presence of sulphur is undesirable.
2. *Argillaceous*, i.e., those which contain silica, alumina and iron oxide. Mostly siliceous clay is used for this purpose. Bauxite is added if the proportion of alumina is not sufficient. Blast furnace slag is also used.
3. *Gypsum* is added during final grinding. It controls the ratio during setting and hardening.

The solid state reactions taking place between the raw materials during the manufacture of cement are promoted by iron oxide which chiefly acts as a flux. The reactions between the limestone and the clay would take place at a much higher temperature without the fluxing action of iron oxide. Therefore, presence of iron oxide in the raw materials is desirable.

For every tonne of iron 750 to 800 kg of slag is produced. Accumulation of this slag produces a serious disposal problem, hence manufacture of slag cement has been started. The molten slag from the blast furnace is poured in water when it gets granulated, i.e. broken into small porous pieces. Generally, basic slag containing about 50% CaO, 30% SiO_2 and remaining alumina plus iron oxide with less than 3% of MgO is used. Acid slags are used to a lesser extent. The Associated Cement Company is making cement at Jamshedpur using blast furnace slag. A cement factory with a capacity of 500,000 tonnes per annum is working under Bhilai plant of Hindustan Steel Ltd. using the blast furnace slag. As the slag from the Rourkela plant contains manganese, it was considered unsuitable for cement manufacture, but a scientific team of Geological Survey of India has reported its feasiblty. For making slag cement both the granulated slag and limestone are ground and calcined in the cement kiln or the slag and Portland cement clinkers are ground separately and mixed. For instance, at Durgapur 0.7 million tones of slag cement is manufactured by mixing nearly equal quantities of granulated blast furnace slag and Portland cement clinkers and grinding to fine powder in the usual way.

The proportion of calcareous and argillaceous material to be used depends upon their composition and the properties desired in the final product. It may be said that on an average for each tonne of cement produced about 0.2 tonnes of gypsum, 0.35 tonnes of coal and 3 tonnes of water and 80 KWH of electricity are needed besides limes and clay minerals. There are two processes of manufacture.

The *dry process* is used to a considerable extent in the U S A. but not in India or in Europe. In India wet process is most widely used. A few plants using dry process have been installed and the recent trend is in favour of the dry process mainly due to the economy in the fuel costs. In this process the raw materials are crushed, dried, ground and mixed in the proper proportion. The mixture is then finely powdered and fed into a rotary kiln 50 to 80 metres long and 3 m in diameter. The kiln is inclined and is rotated at the rate of about one revolution per minute. Powdered coal is blown from the lower end and combustion gases escape from the upper end. In the first half portion of the kiln limestone decomposes. Alumina and silica combine at 1400 to 1500^0 with lime giving the fused product. The clinkers are removed from the lower end, cooled, ground with about 4 to 5% gypsum and marketed. In the dry process the cost of fuel is less than in the wet process.

The *wet process* for manufacturing Portland cement is used to a very much larger extent than the dry process. In this process greater control is obtained over the composition and the product is more homogenous. The cost of grinding the raw materials is less than in the dry process but the kiln has to be longer and more water has to be removed, therefore fuel consumption is greater. The choice between dry and wet process depends mainly on the cost of fuel and the cost of grinding. If the raw materials available are hard, the wet process will cost less to grind the materials than the dry process. In many cases the overall advantages of the wet process are greater than those of the dry process.

The raw materials are carefully proportioned to give the required amount of lime, silica, alumina and iron oxide in the final product. Iron oxide and alkali oxides reduce the reaction temperature. Limestone after excavation from quarries is crushed and the size reduced to below 2 cm. It is fed along with washed clay in a grinding mill, water is added, and the materials are ground while wet. This gives a mud-like mixture, called slurry, containing 30 to 50% water. The slurry is collected in a tank and if necessary its composition is adjusted

As the temperature of the kiln is below the fusion temperature of the products, it is necessary to grind and mix the materials well so that chemical reactions may take place more easily. The slurry is fed directly into the cement kiln and calcined up to a temperature of about 1400 to 1500^0 when the materials react chemically and hard lumps called clinkers are obtained.

Rotary kilns (Fig. 8.1) are used for burning cement. A rotary kiln is cylindrical in shape, 2 to 4 metres in diameter and 70 to 150 m long. It is made of steel plates lined with firebricks. It is inclined (inclination 1 in 25, and mounted on heavy friction rollers and rotated at a speed of about one revolution per minute. At the upper end of the kiln is a flue through which the combustion gases escape after the removal of dust. At the lower end there is an opening through which the clinkers fall. There is a firing pipe at the lower end for blowing powdered coal or oil along with air. The hot gases leave through the flue at the upper end at a temperature of about 700° and the clinkers fall from the lower end at about 1200°.

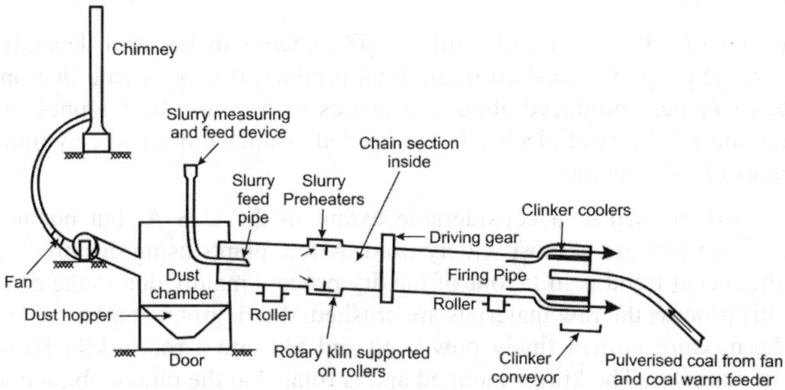

Fig. 8.1 A cement kiln.

The slurry is stored in a tank situated at a height so that it can be fed in the kiln by gravity. It passes through a measuring and feed device and a pipe (which is sloping towards the kiln because the slurry is viscous) leads it into the upper end of the cement kiln. At this end there are slurry pre-heaters. These are chambers riveted on flanges on the kiln body. These pre-heaters have inlet and outlet passages into the shell of the kiln with grates at the inlet and outlet ends. The slurry enters from the upper end and hot gases enter from the lower end. They come in contact inside the preheaters with the result that heat of the hot waste gases is transferred to the slurry. Water is thereby evaporated and the material gets heated. As the materials travel down the kiln they meet gases at higher and higher temperature, moving in the opposite direction. Decomposition of the carbonates takes place at about 900°. As the mass moves downwards, formation of the silicates and aluminates takes place above 1100°. To ensure complete reactions temperature is kept to about 1500° at the lower or firing end. The fused mass solidifies into small lumps, greenish black in colour. These lumps are called cement clinkers. At the lower end of the kiln 16 clinker coolers are attached to the kiln shell, all round the periphery. From the outlet of the kiln the clinkers enter the coolers. There are chains hanging inside the coolers. The clinkers drop down from the lower end of the coolers through which some air enters. Air helps to cool the clinkers and then enters the main kiln and there takes part in combustion reactions. Only partial cooling is accomplished in the clinker coolers. The clinkers drop on a shaker conveyer on which they cool further and are taken to ball mills for pulverising. The materials take a few hours to move from the feed to the discharge end of the kiln.

High efficiency dust filters need to be installed before discharging the gases into the atmosphere so as to control air pollution

The powdered clinkers on mixing with water will set very quickly and the cement mortar will unworkable. Therefore it is necessary to add some retarders to slow down the setting process, allowing time for placing cement mortar in the desired place. Gypsum, $CaSO_4.2H_2O$, is commonly used as the retarder. Usually about 4 to 5% of gypsum is mixed with the clinkers and the product is ground finely and stored.

Powdered coal is mostly used for heating but oil is also used. The coal received from the mines is crushed and taken by belt conveyer to ball mills and pulverised. This powdered coal is stored in coal bins from which it is fed by a worm conveyer into a pipe. Air is also forced into this pipe by a fan. This pipe, called firing pipe, feeds the air draft mixed with powdered coal in the kiln from the lower end. The flames and hot gases move upwards and supply the heat necessary for the reactions.

The reactions taking place during firing are complicated but may be represented thus :

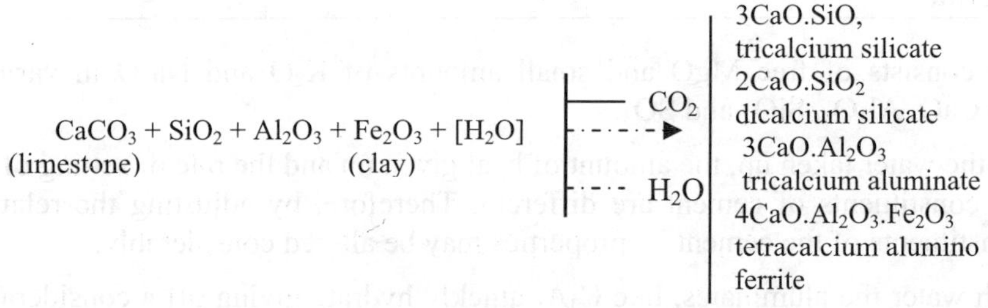

Cement for common purposes contains about 63% CaO; 22%, SiO_2; 6% Al_2O_3; 3% Fe_2O_3; 3% MgO, the balance being other constituents. As will be discussed presently, the composition may be widely different for certain uses. According to Meade, the ratio of the silica to the alumina content in the cement should be between 2.5 and 4 to 1 and iron should not be more than alumina. Magnesia and alkalis should each be less than 3%.

8.1.2 The Setting and Hardening of Portland Cement

Portland cement on mixing with about one third its own weight of water is converted into a plastic mass, called cement paste, which slowly loses its plasticity and becomes stiff and ultimately a rock-like mass is obtained. The stages in the change are indicated by the terms *initial set* and *final set*. Initial set is said to have taken place when a standard needle does not penetrate the paste, made in standard manner, beyond a certain depth. When the needle does not penetrate at tall, the cement is said to have reached the final set. The time for the initial set should not be less than 45 minutes and for the final set not more than 10 hours. It is necessary that in the beginning the cement should set slowly to allow time for mixing and placing it in position and to ensure that it adheres well with adjacent batches of concrete. At the same time it is necessary that the concrete should quickly become hard after the initial set so that it may soon be loaded. Setting of cement is completed in a matter of hours, while hardening is not completed for months. As a matter of fact, it has been found that hardness and strength continue to increase for years. It may be added that there is no fundamental difference between setting and hardening.

The hydration reactions begin immediately on mixing water and continue till the anhydrous compounds that constituted the cement disappear. Hydrolysis reactions also take place. The reaction products, energy changes and speeds of reactions during setting and hardening of Portland cement have been studied for over half a century. The main constituents of the common cement are given in Table 8.1.

Table 8.1

Name	Abbreviation	Formula	Percentage
Tricalciun Silicate	C_3S	$3CaO.SiO_2$	45
Dicalcium silicate	C_2S	$2CaO.SiO_2$	25
Tricalcium aluminate	C_3A	$3CaO.Al_2O_3$	10
Tetracalcium Aluminoferrite	C_4AF	$4CaO.Al_2O_3..Fe_2O$	10

The remainder consists of free MgO and small amounts of K_2O and Na_2O in various combinations with CaO, Al_2O_3, SiO_2 and SO_3.

The amount of the water taken up, the amount of heat given up and the rate of setting of all the four principal constituents of cement are different. Therefore, by adjusting the relative amounts of the constituents of the cement its properties may be altered considerably.

On mixing with water the aluminates, like C_3A, quickly hydrate giving off a considerable amount of heat and give the initial set. To check the rate of initial set gypsum, $CaSO_4$, $2H_2O$ is added which forms crystals of calcium sulpho-aluminate which retard the initial set and allow time for placing the concrete in position. The exact mechanism of the action of gypsum is not well understood. Calcium sulpho-aluminate has a low specific gravity and occupies a large volume. Therefore, if excess of gypsum is added there is a danger of high stress being produced due to expansion after hardening of the concrete. This may lead to cracking and failure of the structure. The aluminates are mainly responsible for the strength during the first day. These reactions are complete in four weeks. Thus if C_3A content is increased the cement would set quicker and would attain greater strength during the first four weeks but the ultimate strength would be less.

Tricalcium silicate, C_3S reacts more slowly with water but this reaction is the most important reaction and may be represented thus :

$$2(3CaO.SiO_2) + 6H_2O + 3CaO.2SiO_2.3H_2O + 3Ca(OH)_2$$

Tricalcium silicate water Tobermorite gel Calcium hydroxide

(Calcium silicate hydrate crystals resemble a natural mineral tobermorite, hence the name tobermorite gel). This reaction is slower and is almost complete during the first week and is responsible for the early hardness and strength of cement. The large amount of calcium hydroxide produced in this reaction adversely affects corrosion resistance of cement. It is noteworthy that no calcium hydroxide is liberated in the hydration of C_3A.

The reaction of water with dicalcium silicate is similar to that with tricalcium silicate but is very slow and its effect on the strength and hardness continues to be felt for a year. Aluminoferrite also hydrates :

$$2\,(2CaO.SiO_2) + 4H_2O \rightarrow 3CaO.2SiO_2.3H_2O + Ca\,(OH)_2$$

The engineering properties of concrete are mostly due to tobermorite gel (calcium silicate hydrate gel), which is the main cementing agent. This gel has an enormous surface area. It has been calculated that if one gram of a cement clinker is ground to the fineness of cement its surface area will be 3,000 cm^2. One gram of tobermorite gel has a surface area of million cm^2. Therefore, the surface forces of even a small weight of this gel are enormous and cause adhesion of the gel between its own particles as well as with other crystals, grains of sand and pieces of aggregate present in the concrete. Calcium hydroxide has greater surface energy than tobermorite has, but its surface area is much smaller; therefore its binding effect is not felt to such an extent.

In addition to surface forces, the strength and dimensional stability of the concrete depends upon factors like porosity. The pores do not contribute strength and hardness and are, therefore, undesirable. Further, if water enters the pores and solidifies, as happens in colder climates, it exerts great pressure because ice occupies a greater volume than water, and cracks may form. Airentrained concrete, i.e., concrete having a very large number of tiny air bubbles, is used to relieve this internal stress. Changes in atmospheric humidity also affect the amount of water in the concrete and cause volume changes. Therefore, if pure cement is used for construction, besides the cost factor, the volume-changes will be great and the mass will crack and will be weak. Inert materials like sand (called fine aggregate) and gravel or coarse stone (called coarse aggregate) are mixed with cement and water so that the volume-changes, and also heat-changes may be minimised. Such a composition, which solidifies into a rock-like mass, is known as *concrete*. A mixture of cement, water and sand is called *mortar*. An inert material, like sand, gravel and stone pieces, is called *aggregate*. The voids, or the space, between the larger particles is filled by smaller particles. The remaining voids are filled by cement paste. The entire aggregate is completely embedded in cement paste which binds it firmly.

Cement concrete is used for structural purposes under compressive loading. It has very poor tensile strength.

The pozzolans are effective because of the presence of reactive silica and alumina which slowly react with lime at ordinary temperatures forming calcium silicate and aluminate hydrates. They consist only partly of such active materials and the remaining constituents are inactive. The reaction being slow, it is necessary to grind well lime and *surkhi*, so that they are in a finely divided state. Concrete made by using pozzolanic cement, for example from the surkhi mortar is more resistant to corrosive enviroments than Portland cement. The structures have been found to be quite resistant to sea water, which corrodes Portland cement. As the heat of hydration is small and the reaction proceeds slowly, the pozzolanic cement is especially suited for making massive structures like dams. Mixture of pozzolans and Portland cement have been used with beneficial effects.

Pozzolan cement is made from clinkers produced either in the rotary kiln or vertical shaft kiln. The clinkers are mixed with 10 to 25% of pozzolanic material and ground \with 4 to 5% of gypsum. Thus the proportion of clinker: pozzolanic material: gypsum is about 80 : 15 : 5. The setting times, strengths and other characteristics of pozzolan cements are similar to ordinary Portland cement, excepting that in the case of pozz6lan cement the 3.days strengths are not specified. However 7-days strengths for pozzolan cements are the same as that of ordinary Portland cement.

8.1.3 Decay of Cement Concrete

Cement concrete, though mechanically very strong, is highly susceptible to chemical attack. Concrete contains some free lime. In acidic water lime dissolves making the concrete weak. Most of the natural waters have a pH value slightly above 7 (i.e., are slightly alkaline) and do not have any marked effect on concrete but as the pH value decreases (i.e., as the acidity increases) deterioration of concrete increases. Lime is more soluble in soft water than in hard water. Therefore, concrete deteriorates more quickly in contact with soft water because the lime dissolves.

Various sulphates and chlorides present in water also remove lime. Sodium carbonate removes silica. All these weaken the concrete. If it is soaked in mineral oil for some time, its resistance to abrasion decreases. Some cases of failure have been reported, the cause of which has been traced to the use of old sugar sacks for carrying sand or for covering the freshly laid concrete. It has been reported that even as low as 0.1% sugar when added to cement delays its setting time and greatly reduces its strength during the first four weeks.

The most serious type of damage to concrete takes place when a new compound which occupies more volume is formed. The common example of this type is in the presence of sulphates. Life of concrete is greatly reduced in the presence of sulphates because sulpho-aluminates are formed. This causes expansion and results in the destruction of the concrete structure. Such a danger is avoided by eliminating tricalcium aluminate from the cement composition. In cement containing tetra-calcium alumino ferrite, instead of tricalcium aluminate is formed. It is resistant to attack by sulphates.

A very inexpensive method of protecting cement concrete against corrosion is to give a coating of a bituminous material which will prevent contact between water and concrete. Concrete can also be protected by coating the surface with silicon fluoride in a soluble form with oxides of magnesium, zinc or aluminium. This precipitates insoluble calcium fluorides in the capillaries and prevents dissolution of lime.

8.2 Refractories

8.2.1 Definition and Functions of Refractories

Any substance that is difficult to fuse is a refractory, but when this term is used for engineering materials, several other properties are implied. Refractories retain their chemical and physical identity and also their mechanical function at elevated temperatures. Many engineering, more particularly metallurgical and ceramic, operations require the use of high temperatures for which materials are needed which can withstand the high temperatures. Common building materials like the ordinary brick, if exposed to such high temperatures, would fuse and the structure would fail. Besides resistance to high temperatures, the material is required to resist other destructive factors like heavy load, fluxes, corrosive fluids like slags, thermal shock due to rapid changes in temperature, abrasion (rubbing) due to movement of materials or dust-carrying gases. In addition, a refractory should be satisfactory for use as a lining for a furnace or equipment handling hot metals. Accordingly, a number of properties need to be considered in a refractory material. In 1981 about 9 lakh tonnes of refractories were made in India.

Properties

The following properties need to be considered while selecting a refractory material.

1. *Refractoriness :* This is a measure of the fusibility of the refractory. It indicates the temperature at which the refractory softens and is generally indicated in terms of pyrometric cone equivalent (P. C. E.) value. As the primary function of a refractory is to withstand high temperatures, it is of prime importance that the softening (or melting) temperature should be well above the operating temperature. It is noteworthy that in a furnace the inner end of the refractory wall is at a much higher temperature than the outer. Therefore, unless the brick melts away completely, it can often be is used to withstand a temperature higher than its softening temperature. The outer end will be at a lower temperature and still in a solid state, thus giving strength.
2. *Strength* : It must be strong enough to resist physical wearing away and to take load. *Refractoriness under load (RVL)* is an important consideration because usually a refractory fails at a lower temperature when subjected to load (Std load 1.75 kg/cm^2). This is supposed to be due to the fact that some crystals which have become fluidic act as a lubricant and deformation becomes easier when subjected to load. In many cases, especially in taller furnaces, the refractory has to support a heavy load, hence strength under the combined effect of temperature and standard load is an important criterion.
3. *Specific gravity :* The specific gravity is an important routine test for refractories as it gives an idea of its true porosity. Bulk density is represented by weight per unit volume of the refractory, the volume includes pore also. It is also a criterion while buying the raw material for making bricks or for lining, etc. Raw materials of higher specific gravity will give fewer bricks per unit weight. This means that the cost of a brick of higher specific gravity is more than that of lower specific gravity (taking the price of the raw material to be the same per unit weight). While using a refractory material in powdered form for making furnace lining or for repairs, lesser weight of a

material with lower specific gravity will be required. Materials of lower specific gravity are also preferable because bricks in the lower part of a full structure will not be subjected to a heavy load.

4. *Thermal expansion :* In furnace design, allowance has to be made for thermal expansion. This becomes very important factor if the refractory is subjected to rapid changes in temperature as a furnace door is, which on opening is suddenly exposed to air at room temperature. Within a single brick in a wall there may be a temperature drop of a few hundred degrees from the hot face to the cold face. This causes differences in thermal expansion in different sections of the brick. There may be changes in the internal structures of the material with the result that there are changes in volume, as will be explained under properties of silica refractories. All these result in the development of high internal stresses.
5. *Thermal conductivity :* In most cases the thermal conductivity should be low so that the heat of the furnace may not be lost. Exceptions to this rule are not many, e.g. coke oven walls, muffle furnaces and retorts are heated from outside, and should be made of refractories of high thermal conductivity. To prevent heat loss sometimes a refractory is backed by an insulating material like asbestos. It is to be noted that under these conditions the refractory will be maintained at a higher temperature and consequently its life will be shorter.
6. *Porosity :* If the refractory has open pores the gases and slags will enter the material more easily and to a greater depth and may react and reduce the life of the refractory.
7. *Electrical conductivity :* For electrical furnaces the electrical conductivity of the refractory should be low. Except graphite, all other refractories are bad conductors of electricity. The electrical resistivities of the refractories drop rapidly with rise in temperature.
8. *Chemical composition :* It gives the true identity of the particular type of the refractory. The chemical composition of the refractory should be such that the surroundings do not chemically attack the refractory and corrode it.

8.2.2 Classification

The most commonly used classification of refractories is on the basis of the chemical nature of the principal oxides (or compounds) present in them. Accordingly, the refractories are classified.

1. *Acid Refractories :* Acid refractories are those which consist of acidic materials and therefore are not attacked by acidic materials present in the furnace but are easily attacked by basic materials. These are of two types : (a) Siliceous materials have silica as the principal constituent as in sand, ganister, quartz, etc.;

(b) Alumino-silicates consist of Al_2O_3 and SiO_2. Fire clays are the most important refractories of this group.

2. *Neutral Refractories :* Neutral refractories are not neutral in the strictly chemical sense. These consist of feebly acidic or feebly basic materials with the result that they resist the action of acidic as well as basic materials. Examples of this group are chromite, graphite, silicon carbide, etc.

3. *Basic Refractories :* Basic refractories are those which consist of basic materials, therefore they are not attacked by other basic materials but are easily attacked by acidic materials. Magnesite, chrome-magnesite, dolomite, etc. are the most widely used refractories of this class.

Some of the common refractories are described below :

8.2.3 Silica Refractories

The raw materials for silica refractories are the various forms of natural deposits of SiO_2 which contain only small amounts of other oxides. Such deposits are quartz, quartzite, ganister, sand, sandstone, etc. The presence of an excess of basic oxides is particularly undesirable as they lower the fusion temperature. One great difficulty with silica refractories is that silica occurs in several allotropic forms which are stable at different temperatures and have different specific gravities. These changes will be discussed presently. Therefore, a raw material is chosen which can be easily converted into the more stable form. For this reason common sands and pure quartzite are not very satisfactory, the raw material should have some binding material preferably fine-grained silica particles themselves. About 4 lakh tonnes of quartz and quartzite are produced annually in India.

For the manufacture of silica bricks the rock is crushed and ground with 2% lime and water. At higher temperatures lime forms fusible calcium silicate which acts as a binding material. The mixture is then made into bricks by hand moulding or by machine pressing. The bricks are dried either in the sun or in heated rooms. The dried bricks need to be handled with care, otherwise they can crumble. They are then transferred to kilns which are heated by burning coal or gas. In the manufacture of refractory bricks on heating, mechanically and chemically held water is first removed. On heating further very often mineral transformation and volume changes take place. In the case of silica bricks another very important aim of heating is to convert quartz into cristobalite or tridymite which are the more stable allotropes of silica. If this change does not take place, while firing, the volume will change during use and the structure will become weak. The temperature is slowly raised, in about 21 hours, to say 1500^0 and maintained at this temperature for nearly 12 hours so as to allow quartzite to be converted into cristobalite which slowly changes to tridymite during cooling. Maintaining the temperature at 1500^0 for a longer period is desirable but is not economical. Cooling has to be done still more carefully and takcs 1 to 2 weeks. This results in the formation of a mixture of cristobalite, and excess of tridymite.

Properties

Silica bricks are slightly yellowish in colour with brown specks distributed throughout the mass. The porosity is about 25% and the specific gravity is 2.3 to 2.4.

An important property is that quartzite is converted into tridymite at 870^0 which changes to cristobalite at 1470^0, each of which has α and β forms and different specific gravities. The changes may be summarized as follows :

		870^0		1470^0	
Allotrope	Quartzite	$\rightleftharpoons$	tridymite	$\rightleftharpoons$	cristobalite
Sp. Gr.	2.65		2.26		2.32

Hence the transformation of quartz to tridymite causes a volume increase of 17.2 per cent; of quartz to cristobalite, 14.3 per cent. The linear expansion of silica bricks during firing, due to these transformations, is about 3.5 per cent. Therefore, if during firing of the bricks, quartzite is not converted into tridymite and cristobalite the bricks will expand on being heated in the furnace and the structure will fail. The reverse reaction, i.e. the formation of quartzite, takes place so very slowly that for all practical purposes tridymite and cristobalite may be taken as stable forms. That is why it is important that these forms should be produced while making the bricks.

Silica bricks have high refractoriness under load and can stand a load of about 3 kg/cm^2 at about 1600^0. No change in shape is noticeable up to 1350^0 under load. Bricks almost free from alumina do not deform under load even at 1680°. Below 600^0 silica bricks cannot withstand thermal shocks, (i.e. rapid changes in temperature). These properties make silica bricks very satisfactory for use in steel melting furnaces and for coke ovens. The thermal conductivity is better than that of fire clay bricks but not as good as that of the basic refractories.

Silica being an acid oxide, the bricks cannot be used in contact with basic fluxes and slags because of the chemical reactions between them. Silica bricks are relatively light and have high mechanical strength and rigidity.

Uses

The largest consumer of silica bricks is the iron and steel industry. They are widely used wherever the material in contact with them is not basic and there is a steady high temperature, i.e. the bricks are not exposed to temperature fluctuations. In addition, they are used in coke ovens, glass furnaces, etc. Silica bricks are used more than any other type of refractory except firebricks.

8.2.4 Alumina-Silica Refractories

In the alumino-slilicate refractories, as the name suggests, the two principal oxides present are alumina and silica. Fe_2O_3, TiO_2, CaO, MgO, Na_2O and K.O are the main impurities present and amount to 2 to 5 percent. Alumina percentage in the semi-silica type varies from 15 to 20% and in corundum type from 99 to 99.7%. In the silica refractories alumina is avoided because it may form an eutectic, which has a very low fusion temperature. A combination of 94.5% SiO_2 and 5.5% Al_2O_3 melts at 1545^0. The most widely used refractories of this group are the fire clay refractories.

Fire clays are hydrated aluminium silicates and their chemical composition may be represented by $Al_2O_3.2SiO_2.2H_2O$. At higher temperatures water is driven out. The common fire clays contain 18 to 47.5 per cent Al_2O_3 and 80 to 50 per cent SiO_2. Clays with higher Al_2O_3 contents are usually called high-alumina refractories. Fire clays are classified in several kinds of clays e.g. flint (hard clay), plastic clays, kaolin, etc. The term 'firebricks' refers to fire clay bricks.

In India most of the fire clay is obtained from deposits under the coal seams of Bihar, Bengal, Madhya Pradesh and Andhra Pradesh. Good deposits are also found near Jabalpur.

The fire clays can hold water mechanically in addition to the chemically held water. During firing, due to the expulsion of all this water and the fluxing effect of the impurities, the product shrinks considerably and becomes out of shape and size. To overcome this difficulty the fire clay is mixed with materials like silica that expand and/or large amounts of materials which do not shrink. This latter material is known as grog. Grog is fire clay which has already been heated to, say, 1300^0 so that no further contraction in volume takes place. Broken and used fireclay bricks are also used as grog. A small amount of plastic clay is added to facilitate moulding. A mixture is made by grinding raw fire clay, grog and plastic clay with water. There are several ways of giving the desired shape to the mixture. The bricks are mostly sun-dried. In colder and more humid climates they are dried in driers.

The bricks are then fired in a kiln. At the lower temperature water is expelled. At 980^0 an exothermic reaction takes place due to the formation of mullite ($3Al_2O_3.2SiO_2$) and the excess silica crystallizes as tridymite. Usually the temperature is raised to 1400^0 and maintained for a few hours to obtain uniform properties. For higher temperature use, the firing temperature may be 1500^0. The furnace is then closed and allowed to cool slowly. The total time taken is about a week.

Properties

Fire clay bricks of several grades are manufactured and their properties and prices differ to some extent. The bricks are light yellow in colour and depending on the content of iron oxide may be reddish in colour. While selecting the bricks for a particular use it is to be noted that fine-textured bricks are more resistant to corrosion by slags and gases because these cannot penetrate the bricks. The open-textured bricks arc more resistant to thermal shock, i.e. changes in temperatures, and are to be preferred where such conditions exist.

Firebricks are slightly acidic in character. The extent to which they are acidic depends upon the SiO content. They cannot therefore resist highly basic slags. They are less refractory than silica bricks though bricks fired up to a high temperature can be used up to 1700^0 but ordinarily they can be used up to 1400^0. The refractoriness under load is much poorer than that of silica bricks and most grades begin to fuse at 1350^0 under a load of 2 kg/sq.cm. The crushing strength at atmospheric temperature is high. It is about 200 kg/sq. cm.

A very useful property of firebricks is resistance to thermal shocks which makes them very suitable for making a checker work of regenerative furnaces and those parts of furnaces, like the charging doors, which are subjected to temperature fluctuations.

Firebricks are cheaper than silica bricks.

In blast furnaces, for which firebricks are very extensively used, the disintegration of firebricks takes place by the deposition of carbon from carbon monoxide and is believed by some authorities to be retarded by the presence of iron oxide.

Uses

Fire clay bricks are more widely used than any other refractory bricks because of the cost factor and as they are satisfactory for most purposes. Under extreme conditions the guiding principle is that they are satisfactory if the temperature is not very high and the slag is not highly basic. As they can resist thermal shocks, furnaces like crucible furnaces, or parts which are frequently cooled are lined with firebricks. They are used in large numbers in lining blast furnaces, regenerators, boiler furnaces and for the general lining given to varied equipments like chimneys, flues, etc.

8.2.5 High-Refractory Materials

There are many specialized service conditions for which, though the quantity of the material needed is small, resistance to very high temperatures is needed. Demand for such materials is increasing for use in gas turbines, nuclear reactors, rockets and missiles besides small furnaces for melting certain materials. Materials for this group come from pure metallic oxides, carbides, nitrides, silicides and sulphides. (In Fig. 8.2 are given the melting points of high melting materials. The melting points of copper and iron have been included for giving a comparative idea of the melting points). It is a matter of common knowledge that impurities lower the melting points. Therefore, great care should be taken to use pure materials and not to use any low melting binding agent. The shape is to be given by special techniques, like the combined effect of heat and pressure.

Oxides

The highest melting oxide is thoria (m.p. 3070^0) but its use is very limited due to its cost and difficulty in fabricating it. Widely used oxides are magnesia (2800^0), zirconia (2580^0), beryllia (2570^0) and alumina (2010^0). The pure oxide is finely ground with water and a temporary organic binder such as starch, wax, a high boiling alcohol, etc., is added. On firing, the temporary binder is destroyed and the fine particles sinter together. Solid surface reactions take place and gradually new tiny crystals are produced which hold the shape. As no material is used, the fusion temperature of the final shape is practically the same as that of the pure material.

Each of the oxides mentioned has some special property. For example, alumina is the hardest of all and is useful where wear resistance is required. Beryllia has high resistance to thermal shocks and is not reduced by carbon or carbon monoxide up to 2000^0. In spite of the high melting point of magnesia it cannot be used above 1700^0 in a reducing atmosphere and 2200^0 in an oxidizing atmosphere. As zirconia undergoes polymorphic changes, its resistance to thermal shock is low, above 1250^0. However, it can be stabilized by adding calcium or magnesium oxide.

Carbides

The carbides have higher melting points than the oxides but they are not resistant to oxidation at high temperatures. Carbides of silicon (decomposes above 2200^0)), boron (m.p. 2450^0), zirconium (m.p. 3750^0) and titanium (3160^0) are the more widely used carbides.

Silicon carbide, also known as carborundum, is the most commonly used carbide. It is made by heating, in an electric furnace, sand with coke and small amounts of sawdust and salt. The gases evolving from the sawdust make the mass porous. Salt forms volatile chlorides of iron and other impurities thus increasing the purity of the product. Several types of bonds have been used for giving the final shape. Silicon carbide decomposes before melting. Above 1000^0 it begins to oxidize in air. It has high thermal conductivity and low thermal expansion.

Boron carbide, B_4C, needs special reference due to its ability to absorb thermal neutrons (i.e. slow moving neutrons) which makes it a useful material for nuclear power plants. It is very hard and has high wear resistance. Above 1000^0 it readily gets oxidized in air and reacts with molten iron. This restricts its use in many furnaces. Other carbides also get oxidized in air much below their melting points and are very satisfactory for use in an atmosphere of nitrogen or in reducing atmosphere.

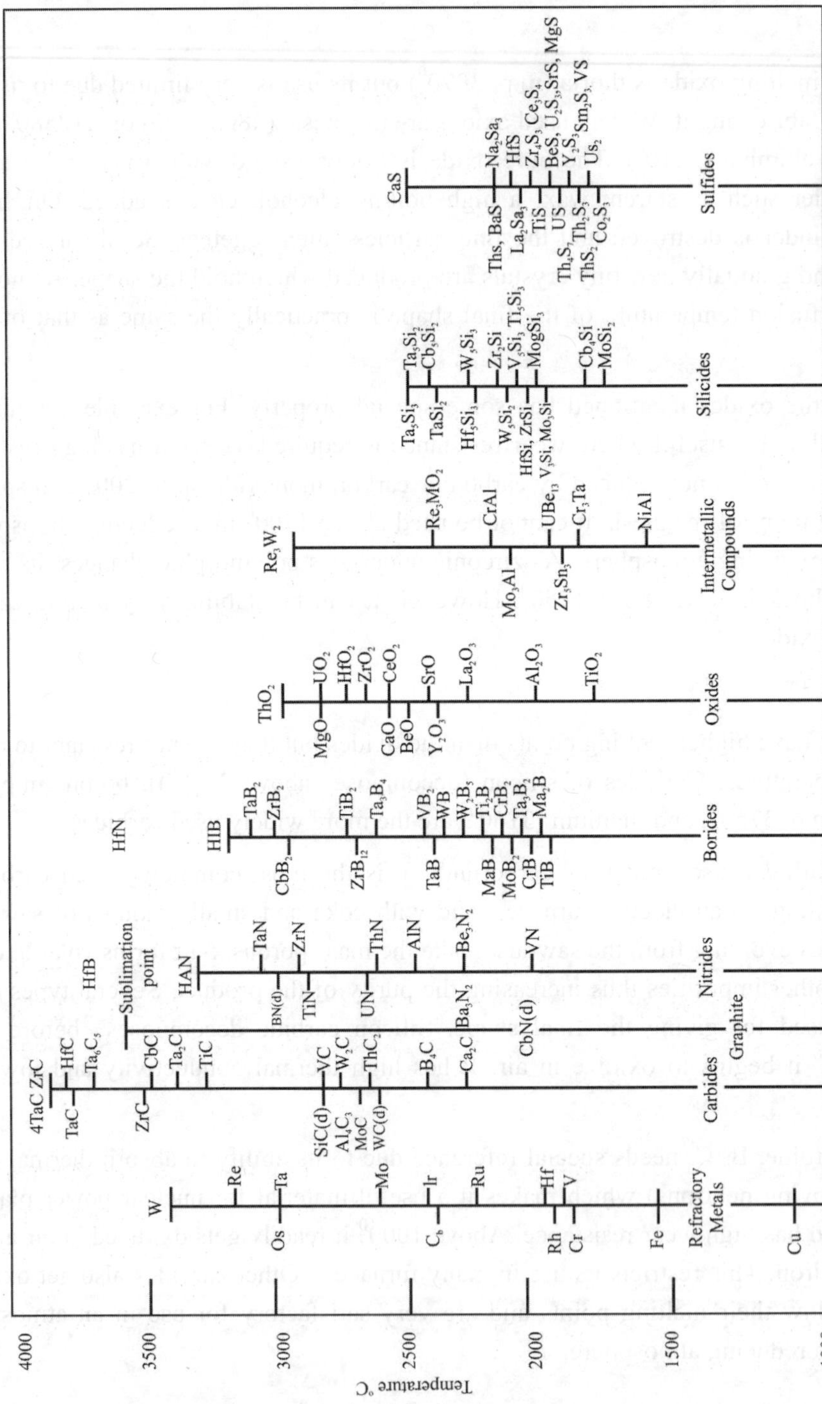

Fig. 8.2 Melting points of high refractory material (melting point of Cu and Fe are given for comparison.

Borides

The industrially important compounds of this group are the borides of titanium (TiB_2 m.p. 2890^0), zirconium (ZrB_2 3040^0), chromium (CrB_2 2760^0), molybdenum (MoB_2, 2100^0) and tungsten (WB, 2860^0). The borides are very hard, more resistant to oxidation than the carbides and have good resistance to chemicals.

Nitrides

As a group the nitrides have somewhat higher melting points than the borides but have poorer resistance to oxidation (Boron nitride is soft like graphite and sublimes at 2730^0).

The *silicides* and *sulphides* have relatively much lower melting points. Silicides resist oxidation on account of the formation of a glass-like coating of silica on the surface.

8.2.6 Super-Refractories

In some cases like very high operating temperatures, severe corrosion due to turbulent flow of molten material, the conventional refractories prove unsatisfactory. For such applications super-refractories are used. The term is used to indicate the superior properties of the refractories.

Very pure raw materials like silicon carbide, mullite, alumina, zirconia, etc. are used. These are fused in an electric furnace when the porosity becomes negligible. The fused mass is broken into small grains, called fused grains. The fused grains of different sizes are taken so that the packing characteristics of the product are improved. These are then made into the required shape by the conventional methods of manufacture.

8.2.7 Failure of Refractories

The most common cause of failure of a refractory is chemical reaction with the environment in which it is working. It should certainly not be in a chemically dissimilar environment, e.g. an acid refractory should not be used in a furnace using basic fluxes and slags and vice versa. In spite of using proper fluxes there may be impurities which may react with the lining. In the chemical reaction the porosity of the refractory plays an important part. The more porous it is, the greater will be the depth to which the slag will penetrate and destroy the refractory. Porosity also affects the reactions with the gases. The deposition of carbon from carbon monoxide in fire clay refractories in a blast furnace is an important cause of its failure (Deposition of carbon sets up high internal stresses resulting in the disruption of the brick). The higher the temperature the faster does the chemical reaction take place. Sometimes rise in temperature beyond the safe limit quickly brings about the destruction of the refractory.

Lastly, the agitation or movement of the materials is important. If there is agitation on the surface, new surface is quickly exposed to chemical attack and is destroyed.

The second important cause is "spalling" which has been defined as 'the loss of fragments or spalls from the face of a refractory brick or structure through cracking and rupture, with exposure of inner portions of the original refractory mass'. The cause of spalling may be thermal, mechanical, or structural. Thermal spalling may be due to unequal expansion or contraction caused by the difference in temperature at different parts or by rapid changes in temperature. Mechanical spalling is mostly due to carelessness in loading the furnace or in the removal of materials from the furnace thereby damaging the refractory. Structural spalling takes place because of the change in composition of the refractory due to reactions with slags, etc., with the result that its coefficient of expansion changes. Thus different parts expand and contract to a different extent. Spalling of a refractory can be reduced by proper design, construction and operation of the furnace.

Abrasion may be listed as the third factor. Abrasion means mechanical rubbing away of the material. In the furnace there is movement of materials and often hard and solid materials slide along the refractory. There are heavy molten slags and metals flowing. Dust-bearing gases often move at high speeds. All these rub off the refractories.

8.2.8 Selection of Refractories

The choice of a suitable refractory depends on a large number of factors. These factors may be classified into the following groups :

1. *Factors related to refractories :* The properties of the refractories available in the market need to be studied. The properties and their significances have already been discussed in this chapter.
2. *Factors related to the furnace design :* The choice of the refractory will depend upon the type of the furnace, conditions of heating (e.g., a coke oven is kept continuously at a high temperature for years but a cupola is daily cooled) and loading, degree of insulation, provision for thermal expansion made in the furnace, whether there are moving parts or not (as are in rotary kilns and in tunnel kiln cars).
3. *Factors related to operation:* The most important factor in this group is the chemical nature of the materials (like the ores, fluxes, fuels, etc.) in contact with the refractory, temperature in different parts of the furnace and temperature fluctuations, fluidity of the metals and slags, velocity of gases in contact with the furnace lining and abrasion due to moving charge, gases and the molten metals.
4. *Cost factors* : While considering the initial cost of different refractories their life during use should not be overlooked.

It may, therefore, be seen that selection of a suitable refractory is quite complicated. Much data is available from the working of the refractories in similar furnaces elsewhere. These will help in actual selection. Out of the large number of factors only a few are so essential that by comparision other factors are of minor importance. For example, in some cases exposure to high temperature is the chief problem. In other cases high load has to be supported, in only a few cases the refractory has to withstand both high temperature and high load. In some cases high insulating property is needed, in others, high thermal conductivity. Greatest attention, in the selection, should be given to the more essential properties for the particular use.

The cost of handling the refractory, for example, transportation, brick-laying cost, are the same whether superior or inferior bricks are used. If an inferior refractory is used and it has a shorter life, it will have to be replaced more often. This will involve additional handling costs, in addition to loss in production. Therefore the true cost of refractories should be considered on the basis of their effect upon the overall cost of production per unit output e.g., cost per tonne of steel produced.

8.2.9 Characteristic Properties of Refractory Bricks

Type	Characteristics
High-Alumina	High refractoriness, which increases with increasing alumina content. High mechanical strength at high temperatures. Excellent to fair resistance to spalling. Greater resistance than fire clay bricks to corrosion by most basic slags and fluxes.
Fire clay	The chemical and physical properties of fire clays vary between wide limits; hence fireclay bricks with widely varying combinations of properties are available. This fact accounts for their suitability for service under widely different operating conditions. Most fireclay bricks have relatively good spalling resistance and thermal insulation value. Fair resistance to acid slags and fluxes; lower resistance to basic slags and fluxes.
Silica	High refractoriness and resistance to abrasion. High mechanical strength at high temperatures. Thermal conductivity at high temperatures is about 25 per cent greater than that of fireclay bricks. High resistance to corrosion by acid slags. Fair resistance to attack by lime, magnesia and iron oxide. Readily attacked by basic slags and fluorine. Not subjected to thermal spalling at temperature above 650^0; poor resistance to thermal spalling at low temperatures.

Contd...

Type	Characteristics
Magnesite	Extremely high refractoriness and high thermal conductivity. Great resistance to corrosion by basic slags; poor resistance to slags containing high percentages of silica. The chemically bonded and metal encased bricks have marked resistance to spalling.
Chrome	High resistance to corrosion by basic and moderately acid slags and fluxes. In general basic slags not adhere to chrome bricks. Under certain unusual conditions iron oxide is absorbed and causes a damaging expansion. Thermal conductivity is lower than that of magnesite bricks but higher than that of fireclay bricks.
Magnesite Chrome	Mechanical strength and stability of volume at high temperature. Excellent resistance to spalling. High resistance to corrosion by basic slags.
Chrome Magnesite	Mechanical strength and stability of volume at high high temperatures. Excellent to good resistance to spalling. High resistance to corrosion by basic slags.
Forsterite $2MgO.SiO_2$	High refractoriness. Excellent strength at high temperatures. Marked resistance to corrosion by alkali compounds. Fair resistance to most basic slags; attacked by acid slags.

(Based on *Modern Refractory Practice*, *Courtesy*, Harbison Walker Refractories Co., Pittsburg).

Note : Basic refractories in general are less resistant to thermal shock than fireclay and high alumina refractories. However, resistance to spalling of some basic bricks or special composition is equal to that of super-duty fireclay bricks.

Basic bricks of any class may be metal-encased and internally plated. Such bricks have much better resistance to spalling than bricks without the casings. Most metal-encased bricks are magnesite chrome compositions.

8.3 Questions

1. Explain setting and hardening of Portland cement.
2. What are the types of cement? Describe the manufaction of cement by any one method and give advantages of it.
3. What are the types of Lime? Describe the manufacture of lime?
4. What is plaster of Paris? Describe its manufacture and setting and hardening along with uses.
5. Mention the ways to reduce thermal spelling of a refractory in furnace
6. Enumerate the important properties a material should possess in order to be a good refractory material
7. What are fire clay bricks? Describe the properties and uses of these bricks
8. What are ceramics? How are they classified? Give an example in each case
9. Why dolomite bricks are rarely used as direct refractories?
10. How do earthenware differ from stoneware?
11. What is a refractory material? Explain the differences between acidic and basic refractories?
12. What is meant by glazing? Explain why the ceramic white wares are essentially glazed.
13. What are insulators? How are they classified? Give examples
14. Describe the manufacture and uses of silica bricks
15. Explain the following properties as applied to refractory materials.

    (i) Refractoriness (ii) Refractoriness under load (iii) Thermal spalling
16. Explain how the dimensional stability and porosity effect the properties of refractories?
17. What is meant by pyrometric cone equivalent of a refractory?
18. Examine the truth in the statement "silica refractories can be used in basic media".
19. What is meant by white wares? Give their two uses. Explain vitrification.
20. Describe the chemical conversion that take place in the manufacture of ceramic products
21. Write note on polymorphism in ceramic materials.
22. What type material are used as transformer liquids?
23. Define porosity of a refractory.

8.4 Objective Questions

1. For initial setting the compared present in cement is

(a) C_3S (b) C_2S (c) C_3A (d) C_4AF

2. To delay the quick setting of cement the compound added to cement is

(a) MgO (b) CaO (c) $MgSO_4.2H_2O$ (d) $CaSO_4.2H_2O$

3. The cement containing the following compound is resistant to sea water

(a) High calcium (b) Low Calcium

(c) High alumina (d) Low alumina

4. Quicklime on exposure to air slowly absorbs moisture, swells and disintegrates to give fine power of

(a) CaO (b) $CaCO_3$ (c) $Ca(OH)_2$ (d) $CaSO_4$

5. The formula of plaster of paris is

(a) $CaSO_4$ (b) $CaSO_4$ ½ H_2O

(c) $CaSO_4.H_2O$ (d) $CaSO_4.2H_2O$

6. Graphite bricks are widely used as material for construction of _____.

(a) Atomic reactors (b) Open hearth furnace

(c) Coke oven (d) Blast furnaces

7. Porasity in an important property of refractory bricks and it increases

(a) Chemical stability (b) resistance to corrosion

(c) Thermal conductivity (d) Resistance to thermal spalling

8. High refractoriness under load is shown by

(a) Dolamite (b) Magnesia

(c) Alumina (d) Silica

9. Sewage (drainage) pipes are made of

(a) earthen ware (b) stone ware

(c) white ware (d) GI pipes

10. Dolamite is an example of _____ refractory

(a) Acidic (b) Basic

(c) Neutral (d) Amphoteric

11. Clays are formed by weathering of

    (a) Igneous rocks (b) Feldspathic rocks

    (c) Dolomitic rocks (d) Both a & b

12. Ceramic obtained from china-clay, feldspar and flink is –

    (a) Stone-ware (b) White ware

    (c) Brick (d) Tile

13. Machines operating under high temperatures and loads are lubricated by ....

    (a) grease (b) mineral oil

    (c) solid lubricant (d) synthetic oil

14. Molybdenum disulphide is a ....

    (a) solid, fuel (b) liquid, lubricant

    (c) solid, lubricant (d) semisolid, lubricant

15. Mineral oils are

    (a) suitable at low T (b) unstable

    (c) thickened in use (d) oxidation resistant

8.5 Fill in the Blanks

1. Permeability of a refractory depends upon_____.
2. Basic refractories are _____.
3. More pores in refractory _____is the thermal spalling.
4. _____ are gaseous insulators.
5. Refractoriness is the ability of material to withstand _____ without appreciable deformation.
6. In porous bricks supplied air act as _____.